THE IoT ARCHITECT'S GUIDE
TO ATTAINABLE
SECURITY & PRIVACY

THE IoT ARCHITECT'S GUIDE
TO ATTAINABLE
SECURITY & PRIVACY

DAVID M.
WHEELER

DAMILARE D.
FAGBEMI

JC
WHEELER

CRC Press
Taylor & Francis Group
Boca Raton London New York

CRC Press is an imprint of the
Taylor & Francis Group, an **informa** business

AN AUERBACH BOOK

CRC Press
Taylor & Francis Group
6000 Broken Sound Parkway NW, Suite 300
Boca Raton, FL 33487-2742

First issued in paperback 2022

© 2020 by David M. Wheeler and Damilare D. Fagbemi
CRC Press is an imprint of Taylor & Francis Group, an Informa business

No claim to original U.S. Government works

ISBN 13: 978-1-03-247523-3 (pbk)
ISBN 13: 978-0-8153-6816-8 (hbk)

DOI: 10.1201/9780367440930

Visit the Taylor & Francis Web site at
http://www.taylorandfrancis.com

and the CRC Press Web site at
http://www.crcpress.com

2lemetry™ is a trademark of 2lemetry, Inc.

Adobe® and Adobe® Sensei™ are either registered trademarks or trademarks of Adobe Systems Incorporated in the United States and/or other countries.

Akamai® is a trademark of Akamai Technologies, Inc.

Amazon®, Amazon API Gateway™, Amazon Cognito™, Amazon CloudWatch™, Amazon EC2™, Amazon Elastic Compute Cloud™, Amazon Elasticsearch Service™, Amazon IoT Core™, Amazon Simple Notification Service™, SNS™, Amazon Virtual Private Cloud™ (Amazon VPC™), Amazon Web Services™, AWS™, AWS CloudTrail™, AWS EBS™, AWS Elastic Container Service™, AWS IoT™, AWS Lambda™, AWS S3™, API Gateway™, and Amazon Webstore™ are trademarks of Amazon.com, Inc.

Anomali® is a registered trademark of Anomali, Inc.

Apple® and Apple® IoS® are registered trademarks of Apple, Inc., in the United States and other countries.

Arduino® is a registered trademark of Arduino, LLC.

Arm®, Arm® Mbed™, and Arm Trustzone® are trademarks or registered trademarks of Arm Limited (or its subsidiaries) in the US and/or elsewhere.

Auth0® is a registered trademark of Auth0, Inc.

Autonomic℠ is a service mark of Autonomic, LLC.

Bitcoin® is a trademark of A.B.C. IPHoldings South West, LLC, a UK-based company and subsidiary of Monopolip LTD.

Black Hat® is a registered trademark of UBM LLC.

Bluetooth® is a registered trademark of the Bluetooth SIG, Inc.

Checkmarx® is a registered trademark of Checkmarx Ltd.

CIP™ is a trademark of Flextronics International, LTD.

Civilization™ and CIVILIZATION IV are registered trademarks of Take-Two Interactive Software, Inc.

CoreOS® is a registered trademark of CoreOS, Inc.

Cumulocity™ is a trademark of Software AG, Inc.

Docker™ is a trademark of Docker, Inc.

Dyn™ is a trademark of Dyn, Inc.

Elasticsearch® is a trademark of Elasticsearch B.V., registered in the U.S. and in other countries.

Facebook™ is a trademark of Facebook, Inc.

Flashpoint® is a registered trademark of EJ2 Communications, Inc.

Ford® and Ford® Model T™ are trademarks and registered trademarks of the Ford Motor Company.

FOUNDATION™ Fieldbus and HART-IP™ are trademarks and HART® and WirelessHART® are registered trademarks of FieldComm Group, Austin, Texas, USA.

Gartner® is a registered trademark of Gartner, Inc. or its affiliates.

GitHub®is a registered trademark of GitHub, Inc.

Google Nest® is a registered trademark of Google LLC.

Google®, Google® Android™, Google Cloud™, Google Cloud Functions™, Google Cloud IoT Core™, Google Cloud Platform™, Google Cloud Storage™, Google Compute Engine™, Google Home™, Google Persistent Disk™, Google VPC™, TensorFlow®, the TensorFlow logo, GCP Firebase™, and any related marks are trademarks or registered trademarks of Google LLC.

HART® is a registered trademark of Rosemount Inc.

HART-IP™ is a registered trademark of Fieldcomm Group, Inc.

Dedication

To everyone who truly loves people, to everyone who uses their skills to serve others,
to everyone who seeks and upholds truth, I dedicate this book to you. For all
the technology in the world could never serve a better purpose.

— Damilare D. Fagbemi

To my Father in Heaven, may you bless the work of our hands;
To my family, may you always see Him in all that I do;
To my mentors and teachers, may I carry forward all that you
have given me with integrity and honor.

— David M. Wheeler

Contents

Foreword

by Dr. James Ransome

A Little about IoT and Security

The Internet of Things (IoT) involves adding internet connectivity to a system of interrelated computing devices, mechanical and digital machines, objects, people, and/or animals. Virtually every aspect of global civilization now depends on or is affected by interconnected cyber systems. Each "device" is provided a unique identifier and the ability to automatically transfer data over a network. IoT security architecture concerns itself with designing and safeguarding connected devices and networks in the IoT. The current explosive growth of the IoT ecosystem has resulted in the complexity of networked systems as a result of the heterogeneity of platforms and operating systems, multifunction protocols, and ubiquity of network access, resulting in an ever-expanding diversity of systems and software. This is giving rise to the attack surface and risks exponentially increasing in IoT systems. Those risks can be greatly reduced with secure system design and development. Allowing devices to connect to the Internet opens them up to a number of serious potential attacks and risks if they are not properly protected. This is due to factors such as their connectivity, the sensitive data they process, and the compute power that is available via IoT devices.

As the IoT evolves, cybersecurity and the issues associated with it will affect everyone on the planet in some way, whether it is cyber-crime, cyber-fraud, or cyber-war. It will be important for this industry to build out substantial product protection and security systems. You must understand the adversary and weakness in the systems you design by taking a closer look at the IoT attack surfaces, attack vectors, and breach consequences. Security design and architecture will be a critical part of this effort.

The proliferation of IoT devices running various kinds of software highlights the concerns that we can no longer ignore the threats of insecure software because software has become the lifeblood of the modern world. Core software and hardware security development practices can significantly eliminate the core factors of risk in IoT security. Reliable IoT applications are not impossible to achieve; however, it will require that all connected networks remain equally secure. This can only happen if comprehensive security measures are taken during the initial stages of architecting, designing, and developing the product and/or application. Hence, the need for experienced and qualified security architects who, needless to say, are in short supply and the gap for these professionals grow on a daily basis. *The IoT Architect's Guide to Attainable Security and Privacy* was written to help fill this gap. The responsibility falls on all of us who are security professionals associated with IoT to rethink how we design and build secure products that respond to and defend our IoT infrastructure against aggressive adversaries, as well as preserve the privacy and build the kind of life we want in our society.

Why I Care about IoT Security

I have spent the last 23 years as a cybersecurity executive since retiring from the US government in 1997. In addition to my CISO/CSO roles, I have had several technical leadership roles, which include Vice President of Integrated Security and Director of Industrial Cyber Security for CH2M HILL where I was responsible for developing models for converged wired/wireless network security. In a subsequent position as the Senior Director, Secure Unified Wireless and Mobility Solutions at Cisco, I co-architected, developed, and marketed the Getronics-Cisco "Wireless Integrated Security, Design, Operations & Management (WISDOM)" reference framework that was a productization of the original WISDOM framework developed as part of my doctoral dissertation titled "Wireless Integrated Secure Data Options Model (WISDOM) for Converged Network Security." This resulted in an increase of over $45 million USD revenue for Getronics and Cisco within a two-year period. I was also an adjunct professor for an NSA/DHS Center of Academic Excellence in the Information Assurance Education program as well as the author of eleven books covering various areas of cybersecurity. Most of my roles and books have centered around the various networks, systems, and technologies that make up what is now called the Internet of Things. Since the year before I retired from the US DOE-Lawrence Livermore National Laboratory, I have had a keen interest in Internet security, particularly the devices and systems that are connected to the Internet.

Why I Believe You Should Read This Book

I have a great passion for the topic of this book and have anticipated its publication since discussing it with the authors. I have known Damilare and David for several years now as co-workers and co-patriots in this journey we call cybersecurity. I have great respect for their work and reputations at both Intel and in the industry at large. Given the technical respect, knowledge, and integrity they have, when they talk . . . people listen, and when they write . . . people read with anticipation. They also have a rare and critical skill necessary in the security industry today in that they know how to use a variation of the KISS methodology, which I call KYSS (Keep Your Security Simple), while also making it relevant, economical, efficient, and scalable. Although security is complex in the background, the use experience must be simple to the user. Keep your security simple or it will be ignored. As an added bonus, David's wife, JC Wheeler, joined the team to write Chapter 8, which combined with Chapter 9 would be an outstanding book on privacy and its relation to IoT security all to itself.

Although this book includes the requisite technical specifications, processes, reference charts, tables, and diagrams, the storytelling pulls it all together. Most importantly, it is filled with practical and relevant examples based on years of experience with the technical specs, design requirements, and architectural principles broken up by lively discussions and story-telling surrounding issues related to IoT security design flaws and architectural issues based on real-world experiences. This book doesn't bog you down in complex theories and descriptions but is rather practical and just "what you need to know." It provides both general and detailed overviews of IoT security design and architecture principles but also leads the reader to other books and reference materials for deep dives into complex areas. If you only had one space on your shelf for a practical reference book for IoT Security Design, this is the book you need to have.

I thoroughly enjoyed the authors' holistic and entertaining approach to educating the reader on the principles of designing secure IoT. Some of the key areas covered that I found of particular interest that will likely be of interest to you as the reader include:

- The evolution and history of the Internet and IoT.
- Using the castle model as an analogy to describe the various inherent vulnerabilities and threat vectors for IoT systems and infrastructure.
- Business drivers for both adversaries and security professionals. Philosophical and business issues are discussed. Since cybercrime is driven by human motives, it is dynamic and evolving. This means that defenders who build and deploy IoT systems must be just as nimble and shrewd in their methods of defense.
- Cost-effective, efficient, and scalable design principles for IoT systems focused on the device, gateway, and cloud layers, with a particular emphasis on how it all works together to protect and secure edge devices.
- Design principles of IoT security architecture, practical secure system architectural models, and processes that include threat modeling, analysis, and disposition, threat mitigation, longevity issues, and focused discussion on Industrial IoT system (IIoT) architectural and design challenges, including the importance of control loops.
- Architecting for manageability, device trust, end-to-end encryption, longevity, inclusion of intelligence, and scalability.
- Design principles to secure the cloud to include how the cloud enables and scales to IoT security, data management and analytics, device management, threat modeling, and analysis as well as a case study providing practical examples and applications of the principles described.
- Weakest links that are the things that are accepted threats in the system.
- Pushing security out to the edge and Real-Time Operating Systems (RTOS).
- Connectivity design issues and solutions to include communications, wired versus wireless, network security, and protocols. The art and a science to secure protocol design and the case of why you shouldn't create your own security protocols but instead use protocols that are well published and standardized.
- Privacy risk, design principles, encryption, and legislation for IoT. Moral and ethical consequences of risks as a result of the technology related to security and privacy are covered without any judgment as to whether these issues are good or bad but rather where the vulnerabilities are and how to mitigate or fix them. Philosophical issues are also addressed.
- How data becomes personal information, and how personal information grows to become something more dangerous, known as personal knowledge. The combination of

multiple data sets of personal information and PII together to represent a different class of information known as personal knowledge. The two chapters on privacy could be a privacy book in itself.

- Security design principles as they relate to usability. There is an existential tension between usability and good security. The purpose of security usability is the alleviation of that tension. If we strive for the utmost security, the system becomes unusable. If we are too lax with security for the benefit of the greatest usability, the system is soon compromised and everyone is up in arms. We must find a balance, ensuring that the security that is implemented is usable.
- Challenges in the provision of usable IoT security controls.
- Eight principles that help you address the security usability challenges that they have outlined. The principles are not an exhaustive list, but rather a foundation that provides you with the necessary insight and starter fuel required to build usable IoT security. An invitation to the reader to extend their list is included.
- How to future proof your security designs in IoT with some forecasted views for some key sectors such as healthcare, agriculture, smart cities, energy, and transportation as well as projected future advances and challenges as a result of the evolution of cloud computing, serverless architecture, infrastructure as code, and autoscaling.

This is done by leveraging the evidence outlined by current trends and technological advancements to envision future trends and their impact on IoT solutions and their users. The expectation that artificial intelligence (AI) and machine learning (ML) systems will emerge as a new form of attacker is also discussed.

I will certainly be recommending this book to all I meet in the technology sector, whether it be on advisory boards, speaking at conferences, or to clients. *The IoT Architect's Guide to Attainable Security and Privacy* is destined to be a "have-to-have" reference book for all who are interested in this topic.

— Dr. James Ransome

About Dr. James Ransome

Dr. James Ransome was most recently the Senior Director of Security Development Lifecycle (SDL) Engineering and the Senior Director of Product Security at Intel and McAfee/Intel Security, where he was responsible for building and managing product security programs at both companies. His career has been marked by leadership positions in private and public industries, including three chief information security officer (CISO) and four chief security officer (CSO) roles as well as other technical leadership roles. Prior to entering the corporate world, James had 23 years of government service in various roles supporting the US intelligence community, federal law enforcement, and the Department of Defense.

James holds a PhD in Information Systems, specializing in Information Security, and graduate certificates in International Business and International Affairs. He developed/tested a security model, architecture, and provided leading practices for converged wired/wireless network security for his doctoral dissertation as part of a NSA/DHS Center of Academic Excellence in Information Assurance Education program. James is a member of Upsilon Pi

Epsilon, the International Honor Society for the Computing and Information Disciplines, and he is a Certified Information Security Manager (CISM), a Certified Information Systems Security Professional (CISSP), and a Ponemon Institute Distinguished Fellow.

James is the author of several published books, including *Wireless Operational Security, VoIP Security, Instant Messaging (IM) Security, Business Continuity Planning and Disaster Recovery Guide for Information Security Managers; Wireless Security: Know It All*; *Cloud Computing: Implementation, Management, and Security*; *Defending the Cloud: Waging Warfare in Cyberspace*; and *Core Software Security: Security at the Source*. He also developed the initial wireless network architecture, SCADA, Cryptography, and VoIP security leading practices for the Federal Communications Commission Network Reliability and Interoperability Council Focus Group on Cybersecurity—Homeland Defense.

Foreword

by Erv Comer, Zebra Technologies

I first became aware of the quality of Dave's work in security in the mid-1990s. I was leading a security analysis team at Motorola's Government Electronics Group in Scottsdale (AZ) developing cutting-edge "stuff" for our three-lettered friends when we first met. The community of security practitioners and designers was a small crowd, and we followed strict need-to-know protocols, which limited interaction, but things always seemed to be happening when Dave was involved. Happening for the better. A few years later designing the Internet in the Sky, I had the immense pleasure of working alongside a remarkable engineer with a devout passion for security—Mrs. JC Wheeler. I'd connected at a professional level with a couple holding a passion for security that I thought only I possessed. Our friendship has never waned over the years. We continue to pursue the engineering discipline of security that we so intensely enjoy.

Although I was honored to be asked to write this Foreword, I honestly didn't believe I'd learn much from the book. However, experience had taught me that Dave and JC had something robust otherwise they would not be seeking the scrutiny of a security practitioner. I've worked with the National Security Agency (NSA) in the development of their System Security Engineering CMM, obtained my Master's Degree in Engineering with a thesis on security for product development, reduced all the academia and theory to practice at Motorola across nine global business units within the past 15 years, and recently reincarnated the whole thing at Zebra Technologies—a leader in the IoT mobile computing space.

This book conveys to the reader a great understanding of IoT security methods and practices that must occur in order to properly account for security within the IoT operational environment. The IoT space is large, but the concepts and methods revealed are applicable across the entire space. End points, edge services, and clouds all come into play. The architectural views, in combination with the iterative security/privacy threat analyses, reveal abuse and misuse cases not easily recognized when analyzing from a single point of view. The robustness of the design is strengthened as the process builds upon itself, with designers becoming security conscious of their own designs. Toss in concerns over privacy, regulatory constraints, and analytic feedback channels and the security analyses will take a different route, yielding more insight. Change a few technology or environmental variables and completely different results emerge.

There is an absolute treasure trove of information within this book that will benefit anyone, not just the engineering community. This book has earned a permanent spot on my office bookshelf because of the wealth of content it provides.

— Erv Comer, Fellow of Engineering, Office of Chief Architect
Zebra Technologies

Preface

This book describes how to architect and design Internet of Things (IoT) solutions that provide end-to-end security and privacy at scale. It is unique in its detailed coverage of threat analysis, protocol analysis, secure design principles, intelligent IoT's impact on privacy, and the effect of usability on security. The book also unveils the impact of digital currency and the dark web on the IoT security economy.

We wrote this book not only to be a valuable security and IoT architectural resource for you but also to be an enjoyable read. With that in mind, we've added personal experiences, scenarios, and examples designed to draw you in and keep you engaged.

Why This Book?

When we purchase a book, we are looking for new insights. Being security architects at Intel, in addition to our combined experiences at McAfee, Honeywell, Motorola, and General Dynamics, our work has afforded us the unique opportunity to be involved in hundreds of diverse IoT and other Internet-related product releases. This depth and breadth of exposure has provided us with a deep understanding into what works, what doesn't, and why. It's provided us with insights into the future of technologies that are shaping IoT, and insights into the security of IoT and networked systems. It's given us a knowledge of the security solutions that scale well, the vulnerabilities that are likely to creep into IoT systems, and the assurance practices that significantly reduce risks without overburdening teams. Its these insights and design practices that we have put into this book. Your success as an IoT innovator is important to society, so you bet it's important to us.

Please visit our blogs, listed in our biographies below, for additional examples, detailed analyses, and discussions.

Who Would Benefit Most from This Book?

If you have a vested interest in IoT systems that preserve security and uphold privacy, this book is for you. The primary beneficiaries are IoT architects (security, system, software, or solution architects), and IoT strategic marketing. Executive managers and policy makers will also benefit from its insights.

Topics at a Glance

- The **fundamental components of an IoT system**.
- The **architecture of IoT systems**, how they differ from other client-server and cloud-based systems, and why are they architected the way they are.
- The **security-mindset** and **secure design principles** required to build secure systems and communicate effectively with customers, architects, and designers.
- The motivation and methods of **cybercriminals**.
- The IoT Security Economy, **dark web, dark money**, and **digital currency**.
- **IoT attack vectors**—how IoT systems are attacked.
- How to perform a **threat analysis** and **construct countermeasures** to protect a system, whether that is an IoT, cloud, or some other system—including detailed examples.
- A broad survey of common and not-so-common **IoT communications protocols**, and the roles of wired and wireless communications in IoT system.
- How to perform a **protocol analysis**—analyzing a protocol for security and finding security vulnerabilities in protocols, including a detailed example.
- **Artificial Intelligence and Digital Privacy**—the privacy impacts of combining AI and IoT; featuring scenarios for an autonomous vehicle ecosystem and smart refrigerator.
- Digital **privacy laws** and **regulations** and how they impact IoT architectures.
- A **Privacy Playbook** to mitigate unnecessary exposure of personal data.
- **Designing Usable IoT Security**—principles for building user-friendly security controls into IoT systems.
- The **future evolution of the Internet of Things and AI**, the impact on our lives, the security consequences we must prevent . . . starting now, and the responsibilities we all share.

Acknowledgments

We humbly acknowledge that we could not have written this book alone. Writing is a journey that can often be fun, and at other times, it can be grueling. It is a journey that is only possible with the help and support of family, friends, mentors, and colleagues.

First and in every sense above all, both authors would like to start by thanking God, without whom we truly believe we would not be.

We would like to thank our editor John Wyzalek at Taylor & Francis, who shared our vision for this book and was a patient and supportive advocate throughout the project. We would also like to thank the stellar production team at DerryField Publishing Services who have worked tirelessly to make our vision a reality: Theron Shreve and Marje Pollack.

Many thanks to James Ransome (Senior Director of Security Development Lifecycle [SDL] Engineering at Intel), Brook Schoenfield (Director of Advisory Services at IOActive, previously Master Security Architect at McAfee, and author of *Securing Systems: Applied Security Architecture and Threat Models*), and Erv Comer (Fellow of Engineering, Office of Chief Architect at Zebra Technologies), for their support with this project. We are also very grateful to JC Wheeler who through a fascinating chapter contribution, shed new light on the digital privacy debate and the privacy risks of a world powered by the IoT, as well acted as a technical reviewer and editor for many other chapters.

Finally, we would like to thank our mentors and peers in the engineering and security communities who we so deeply appreciate. As that wonderful saying teaches, "If you want to go fast, go alone. If you want to go far, go together."

— Damilare D. Fagbemi and David M. Wheeler

It is with a joyful grin that I say these heartfelt thanks to my beloved wife, AtTIyah, Papa and Mama, and the famous five: AtTIyah, you were a loving stalwart even before you were mine. You said, "I'm waiting for Chapter 1," and with that, we were on our way. Papa and Mama, somehow you kept me alive, continually investing in my education so I could learn to read and . . . write. The famous five, our famous five, we could scarcely have known that all those novels we shared and all those precious moments we enjoyed would take us down these paths, and into the stories and tales of far far away.

A special thanks to James Ransome, Brook Schoenfield, Rotimi Akinyele, Antonio Martin, and Marcus Lindholm. It is because of priceless people like you that I find myself believing that all security professionals must be wonderful people.

— Damilare D. Fagbemi

Thanks to my family for their support and patience while I was consumed with this long and, at times, arduous project! To my wife, you always bring out the best in me and enable me to achieve more than I could have imagined—I am so blessed to "do life" with you.

Thank you to my mentors at Intel—Brendan Traw, Baiju Patel, Lori Wigle. I value the time you gift to me and your examples of integrity and selfless leadership. I have learned as much from watching you as I have from our conversations. And to Dr. Jesse Walker, thank you for imparting your invaluable insights in protocol analysis and cryptography to me; you may never really know how that reignited the researcher in me.

A heartfelt thanks to all my technical peers and colleagues at Intel. Through our inter-actions and work each day, I learn so much. There is insufficient space to adequately thank each of you. Among these, I must acknowledge Geoffrey Cooper, Tony Martin, Brent Sherman, and the late and beloved George Cox.

To my friend (and JC's mentor) Dr. William T. Scott, thank you for the many lunches and, of course, evenings shared over good wine with you and Sue, talking about security and com-munications. That has shaped this book in ways you could not have imagined.

— David M. Wheeler

About the Authors

David M. Wheeler, CISSP, CSSLP, GSLC, GREM, is a Senior Principal Engineer in the Platform Security Division of the Architecture Graphics and Software group at Intel Corporation and has thirty years' experience in software, security, and networking for both commercial and government systems. In his current role, Dave is responsible for the research and development of new cryptographic algorithms and protocols, several security APIs, and libraries across Intel including for IoT platforms. He performs security reviews for both Intel's IoT and cryptographic implementations and represents Intel at the IETF.

Within the Internet of Things, Dave has contributed to Intel's Software-Defined Industrial Systems architecture and Intel's Internet of Things group's Health Application Platform. Prior to Intel, Dave held various lead software and systems architecture positions at Motorola, Honeywell Bull, General Dynamics, as well as his own firm. Dave has designed and built several hardware security engines, including a Type-2 security coprocessor for a software-defined radio, and the Intel Wireless Trust Module—a hardware cryptographic coprocessor on the Intel XScale processor. He has implemented several cryptographic libraries and protocol layers, including an IPSec-type implementation for an SDR radio; header compression protocol layers for IP, TCP, and UDP over multicast; a connectionless network layer protocol; two-factor authentication verification over RADIUS for a firewall VPN; PPP for serial; an instant messaging protocol over Bluetooth; and many others.

Dave has been a key contributor to other full-stack product implementations, including Intel's Blue River Network appliance, several complete public Internet applications in PHP, JavaScript/Sails, and even VBScript. Dave has also worked on smartcard security for banking and gaming applications at a startup, Touch Technology. While at Motorola in 1992, Dave authored the "Security Association Management Protocol" for the National Security Agency and subsequently spoke nationally about key management and key management protocols. He has led clean-room implementations for ISAKMP, IKEv2, and a custom network-keying protocol. Dave's extensive experience in security, networking, software and hardware is leveraged across a broad segment of Intel's Internet of Things to make Intel's products and software projects secure.

Blog: http://crypto-corner.typepad.com
Twitter: @dmwheel1
LinkedIn: https://www.linkedin.com/in/davidmwheeler/

Damilare D. Fagbemi CISSP, GXPN, had what might be considered the best possible introduction to the field of information security. An innovative software system that he built, the first of its kind in Nigeria at the time, was hacked minutes before a highly publicized deployment. After that, needless to say, Damilare got interested in information security fairly quickly. He began learning about the security of data and networks, then took and passed the CISSP. Considering his background in software development, he wondered where the intersection might be between the vast disciplines of software and security. A few years later, in Ireland, he stumbled upon a job advertisement for product security engineering. The rest as they say, is history.

Since then, Damilare has had the opportunities to serve as an engineer, architect, and technical leader at high-tech firms such as Intel Corporation and McAfee LLC, in the United States and Ireland. In those roles, he has had the pleasure of working with talented product teams to architect and build secure Internet of Things (IoT), web, and mobile solutions. As part of Intel's innovation in Smart Cities, he designed an IoT solution for Intelligent Transportation and contributed to the architecture of an artificial intelligence (AI)–powered platform for rapid decision making at the IoT edge. Damilare leads the Libraries Product Security Expert Center in Intel's Architecture Graphics and Software group, where he has enjoyed creating and leading a cross-organizational and cross-located security engineering team. He has taught security architecture and design across three continents—North America, Africa, and Europe—and served as Chapter leader of the Open Web Application Security Project (OWASP) in Nigeria. He is also a former co-founder of a software development company, with clients spanning private and government sectors.

Damilare holds a Master's Degree in Advanced Software Engineering from the University College of Dublin in Ireland, a Bachelor's Degree in Computer Science from the University of Ibadan in Nigeria, is a Certified Information Systems Security Professional (CISSP), and a GIAC—Global Information Assurance Certified—Exploit Researcher and Advanced Penetration Tester (GXPN). When he is not stuck at a computer, he can be found exploring nature and trying to stay active without a fitness tracker.

Blog: https://tech.edgeofus.com
Twitter: @damilarefagbemi
LinkedIn: https://www.linkedin.com/in/damilarefagbemi/

JC Wheeler began her career at US West Cellular analyzing analog network traffic and contributing to the rollout of one of the first commercial CDMA infrastructures in the nation, where she helped design the metrics and tools for CDMA traffic analysis. She then moved to Motorola to design cellular and satellite network protocols, authentication, crypto key management, and end-user features. She began consulting at General Dynamics in 2005, where she designed and integrated VoIP, header compression, multicast communications protocols, over-the-air provisioning, and IPSec variants for both MANET and satellite SDR waveforms. The small business she co-owned won a DoD SBIR and was a semifinalist in The Arizona Innovation Challenge for its smartphone secure framework; it was also a Navy Phase 2 SBIR subcontractor, building an AI engine to troubleshoot MANET radio configurations. JC is now retired and enjoys researching new technologies and macroeconomic trends.

Part One

Chapter 1

How We Got Here

– The discovery of new truths
– Comfort and companionship
– The rush of excitement
– An increase in efficiency
. . . humanity's lure.

— Damilare D. Fagbemi

1.1 We Forgot Security When Building the Internet

November 2, 1988, was a pivotal day in the history of the Internet. Do you recall what you were doing on that day? Few actually do. For us, as authors of this book on IoT security, it was amusing to think back to where we were, and the stark differences in our personal experiences. Damilare was the happy new arrival to a young Nigerian family and a few months away from his first words. He certainly had no clue as to the day's significance in Internet history. And truth be told, the Internet had hardly arrived in Nigeria or many other parts of the world. Dave, on the other hand, was a college student at Grand Canyon University in the United States, majoring in Computer Science with an important deadline requiring the use of the VAX™ computer system on campus. For him, it is still a clear memory. On that fateful day, a Cornell University student name Robert Morris took down large segments of the Internet by introducing the first self-replicating software that travelled across networked computers.[1] Although Dave and a few others have vivid recollections of the Morris Internet Worm, to the rest of the world, November 2, 1988, was just another day, three decades in the past, that they cannot exactly remember what occupied their time. There is a profound observation to be made by their memory lapses that has little to do with the amount of time that has passed.

Fast forward to October 21, 2016, when large areas of Europe and North America lost access to the Internet for a span of several hours. Investigation by security firms Flashpoint® and Akamai® security showed that a Distributed Denial-of-Service (DDoS) attack had been launched against Dyn™.[2] As a DNS provider, Dyn provides to end users the service of mapping an Internet Domain Name to its corresponding Internet Protocol (IP) address. For example, a web browser uses this mapping service to convert a computer's name entered in the address bar into a number corresponding to that computer's unique address on the Internet. The attackers had sent tens of millions of bogus requests to Dyn, rendering Dyn unable to process valid requests. How had the attackers managed so many requests? They had infected millions of IoT devices such as cameras, printers, and baby monitors with malware, essentially creating a super network of machines obedient to the whims of the attackers. October 21, 2016, is etched in the memory of security professionals and computer scientists as the Mirai Botnet attack. But that day is also etched in the minds of millions of individuals and businesses that will not easily forget the day the Internet, Facebook®, and point-of-sale systems all went dark at the same time. The difference three decades makes on the dependence of networked systems is indeed profound.

Thanks to IoT, the Internet is fast becoming not only the lifeblood of our digital lives, but our physical lives as well, and it is still arguably in its infancy. Imagine what can happen when homes, self-driving cars, power grids, oil tankers, oil refineries, inventory control systems, distribution systems, medical devices, farms, and refrigerators all complete their transition to the Internet of Things? It's a sobering thought.

Security was not a priority for the research students and professors who designed the ARPANET (the precursor to the Internet), and their security decisions have left an indelible mark on the Internet even to this day. In their defense, at the time, few computers were in existence and even less—about sixty thousand at the time Morris launched his worm—were connected to the Internet. Internet-connected computers were largely owned by research institutions and universities. Contrast that with today's IoT ecosystem. Gartner® predicts that there will be 20.8 billion Internet-connected devices by 2020. Malware that is able to infect even a small fraction of such devices could have a major global effect.

The concern that keeps us, the authors, up at night, as well as many of our peers who we are honored to know and work with, is that security, more often than not, is still an afterthought in the development cycle. It is for this reason that we wrote this book. Our hope is to bring security to the forefront of the conversation when designing, building, and integrating IoT systems.

1.2 What's This Book About and Who's It For?

Of course no one book can cover the entirety of IoT security. After all, it is the Internet of Things, or quite literally, all the things connected to the Internet. Covering all of the Internet would make for an exhausting endeavor. We view this book as a tour through IoT security for the intermediate-level engineer or architect, covering IoT foundations, IoT systems architecture, IoT security concerns, threat analysis, and communications security analysis. Along the way, we provide our insights, experiences, and examples. Our goal is to drill down even further into critical IoT security topics in our blog.

If you are an architect seeking to know more about securing IoT systems, an engineer branching into IoT, or a technical marketing person wanting to educate your customers, this book is a great resource that we are confident will remain on your bookshelf for years to come.

1.3 Let's Break Down the Book

This book is divided into three parts:

Part One provides an introduction to IoT systems, their uses, and their vulnerable nature. The chapters in this section are:

1. **How We Got Here:** In this chapter we introduce the Internet of Things, how it came about, why it exists, and its major components.
2. **The Castle and Its Many Gates:** At this juncture, we begin to explore the inherent security concerns in IoT systems and the proper mindset of the security architect and engineer through the castle analogy. We use the castle analogy to discuss the attack surfaces of IoT systems.
3. **The IoT Security Economy:** In this rather hair-raising trip into the dark side, we consider the regular economics of IoT and how cyber criminals subvert that economy to make money by compromising IoT systems. This leads us to consider the question, Why is security considered expensive for many IoT product companies?

Part Two has us rolling up our sleeves and diving into the technical analysis of the secure architecture and design of IoT systems of the future. This chapter is highly relevant for anyone who is building an IoT system.

The chapters in this section are:

4. **Architecting IoT Systems That Scale Securely:** In this chapter, we take a deep dive into the various elements that make up an IoT system, such as the edge device, gateway, and cloud layers. We consider the constraints that are placed on IoT systems and finish up with an explanation of why security is hard in IoT systems.
5. **Security Architecture for Real IoT Systems:** Securing any system requires careful analysis of the system, as well as of the attackers. This chapter reviews the processes and tools a security engineer uses to properly analyze and prepare an end-to-end IoT system to mitigate attacks and then walks through a threat-analysis exercise using an industrial factory example.
6. **Securing the IoT Cloud:** Cloud computing represents a major attack surface for IoT solutions. As described earlier, cloud services process and make sense of inputs from IoT sensors and gateways. They also manage and provide instructions to gateways, sensors, and actuators. In this chapter, we use practical examples and illustrations to explore solutions to cloud security concerns that are particular to IoT use cases.
7. **Securely Connecting the Unconnected:** IoT systems are nothing if they are not interconnected. We look at some of the most common communication protocols and discuss how to perform security analysis on protocols.
8. **Privacy, Pirates, and the Tale of a Smart City:** This chapter takes a unique and captivating look at the digital privacy debate through the development of two realistic scenarios—one taking place in the present, and one in a smart city from the not-so-distant future.
9. **Privacy Controls in an Age of Ultra-Connectedness:** The realities of privacy concerns in an ultra-connected world require workable strategies for designing and building privacy into IoT systems. Having looked at the evolving privacy challenges posed by the IoT, this chapter reviews the algorithms and software techniques used to preserve privacy. This chapter provides a balanced perspective of definitions, policies, legal protections, and controls.

10. **Security Usability: Human, Computer, and Security Interaction:** An IoT system has many pieces, all of which must be securely managed. It isn't enough to design security into a system; the administration of the system must also be done securely. What happens when an IoT system's security features are too convoluted or unintuitive? History shows that system owners bypass or ignore them. How can we design secure access, network protections, and security administration features into IoT systems so that those systems are actually usable?

Part Three is the forward-looking section of the book. While analyzing the current state of IoT is important for today, we as security architects must also look to the future. What does the future hold for IoT, and how are IoT systems likely to change in the next 20 years?

11. **Earth 2040—Peeking at the Future:** We wrote this book so that it can be future proof, at least for a few decades. In Chapter 11, we explore how IoT innovation is changing our lives today. Afterwards, we step into a time machine as we explore the future of IoT and its impact in various sectors, such as healthcare, agriculture, cities, homes, autonomous driving, energy, and transportation.

Epilogue: We provide our opinion concerning . . . "where we go from here."

1.4 What's an IoT System?

IoT stands for Internet of Things. We can describe an IoT system as one that comprises internetworked physical devices (also referred to as "connected devices" and "smart devices") embedded with electronics, software, sensors, actuators, and network connectivity, which enable these objects to collect and exchange data.

Let us start to flesh out what IoT is by describing the components as simply as we can:

- **Sensors:** Electrical components that are able to detect events or changes in their environment, such as temperature, pressure, or motion.
- **Actuators:** Any component of a machine responsible for moving or controlling a mechanism or system.
- **Software:** The compiled computer instructions that control the sensors and actuators, thus determining the final goals or outcomes of the system. Some software could reside on the sensor or actuator devices, but there will often be a major chunk of central software separate from both.
- **Network:** The channel through which the devices communicate. The letter *I* in IoT represents the deployment of many of these systems over the Internet—that is, sensors sending data over the Internet to a central management system, which in turn can send control signals down to physical devices.

How did the Internet of Things come about? What triggered the innovation to connect everything to the Internet? And why has security remained so elusive?

1.4.1 Everyone Needs to Know the Location of the Nearest Pizza

It was 1995 and the juggernauts of Internet business were rapidly gathering steam. At this time, businesses were just beginning to get on the Internet. And most regular, brick-and-mortar businesses were yet to create websites or web presences.

Two brothers, Elon and Kimbal Musk, who lived in the Silicon Valley[3] technology hub, decided they wanted to create an Internet company. They were not sure what kind of company they wished to create but simply knew that it had to be an Internet company. Elon had served an internship in Silicon Valley the previous summer, where he happened to hear a bungled pitch from a Yellow Pages sales representative for an online listing to go along with the printed book of business addresses. Although the representative seemed unsure what he was talking about, Elon recognized that searchable web directories were a quick and easy way to offer value online.

The Musk brothers purchased a local business directory for a few hundred dollars. Afterwards, they negotiated free access to some digital mapping software from a Global Positioning System (GPS) company called Navteq®. Finally, they wrote the code that combined both databases. Regarding his motivation for that venture, Elon often said that everyone should be able to find the closest pizza place and figure out how to get there.[4] Zip2®, the first major location-aware Internet business, was born.

The intrigue of products like Zip2 and many that followed is the augmentation of computers with other services or sensors. In this case, augmentation with GPS location. Location-aware or context-aware computing refers to computing solutions or systems that are able to sense their physical environment, using such input to influence the decisions taken or recommendations provided by a central program or brain in the system. The value provided by Zip2 was the availability of digital maps and city guides,[5] which not only provided directions to select destinations but also included information on businesses in different areas.

The augmentation of computing with location services was just the beginning.

1.4.2 Computing Everywhere

For the last two decades, technologists and scientists have longed to get computers off our desks and into myriad environments in and outside our homes, where they can do our bidding in more natural and less intrusive ways. This all started long before Zip2 was being created. It turns out that before Elon was searching for nearby pizzas, a fellow named Mark Weiser was dreaming up something quite similar, a field of research that seeks to make computing available anywhere and everywhere. Weiser's concept is commonly referred to as *ubiquitous computing*.

Mark Weiser coined the term ubiquitous computing in 1988 during his tenure as Chief Technologist at the Xerox Palo Alto Research Center (PARC), and he is often referred to as the father of ubiquitous computing. In one of his early papers on the subject,[6] he described how the best technology ought to disappear, weaving itself into the fabric of everyday life. In such an environment, humans need not interact with desk-based computers, but rather with a network of smaller computers that are often invisible and embodied in everyday objects, doing everyday things that augment the life of the human. Such computers would also be equipped with sensors that enable them to interact with their environment. For instance, a smart fridge that is aware of its tagged contents is able to provide food recipes capable of being made from its available stock or tell its owner about any food that's going bad.

Ubiquitous computing was the harbinger of the Internet of Things (IoT) as we know it today. The present excitement about connected devices is simply the latest stage of that phase of computing evolution—off our desks and into our lives. We can easily say, as Mark Weiser actually theorized, that present-day advancements in decentralized computing are a step on the trajectory toward practically invisible technology.

1.5 An IoT System's Major Components

IoT systems can also be described using a basic three-layer structure, sometimes referred to as MGC, referring to **EM**bedded devices, **G**ateways, and **C**loud back-end systems.[7,8]

- **EM**bedded devices (or just Devices) layer: The Devices layer of an IOT system is composed of the edge nodes or computing systems that interact with the physical environment, either by mechanically or electromagnetically changing physical systems or by collecting data.
- **G**ateway layer: Gateways are communication and processing hubs that directly service the Devices layer. Gateways provide data relay services back to the cloud, off-loading the cloud of some of the time-critical processing required to properly manage control loops, and they provide network and data security to devices.
- **C**loud layer: The IoT Cloud Layer represents the back-end services required to set up, manage, operate, and extract business value from an IoT system. In the Cloud layer, sophisticated business analytic engines, artificial intelligence (AI), and web applications are used to process and massage the data collected by an IoT system and use that processed data to present sophisticated models of the system to operators and users.

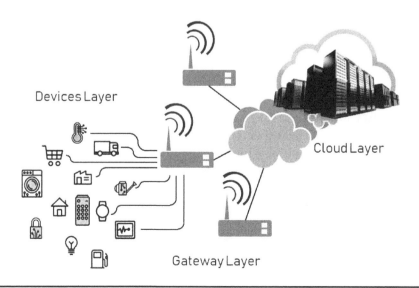

Figure 1.1   Conceptual Representation of an IoT System

Please see Figure 1.1 for a conceptual view of an IoT system.

One way to think about how an IoT system actually works is to think about how our human bodies work. Let us explore this analogy just a bit more.

1.5.1 The Human IoT System

At the conceptual level, the workings of an IoT system, which is designed to sense and respond to the environment, can be compared with the human body, which is wonderfully designed in the same way. For instance, when we enter a cold room, our skin (sensor) detects the cold, breaking out in goosebumps as it sends signals to our brain (central software and actuator). Our brain decides, based on historical evidence biologically programmed in knowledge, that warmer air is better for us. It sends a signal of discomfort to our skin, which often causes us to cover ourselves with our hands and hug ourselves for warmth. Smart home devices, such as the Nest thermostat from Google Home™[9] do the same thing. Such devices are designed to automatically regulate the temperature in an environment. They are usually connected to a central heating, ventilation, and air conditioning (HVAC) system. Their sensors pick up the temperature which is sent to software on a central unit, which might also house the sensor. The central software analyzes the data from the sensors before turning the temperature up or down via the HVAC system. This is, of course, a simplified example. We will describe the architecture and components of IoT systems in much greater detail in Chapter 4.

1.6 Shall We Just Connect Everything?

Powerful computers and advanced sensors have created the perfect environment for the proliferation of IoT systems. The IoT market is growing quickly and the Boston Consulting Group predicts that it will reach a valuation of $267B by 2020.[10]

Unsurprisingly, the promise of IoT business has drawn the attention of entrepreneurs and technologists, who're racing into IoT ventures, conceiving and experimenting with myriad solutions. Indeed, with the proliferation of IoT and connected devices running powerful computers, it can be said that the sphere of IoT is becoming less of a world of computers in everyday things, but rather computers moonlighting as everyday things.

In many cases, the value provided by IoT solutions is obvious. For instance, thermostats that control the temperature in a home or building, door locks that enable remote access control to rental property, sprinklers that intelligently nurture the grass based on their awareness of temperature and humidity, the list is seemingly endless. However, some solutions raise eyebrows regarding their necessity—that is, smart coffee makers or toasters that alert their owners via email or text messages when the brew or the bread is ready.

It is worth pointing out that this book does not intend to outline the viability of IoT solutions, but rather to provide recommendations for designing and implementing such solutions securely. Yet, a security concern with the current flood of IoT solutions and devices is that many of those devices are not thought through. Before IoT, a cheap toaster was just that, a cheap toaster that couldn't last very long. Now, a cheap toaster with poorly written software that's exposed to computer networks or the Internet is not just an unreliable toast-making machine, but a ticking information security time bomb.

At this juncture, it becomes noteworthy that we are in an age of convenient tech. The most popular web businesses and technologies are those that provide instant, seamless gratification in the fulfillment of a given task. If you find yourself wondering, "Isn't technology meant to simplify the execution of tasks that would be otherwise tedious?" you're right. But there's a difference between a tractor that the farm owner or farm hand drives to prepare acres of land for planting, and a smart lock that the owner uses to open or lock their home whilst they are physically 10,000 miles away, possibly in another country.

As we discussed at the outset of this chapter, the Internet was not designed to intrinsically provide security. Hence, software or systems designed to use the Internet must also design and implement secure mechanisms for data transport, system access, user authentication, etc.

By definition, IoT solutions are Internet accessible and thus incorporate remote usage and administration capabilities. For instance, temperature can be remotely regulated for smart homes, power grids can be administered remotely, door locks can be disabled remotely, cars back up data to and download software from the cloud, and nuclear reactors can be administered remotely.

Remote administration systems and interfaces of connected critical infrastructure such as power grids are juicy targets for highly skilled and resourced attackers. There is an assumption that security controls deployed in environments hosting such infrastructure are staffed with highly expert security management personnel paying great attention to detail—at least that's the case in the leading tech countries of the world. Nevertheless, in September 2017, security researchers at Symantec™ reported that attackers had gained direct access to US power grid controls, potentially allowing them to induce blackouts in large areas of the country.[11] We bring this up not to dissect, confirm, or refute the claims of Symantec but to illustrate the lack of invincibility of even the most fortified IoT deployments.

If fortified systems are still vulnerable, how about rapidly thrown together IoT solutions championed as convenience machines for the unsuspecting customer. Toasters, coffee makers, microwaves, fridges, sprinklers, door locks, light bulbs, baby monitors, the list goes on. Many manufacturers of such devices, some of whom are experienced in the creation of more conventional and unconnected versions of those devices, jump on the IoT devices without investing in the proper development and validation of the enhanced, juiced-up devices they're producing. Usually, customers of such companies are not used to expecting security as a feature on their bulbs and toasters, they just expect them to work, much more conveniently and seamlessly, of course.

Since customers aren't demanding security and are likely to balk at expensive bulbs, energy saving and remotely administrable as they might be, many IoT manufacturers are hesitant to invest in the secure development and validation of their creations. This leaves us with a sizable number of insecure and broadly deployed IoT devices that are exposed to the Internet. The website http://shodan.io, touted as the world's first search engine for Internet-connected devices, does a very good job of exposing any such insecure devices to the public, allowing anyone and everyone to find and have a go at cracking an IoT device.[12]

1.7 Wait! We Need to Add Security!

As we look back at history, from the Morris Internet worm, through ubiquitous computing, to Zip2, and all the fabulous IoT magic going on today, we just want to scream, "Wait! We need to add security!" But it seems that too many companies glibly run off to do something wonderful, not thinking about the consequences. Not ever thinking that something like the Mirai botnet could happen.

Born Curious: Damilare Reflects

In Africa and the Caribbean, masquerade ceremonies are religious events where revered masked figures, masquerades, parade as part of ceremonial rites. In southwestern Nigeria, the Egungun masquerades are believed to be connected to ancestors of the Yoruba people. The Egungun are also believed to possess spiritual powers and, as such, elicit fear in a good number of adults. Their elaborate movements and shrieks also terrify most children, who've been regaled with stories of the Egungun's prowess.

I remember attending such an event with Yemi, a niece of mine. She was barely two years old and hobbled most adorably whenever walking took her fancy. Yemi wasn't afraid of the Egungun, she was intrigued! As she saw the first masquerade in the ceremonial procession approach, her eyes grew wide. As the Egungun danced and shrieked in their grand and colorful costumes, she laughed. Before I knew it, she was hobbling toward them. Yemi didn't know about masquerades, and she didn't know they were supposed to be revered powerful forces. She simply saw something interesting and decided to explore.

Now, besides the famed spiritual prowess of the Egungun, which some of us will instinctively dispute, there were also physical dangers to Yemi. As she ran into a group of stampeding men/women who were masked and most likely had some constraints on their vision, she could very easily have been trampled upon.

As it was for Yemi with the Egungun, the vision or promise of IoT can be quite thrilling. Computers combined with sensors can help us to investigate and analyze many interesting scenarios in different environments and provide recommendations that help us make informed decisions. But if such systems are not well architected and designed, they can open up a vast array of security vulnerabilities that cyber attackers will be only too glad to exploit. We explore this concept of how to look at IoT with a security mindset in Chapter 2: The IoT Castle and Its Many Gates.

References

1. Shackelford, S. (2018, November 1). "30 Years Ago, the World's First Cyberattack Set the Stage for Modern Cybersecurity Challenges." The Conversation. Retrieved from http://theconversation.com/30-years-ago-the-worlds-first-cyberattack-set-the-stage-for-modern-cybersecurity-challenges-105449

2. Newman, L. H. (2018, October 21). "What We Know About Friday's Massive East Coast Internet Outage." Retrieved from https://www.wired.com/2016/10/internet-outage-ddos-dns-dyn/

3. Uskali, T. and Nordfors, D. (2007). "The Role of Journalism in Creating the Metaphor of Silicon Valley." Stanford University (CA): Innovation Journalism.

4. Isaacson, W. (2014). The Innovators: How a Group of Hackers, Geniuses, and Geeks Created the Digital Revolution. New York (NY): Simon & Schuster Inc.

5. Patricia, B. "Zip2: American Company. Encyclopedia Britannica." Retrieved from https://www.britannica.com/topic/Zip2

6. Weiser, M. (1991, September). "The Computer for the 21st Century." Scientific American.

7. Levis, P. (2016, September 7). "Rethinking the Secure Internet of Things Architecture." Retrieved from http://iot.stanford.edu/doc/SITP-summary-2016-project.pdf

8. Levis, P. (2018, March 20). "Towards a Secure Internet of Things." Retrieved from http://www.dcs.bbk. ac.uk/~gr/uploads/percom-2018-k1.pdf

9. Google Home. (2018). "Nest Thermostat." Retrieved from https://nest.com/thermostats/

10. Grobman, S. and Cerra, A. (2016). *The Second Economy: The Race for Trust, Treasure and Time in the Cybersecurity War*. New York (NY): Apress Media LLC.

11. Symantec Security Response Attack Investigation Team. (2017, October 20). "Dragonfly: Western Energy Sector Targeted by Sophisticated Attack Group." Retrieved from https://www.symantec.com/blogs/threat-intelligence/dragonfly-energy-sector-cyber-attacks

12. Shodan.io. "The Search Engine for Internet Connected Devices." Retrieved from https://www.shodan.io/

Chapter 2

The IoT Castle and
Its Many Gates

Avoiding danger is no safer in the long run than outright exposure.
The fearful are caught as often as the bold.

— Helen Keller[*]

2.1 And the Internet Got Hacked: Analyzing the Mirai Attack

October 21, 2016, was a regular Saturday, until it was not. A new perspective emerged, at least for tech enthusiasts. The "new" perspective was not birthed from a novel concept, but rather some new light was shed on the condition of security in the burgeoning area of the Internet of Things (IoT). Perhaps one can hope that the Mirai botnet attack also bestowed this new perspective upon regular Internet users as they received a personal introduction to the risks inherent in innocent IoT devices such as baby monitors, cheap surveillance cameras, and other devices in their smart home systems.

The bestowal of this new perspective began sometime around 7:00 am EST on that Saturday. Many major Internet platforms and services in North America and Europe were booted off the air waves, or, shall we say, web waves. For hours, such platforms and services were inaccessible. Affected services spanned multiple industries and sectors, such as national governments, online shopping services, news media, social media platforms, and online payment services, to name only a few.

The outage of the affected websites and web services was caused by a distributed denial-of-service (DDoS) attack. Denial of service (DoS) is typically accomplished by overwhelming the targeted machine or resource with superfluous requests in an attempt to overload systems and

---

[*] Retrieved from https://en.wikiquote.org/wiki/Helen_Keller

13

prevent some or all legitimate requests from being fulfilled. In a DDoS attack, the incoming traffic flooding the victim originates from many different sources, making it impossible to stop the attack simply by blocking a single source or single network segment.

As we briefly discussed in Chapter 1, the victim of the DDOS attack was Dyn™, a company that provides a Domain Name System (DNS) resolving addresses for a considerable portion of the Internet.[1] DNS is the system protocol that is responsible for the resolution of human readable web application or web service names—for example, iot.com—to the unique Internet Protocol (IP) addresses of the machines or networks that host such services.[2] With Dyn under attack, systems were unable to resolve the Internet or IP addresses of a good number of web applications or services, effectively rendering them unreachable.

But how did the DDoS attacks originate? Investigation by Dyn and the United States Department of Homeland Security showed that the troublesome traffic originated from 100,000+ IoT devices around the world, which had been compromised by malware known as Mirai. The infected devices included printers, IP cameras, residential gateways, and baby monitors. See Figure 2.1 for an illustration of how the DDoS attack works.

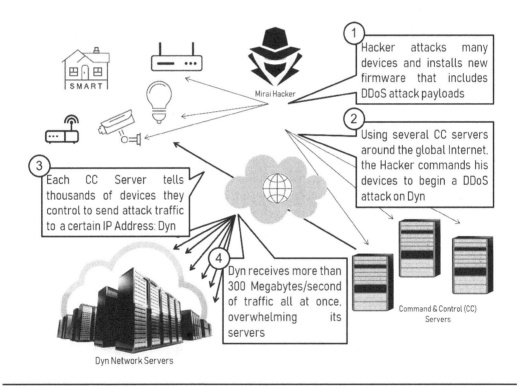

Figure 2.1   Stages of the Mirai–Dyn Attack

Mirai is a simple, yet effective, malware that turns infected networked devices into worker machines—called bots— that can be controlled by a remote attacker. Thus, unfortunately for Dyn and their customers, they found themselves at the mercy of 100,000+ IoT devices that had been instructed to continuously send malformed requests to Dyn.

We know. You are probably thinking, as we did, "That's it? Websites down for just a few hours?" That could be a fair assessment, until one considers how dependent we are on the

Internet and its many offerings. Sure, many hours are blissfully enjoyed on social media, but a great many hours are painstakingly spent buying, selling, mediating, investigating, and the list goes on. One of the concerning aspects of this attack is that websites involved in multiple sectors of the economy, virtual or otherwise, were affected, even if it was only for a few hours. However, how it happened, and the apparent simplicity of such a far-reaching attack, is nearly as intriguing.

2.1.1 Resolution of the Mirai Attack

The Mirai software scans the Internet for certain types of IoT devices, for example, a certain type of baby monitor. When it locates such a device, the Mirai software attempts to log in to the device using the manufacturer's default username and password. This information is contained in the device manufacturer's user manual that a user receives when they purchase the device, but can also be found for free on the Internet. If Mirai is successful in logging in to the device, the Mirai software replicates itself on that new device by replacing its firmware with the Mirai botnet software. Now, there are two instances of Mirai bots both looking for other victim devices. The replication expands geometrically—from two to four to eight . . . to tens of thousands. The Mirai code contains a table of over 50 common factory default usernames and passwords, allowing Mirai to attack many different devices.[3]

The solution to the Mirai–Dyn situation is frustratingly simple:

- Action by End Users:
  - Change default passwords for your devices and set stronger ones.
  - Disable any network ports or interfaces that you do not use.
- Manufacturers:
  - All IoT devices should support the update of software, firmware, and administration credentials, that is, passwords.
  - Make it mandatory for users to change the default username and password during the installation of any device that connects to the Internet.
  - Patch IoT devices with the latest software and firmware updates to mitigate vulnerabilities. The proliferation of IoT devices running various kinds of software highlights the concerns brought to light by James Ransome and Anmol Misra in their book, *Core Software Security: Security at the Source.*[4] They argue convincingly that we can no longer ignore the threats of insecure software, because software has become the lifeblood of the modern world.

Although Mirai is frustrating because the security defect is so simple, in many other cases, malware takes advantage of actual security bugs in the software—operating system and user applications—or hardware of the device. Any one such vulnerability can be used to compromise and control affected devices, steal private information, or monitor the owners or users of the device.

The Dyn attack highlights the potential damage caused by security vulnerabilities in any IoT device that is connected to a network—and most IoT devices are. Successful attacks against such systems are not just a concern for the device owners and manufacturers, but also for anyone and everyone else participating in the Internet ecosystem.

Make no mistake about it; this isn't the only eye-catching IoT attack in recent times. There are many more examples, including pacemakers getting hacked along with cars as well. Take a moment to consider both scenarios. In the first case, someone's heart or heart support system is directly attacked. Unsurprisingly, the attack was such a concern that 465,000 pacemakers were recalled by the manufacturer.[5] You can't make this stuff up . . . "Sorry, the pacemaker we gave you can be hacked remotely, please return." In the second case, how would you like it if someone was kind enough to hack into your car as you drove your kids to school? No, it's not a fancy tale. A practical attack was demonstrated to wide acclaim and fascination, leading to the recall of 1.4 million cars.[6]

As scary as both of those attacks are, they probably aren't even close to being the most worrisome cases we should consider for the future, with the proliferation of IoT deployments. Attacks against critical infrastructure have that esteemed position.

For many of these problems, there isn't an easy answer as to why a vulnerability appears. Sometimes, it is as simple as a software or hardware deficiency in the product of a specific vendor, or even user misconfiguration. But the breadth of products leveraged by the Mirai botnet attack requires us to consider this from a different perspective. The defect should be viewed as a failure in security usability that was common in multiple product types from different vendors, which signals a failure in the common mindset of the security engineering community. Chapter 10 discusses security usability in more detail. In this chapter, we develop the appropriate security mindset and equate it to a common example—the medieval castle.

2.2 "Full Disclosure," Ethics, and "Hacking Buildings for Fun and Profit"

The code for Mirai is publicly available.[7] This was predictable, as the security community frowns upon what is termed as "security by obscurity." In the same way that a home is insecure when its door locks are defective and cannot throw the deadbolt, an IoT system isn't secure just because its vulnerabilities are unknown. In the case of the house, it does not matter how obscure the defect with the locks actually is, since it only takes a single burglar to try to open the door and find they can walk right in. Security is only actually achieved through the implementation of appropriate protections. "Full Disclosure" is the policy that vulnerabilities are completely revealed, but it must be done so responsibly.

Ethical hackers often share software that they've written to uncover security holes in systems, usually at conferences such as Black Hat® and SchmooCon™. However, the Ethics of Responsible Disclosure[8] requires an agreement between the ethical hacker and system owners, for a certain grace period before full disclosure, allowing the owners sufficient time to fix the vulnerability before the hack is published.

However, nonethical hackers or ethical hackers on a "mission" are likely to be bent on disruption and can't be counted on to adopt responsible disclosure. Furthermore, the path to responsible disclosure is less obvious in cases of vulnerabilities such as those exploited by Mirai.

Soon after the discovery that Mirai was the software behind the Dyn DDOS attack, the code was shared as open source on hacker forums. Subsequently, it was also shared on the mainstream open source platform, GitHub®. Almost immediately, copycat attacks began to occur, as eager beta testers began to deploy Mirai in vanilla form or with their own flavors and modifications.

One of the more notable follow-up Mirai attacks occurred in Finland. A month after the Dyn attack, cybercriminals used a Mirai botnet attack to disable systems that controlled heating in two apartment complexes.[9] If you live in Finland, the winters are long and bitterly cold; you do not want anyone messing with the heat . . . it could be deadly.

This attack blocked the Internet connection of the building management systems (BMS). The systems automatically rebooted in an attempt to reset the faulty network connection. But this didn't resolve the problem, so the BMS rebooted again. Eventually, the system got stuck in a reboot loop, leaving the furnaces off and stranding the residents without central heating, in the thick of a cold, Northern European winter. How should a security engineer view systems to account for such types of attacks?

2.3 Defending IoT Castles

The mindset of security engineers should be similar to the mindset of castle builders in the medieval period—protection, fortification, and defense in depth. We develop the security engineering mindset throughout this book, but use the castle analogy here to help those new to security engineering ease into the subject with a vivid and more tangible example. The castle analogy allows us to illustrate security vulnerabilities, attacks, and defenses. The first step is to refresh our memory about castles, what they were, and how they were used.

The medieval period lasted from the 5th century to the 15th century. During that time, castles were fortified mansions built in Europe and the Middle East. They were designed to protect their occupants from both foreign and local invaders, and also served as a base for launching military campaigns.

Often, castles were strategically placed for financial rewards. For instance, the Kronborg Castle in Helsingor, Denmark (near Copenhagen), was well placed at the Oresund, one of the few shipping channels between the Baltic Sea and the North Sea. This allowed the lord of the castle to attack and thus tax all ships passing into or out of the Baltic Sea. In Chapter 4, we'll explore the flow of money in the IoT ecosystem, as well as the underworld of the Internet, known as the Dark Web. Just as castles occupied strategic locations in medieval times to control the flow of money, protection of critical points in an IoT deployment is essential to retain control over the economic benefit that an IoT system brings. Figure 2.2 illustrates a medieval castle, along with its various defenses.

Initially, castles took advantage of natural defenses, such as high ground, but over time, a science of castle building emerged. By the 12th century, newer castles relied on concentric defenses—defense in depth—with several stages of defense within each other that could all function at the same time to maximize the castle's firepower.

Newer castles—12th century to 15th century—often included a common set of defensive features[10]:

- **Moat:** The moat is a deep, broad ditch usually filled with water that is dug around the castle. Its purpose was to stop attack weapons such as siege towers from reaching the curtain wall and to prevent the walls from being undermined. Castle inhabitants crossed moats with the aid of draw bridges.
  - In computer systems, the moat is equated to the network security components that keep hackers away from your computer. Things such as firewalls and NAT* routers

---

* Network Address Translation

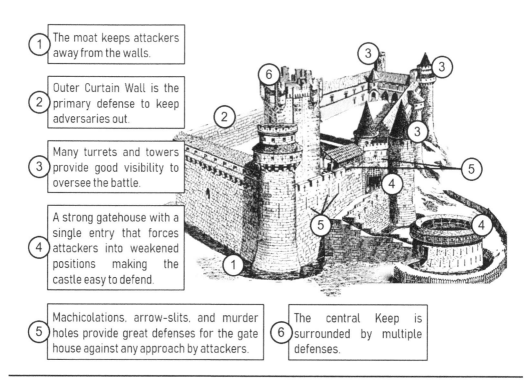

(1)	The moat keeps attackers away from the walls.
(2)	Outer Curtain Wall is the primary defense to keep adversaries out.
(3)	Many turrets and towers provide good visibility to oversee the battle.
(4)	A strong gatehouse with a single entry that forces attackers into weakened positions making the castle easy to defend.
(5)	Machicolations, arrow-slits, and murder holes provide great defenses for the gate house against any approach by attackers.
(6)	The central Keep is surrounded by multiple defenses.

Figure 2.2   Castle Attack Surfaces

ensure that a network adversary cannot even get near your system. That is, of course, the logical view of the system. There is also a physical view, which could be the data center with all the physical controls to keep unauthorized persons away from your systems. But especially in IoT systems, many of our computing resources are not locked in a data center and must rely on other physical controls to prevent theft and destruction. In Chapter 6 we talk about how the cloud is architected to protect itself, and in Chapter 7 we talk about network security and give some general guidance on how to protect your IoT devices from network attacks.

- **Outer curtain wall:** The "curtain wall" was the vast stone wall which wrapped around the outside of a castle. As you might imagine, it's called a "curtain" because it covered everything within. Concentric circles had at least two separate layers of curtain walls, one inside the other. The wall served to hide the inhabitants of the castle so their actions could not be seen, giving them some privacy, as well as provide the last barrier of defense against attacks.
  - In computer systems, the last barrier of defense is the operating system itself. Once an attacker gets into a computer system, it is the operating system's job to contain the attacker. The operating system provides various containment techniques to keep one program from viewing what another program is doing. It also provides layers of privilege, restricting the access that programs and users have to certain resources and capabilities. Just like the wall, containment is the primary purpose of the operating system. Besides just process containment to separate programs, some operating systems support containers, a more advanced type of software separation. Virtualization is another containment feature, which leverages hardware to provide even stronger

separation between programs on a computer system. Finally, some operating systems provide trusted execution environments (TEEs), which are special hardware-enforced separation environments that are used for highly sensitive operations, such as the use of cryptographic keys or processing very sensitive data. In Chapter 5 we discuss containment capabilities, and in Chapters 8 and 9 we discuss privacy—an extremely important and often overlooked topic in IoT systems.

- **Turrets, towers, and lookout points:** Castle towers were designed to give an unobstructed panorama of the countryside around a fortress, so lookouts could spot oncoming attackers.
  - Seeing an enemy coming from far away allows the defenders to mount a more effective defense. In the castle, height and an unobstructed view was paramount to an effective lookout. In computer systems, early warning systems, such as network and host intrusion detection software can identify the early warning signs of a network attack and alert a system administrator or, even better, shut down a network connection preventing the attack altogether. Logging can be used as another early defense system, but only if logging is tied to an anomaly detection system with proper alerting or automated incident response software. In Chapter 5, we discuss system monitoring and anomaly detection as critical parts of the system design.
- **The gatehouse:** Castle entrances were often their most vulnerable features. Doors could not be built out of reinforced stone like the castle walls. Hence, the entrance was vulnerable to battering rams and flaming onslaughts from any approaching attacker. The solution to this was the design of the gatehouse. This was an incredibly strong, fortified entrance building that made it really unattractive to attack this part of the castle. The gatehouse was filled with obstacles—multiple metal portcullis gateways; arrow-slits; many different gates, doors, and drawbridges; and even the infamous "murder holes" or machicolations.
  - Network ports on a computer system are akin to the gates and doors of the castle. The network is the weakest point of a computer system. For this reason, many defenses are built into the network stack and networked applications. Restrictions on the number of open listening ports in a computer system are similar to the design decision to restrict the number of gates on a castle. Too many gates, and there are too many ways an attacker can approach the castle. Virtual private network (VPN) software that requires authentication prior to establishing a network connection is similar to the gatehouse, with guards verifying all parties that want to enter the castle. Other defenses, including software such as iptables* and tcpdump,† which can be used to limit connections, and enforce encryption and integrity on outgoing data. TLS configurations can be set to verify that remote parties are trustworthy by checking for the possession of a trusted cryptographic key. In Chapter 7 we discuss communications and how to secure your network. Chapter 7 also covers some of the common IoT protocols and how to analyze them for security.
- **Arrow-slits:** Arrow-slits are open spaces on the level of each castle tower and the gatehouse—essentially windows—through which archers could fire on anyone nearing the

---

* Retrieved from https://netfilter.org/
† Retrieved from https://www.tcpdump.org/

castle. Of course, attackers could also aim at the defenders in the arrow-slits or attempt to send in flaming arrows that put the castle and its inhabitants in danger.

○ Often when talking about the castle, we focus on defensive properties, and this is fair because a lot of the castle is defensive. And likewise in our computer systems. But we are not without offensive capabilities. Host intrusion and antivirus software can be used to identify malware on a platform and kill it before it does too much damage—just like the archers using the arrow-slits to pick off the soldier bringing siege tools to the wall. Advanced anomaly detection mechanisms are very useful and cost effective in IoT systems because they can leverage the fact that IoT workloads are fairly regular, unlike our laptops. The things our laptops do are closely related to our personal interests and habits. Dave is a technical nerd who uses his computer for all sorts of things. Characterizing what his laptop was running over the last week might include the Civilization™ simulation game one Saturday, web browsing for academic articles on security the next day, and then, all through the week, writing code in VI, compiling and testing that code, and then running various word processing applications at weird hours of the night. IoT system workloads are very regular, and their processing and memory usage signatures can be measured. When an IoT system starts behaving in some way outside of this standard behavior, it is likely malware. This behavior can be reported, or the system can be rebooted from integrity protected software that guarantees the system is running free of viruses and malware. Chapter 6 will discuss how cloud analytics and threat detection services can be used to identify and kill invading malware in an IoT system.

- **Machicolations:** Machicolations is a term referring to a number of different defenses including overhanging holes, or platforms, built into castles. People defending a castle could hurl items—for example, rocks, arrows, boiling oil, or water—through these holes at an attacker below.

○ The machicolations are brutal for attackers. Being covered in boiling oil is bad enough, but then to have a flaming arrow hit you pretty much finishes your day! These are the advanced techniques that protect the castle, but they are really architectural in nature—the machicolations were really a way to design the castle so that other defenses (i.e., arrows and heavy rocks) had increased effectiveness. The machicolations funneled attackers to a position where they were weak and exposed, and it placed defenders in the best position to fend off or destroy the attackers. In IoT systems, there are ways to organize and build your system so that the defenses you have are more effective—for example, organizing the privileges of your software so that no one software component has full admin privileges; separating your features into micro-services running in containers so that they can be started only when they are necessary, thereby reducing the attack surface of the system; utilizing secure initialization only at boot time so that sensitive data is not generally available to anyone with admin rights over the file system, but the sensitive data is given to highly protected services running in TEEs. In Chapters 5 and 6 we talk about architectural designs for IoT and cloud systems, and how to make them more secure.

It goes without saying that it would be practically impossible to defend a castle without in-depth knowledge of its various fortifications or weaknesses. Likewise, it would be silly to attempt to design or build a castle without considering the different ways the finished fortress might be attacked. Throughout the remainder of this book, we present the background necessary to frame the IoT environment and threats, discuss the architecture and design principles required to build secure IoT systems, and walk you through some IoT system examples to share our security design experience and equip you to build and secure your IoT system.

2.3.1 Know Thine Enemy

"If you know the enemy and know yourself, you need not fear the result of a hundred battles"* is a famous quote from the brilliant Chinese general, Sun Tzu. A successful defender must understand the strengths and weaknesses of their castle, but they also must understand their adversary.

As a defender of IoT systems, we must ask the following: What types of cyber attackers exist? And what motivates them to strike? In his book, *Securing Systems: Applied Security Architecture and Threat Models*, Brook Schoenfield does a great job of outlining the attributes of different classes of cyber attackers. He presents a detailed analysis of the different capabilities, goals, methods, and risk tolerance that can be attributed to any class of cyber attacker today.[11] We briefly describe each attacker and their attributes in Table 2.1.

Table 2.1 IoT Attacker Profiles

Threat Agent	Goals	Risk Tolerance	Work Factor	Methods
Cybercriminals	Financial	Low	Low to medium	Known and/or proven
Nation states	Information and disruption	Very low	Extreme	Very sophisticated
Industrial spies	Financial, information, and disruption	Low	High to extreme	Sophisticated and unique
Hacktivists	Information, disruption, and fame	Medium to high	Low to medium	Known and/or proven, system admin errors, and social engineering
Security researchers	Financial (via industry reputation and/or fame)	High	Low to high	Sophisticated and unique
Malicious insider	Financial, information, and disruption	High	Low	None (usually already have easy access to assets)
Script Kiddies	Fun and/or fame	High to extreme	Very low to low	Known and/or proven

It may be helpful to explain a few of the entries in Table 2.1. The goals of an attacker are varied but often fall into only a few categories: financial, information, disruption, or fame. Financial goals include obtaining money, such as credits cards and crypto-currency, as well as gaining access to free computing resources or storage. An unprotected and poorly monitored IoT device may be a great place to store illicit videos or mine for Bitcoin®. Information goals include gaining access to trade secrets, intellectual property, computer code, personal information, or any other digital information that may be used to the advantage of the attacker. The

---

* Retrieved from http://www.history.com/topics/the-art-of-war

goal of disruption is to interfere with the operation of a nation, a company, an organization, or even a single individual. Typically, disruption has an underlying motivation such as revenge or the desire to create economic loss to another entity; the Dyn attack is classified as a disruption. Fame is a motivation for many attackers. This includes the ability to brag to others about the conquest of a particular company's computer network, or to present a paper at a security conference like Black Hat. Of course, some attackers prefer to remain anonymous.

The methods of attackers fall into three basic categories: social engineering, known and proven, and extreme and unique. Social engineering methods are ways an attacker preys on the behaviors of human beings to break into computer systems. This includes guessing a user's password to break into a system or trying simple passwords, such as *Password123*, or calling up a user and pretending to be someone important and convincing the user to give the attacker the password. Known and proven methods are attacks that are documented and have been used in the past, but the systems the attacker is targeting have not been updated appropriately to counter those known attacks. When copycat attackers used Mirai code from Github to attack IoT devices, they were using a known and proven attack method. Extreme and unique methods are attacks that are newly invented by researchers or zero-day vulnerabilities that are found in operating systems or other software. The level of sophistication and amount of work entailed in discovering and using new methods to attack a system in this way is high. However, these types of attacks do occur and, when uncovered, quickly turn into the known and proven type of attack methods.

In future chapters, as we analyze the security of IoT systems, we will refer back to Table 2.1 to reference different attackers, goals, and methods.

2.4 Attacking the IoT Castle

Now that we have established a basic concept of how we can use the castle analogy to represent the defensive and offensive concepts of a computer system, in this section we look a little more closely at IoT systems specifically. IoT systems often have a similar general architecture, even if they are used in widely different IoT ecosystems and highly varied industries. Then, armed with just this IoT architecture, we can map out a series of tables that highlight general areas of security weaknesses. In Chapter 4, we develop this IoT architecture in greater depth. Then, Chapters 5 and 6 immerse us in the design of IoT and cloud security architectures, respectively.

In Chapter 1 we introduced the IoT architecture and described it as the MGC architecture,[12] which is an acronym for the primary component types in an IoT system:

- **EM**bedded devices,
- **G**ateway devices, and
- **C**loud back end systems.

In any system architecture, the attack surface is the collection of interfaces through which data can be entered or exfiltrated from the system. It is worthy to note that by *system*, we refer to the complete set of components and interactions that make up the whole of a technological solution, and the system's entire lifecycle from creation to decommissioning. Said in a slightly different way, we can consider the attack surface as the various entry points that a prospective attacker could use to supply bad inputs to the system in order to modify its behavior or extract

Table 2.2 Attack Surface of a Generic IoT System

High-Level IoT Component	Description	Attack Surface
Embedded device	Dedicated computer system with a specific function within a larger system.	• Network communication protocols and interfaces (i.e., WiFi, Bluetooth®, ZigBee®, Z-Wave®) • Network-accessible software services • Operating systems • Update mechanism
Gateway	Connects the embedded devices in the field (home, farm, car, etc.) to a cloud environment	• Network communication protocols and interfaces (i.e., WiFi, Bluetooth, ZigBee, Z-Wave, TCP/IP, TLS) • Messaging protocols (i.e., MQTT, XMPP, OPC UA, HTTP/S) • Operating systems (i.e., Windows, Linux®) • System update mechanism • Administrator/management interfaces, APIs, or applications • BIOS
Cloud	• Cloud computing is a paradigm that is based on the principle of allowing local devices or clients access to shared Internet-hosted computing resources (e.g., computer networks, servers, storage, applications, and services). Popular cloud platforms are Amazon® Web Services IoT, Microsoft® Azure® IoT Hub, IBM Watson™ IoT platform, Google Cloud™ Platform. • Cloud back ends often host the analytics and administration applications that are used to make decisions or inferences from IoT devices or control gateway and IoT devices.	• Network communication protocols and interfaces (i.e., TCP/IP, TLS) • Messaging protocols (i.e., HTTP/S) • Network infrastructure components: Bastion Host, NAT, load balancer, firewalls • Administrator/management interfaces, APIs, or applications • Code deployment pipeline • Disk storage and databases • Backups

and steal the data the system intends to keep secret. Using our castle analogy, the attack surfaces are the points of entry by which invaders, weapons, or projectiles from weapons can enter the castle or compromise the state of its security or stored treasures. The process for analyzing an architecture to comprehend its weak points and design protections for its attack surface is fully described in Chapter 6. By their nature, IoT systems have many moving parts, and, as such, they usually have a wide and varied attack surface. Table 2.2 outlines a few such attack points for a generic IoT system, organized according to the MGC representation.

2.5 A Closer Look at IoT Attack Surfaces and Breach Consequences

What could go wrong if any of the aforementioned attack points were successfully compromised by an attacker? How might that impact the solution or its users? Those questions will be answered in greater detail in Chapter 5, but let us take a sneak peak for now. Table 2.3 outlines the attack surface name, description, and breach consequence for embedded devices. Table 2.4 accomplishes the same for gateways, while Table 2.5 addresses the cloud.

Table 2.3 Attack Surface of an Embedded Device in an IoT System

Attack Surface	Description	Breach Consequence
Network communication protocols and interfaces	Edge devices such as sensors or actuators must be able to send the data they record to a central system or receive control commands from such a system. This interaction is achieved through the utilization of wireless network communication protocols (i.e., WiFi, Bluetooth®, ZigBee®, or Z-Wave®).	An attacker who successfully intercepts network traffic to or from the edge device can control the device or poison the data sent to the central system. In both cases, the attack compromises the integrity of analytics or decision making based on data from the edge device. If attackers can control the device, they can take it offline—imagine a security camera—or hijack it for ulterior motives, such as a botnet.
Operating system (OS)	Edge devices require operating systems to manage the software and network resources that are required for the device to function.	Security bugs in the OS could lead to the compromise of the entire edge device as well as access to its network traffic.
Update mechanism	It's acclaimed best practice to design a means for updating deployed systems or components. This can happen over the network or in an offline mode.	If an attacker is able to supply a malicious update to the system, which is not properly validated by the system, the entire device can be compromised.
Network-accessible software or services	Includes, but not limited to, administrative web applications running on the device.	Compromise of data, device admin or user authentication credentials, and the device itself.

Table 2.4 Attack Surface of an IoT Gateway

Attack Surface	Description	Breach Consequence
Network communication protocols and interfaces	Gateways utilize wireless network communication protocols (i.e., WiFi, Bluetooth®, ZigBee®, or Z-Wave®) to communicate with sensors and actuators. TCP/IP is often used for communication with the cloud back end.	There are two avenues of attack—the gateway's communication with the sensors/actuators and its communication with the cloud, often over the Internet. In both cases, compromise of network traffic could affect the integrity and confidentiality of the entire system.
Messaging protocols	These protocols are used to define and handle the transfer of messages or data between the gateway and the embedded devices at the very edge of the system. Messaging protocols run over the network communication protocols.	Unauthorized access or modification of data and credentials used by the system.
Operating system	See entry in Table 2.3.	See entry in Table 2.3.
Update mechanism	See entry in Table 2.3.	See entry in Table 2.3.
Administrator/ management interfaces, APIs, or applications	Administrative applications or interfaces exposed (for external access) by the device.	Compromise of data, device admin, or user authentication credentials, and the device itself.
BIOS	Non-volatile firmware used to perform hardware initialization during the booting process (power-on startup).	Physical access to the gateway is usually required for a successful attack. But a successful attack easily grants full control of the gateway to the attacker.

Table 2.5 Attack Surface of the Cloud Component of an IoT System

Attack Surface	Description	Breach Consequence
Network communication protocols and interfaces	The cloud paradigm is entirely based on the concept of software applications and services offered over the Internet. Network connectivity is essential, it's the life source. Hence, it's always connected, and malicious traffic can always be expected.	• For hosted applications on shared or private cloud infrastructure, compromise of host systems, servers, data, or authentication credentials could lead to malicious control of the entire IoT ecosystem for that solution or service . . . cloud applications and infrastructure, gateway, and embedded devices. Sadly, attackers have deleted entire cloud applications, taking companies without good backups out of business overnight. • The stakes are even higher for cloud providers who host the applications on their shared cloud infrastructure. In such cases, compromise of the cloud provider causes mayhem for all companies and businesses that run on that provider. As a result, it takes a lot of skill and resources to provide cloud hosting services, and to do so securely.
Network infrastructure	This includes the networking equipment used to connect, protect, and access the different machines that make up the cloud architecture, such as firewalls, bastion hosts, switches, routers, load balancers, etc.	Same as above.
Messaging protocols	These protocols are used to define and handle the transfer of messages or data between the gateway and the embedded devices at the very edge of the system. Messaging protocols run over the network communication protocols.	Same as above.
Operating System	See entry in Table 2.3.	Same as above.
Update mechanism	See entry in Table 2.3.	Same as above.
Administrator/ management interfaces, APIs, or applications	Administrative applications or interfaces exposed (for external access) by the device.	Same as above.
BIOS	Non-volatile firmware used to perform hardware initialization during the booting process (power-on startup).	Physical access to the physical server that hosts the virtualized servers that cloud computing applications use. An attack here is not as easy to perpetrate or attempt as those previously described, but it is still a concern.

2.6 The Road Ahead

In this chapter, we described recent, rather worrisome attacks against IoT systems. We also established what should be considered the security mindset as that of defending your castle. Then, we explored the different attack points of a generic IoT system and equated those to the castle analogy. It is fair to say that there are many attack points, and they vary widely across different IoT deployments. In this chapter, we only covered the most basic, common types.

With such gloomy talk and concerns of tech attacks in the past, present, and future, what shall we do? Abandon technology altogether and return to the medieval period when life was much simpler? That is never going to happen. The technology cat is already out of the bag, and there is no going back. But we do need to guard our tech castle, or the future of technology is on a doomed trajectory leading to the repetition of past mistakes. But our future does not have to be filled with multiple October 21st events. Truly the responsibility falls on all of us to rethink how we design and build secure products that respond to and defend our IoT castles against aggressive adversaries, as well as preserve the privacy and build the kind of life we want in our society. You have a part to play. It begins with an understanding of the adversary, and the journey begins in the following chapter.

References

1. Schneier, B. (2016, November 8). "Lessons from the Dyn Attack. Schneier on Security." Retrieved from https://www.schneier.com/blog/archives/2016/11/lessons_from_th_5.html
2. The Internet Society. (1987). RFC 1034, Domain Names—Concepts and Facilities.
3. Paganini, P. (2016, September 5). "Linux/Mirai ELF, When Malware Is Recycled Could Be Still Dangerous." Security Affairs. Retrieved from http://securityaffairs.co/wordpress/50929/malware/linux-mirai-elf.html
4. Ransome, J. and Misra, A. (2013). *Core Software Security: Security at the Source.* Boca Raton (FL): CRC Press.
5. Smith, Ms. (2017, September 4). "465,000 Abbott Pacemakers Vulnerable to Hacking, Need a Firmware Fix." CSO from IDG, Retrieved from https://www.csoonline.com/article/3222068/hacking/465000-abbott-pacemakers-vulnerable-to-hacking-need-a-firmware-fix.html
6. Greenberg, A. (2015, July 24). "After Jeep Hack, Chrysler Recalls 1.4M Vehicles for Bug Fix." Wired. Retrieved from https://www.wired.com/2015/07/jeep-hack-chrysler-recalls-1-4m-vehicles-bug-fix/
7. Moffitt, T. (2016, October 10). "Source Code for Mirai IoT Malware Released." WebRoot. Retrieved from https://www.webroot.com/blog/2016/10/10/source-code-mirai-iot-malware-released/
8. Bugcrowd. (2018). "What Is Responsible Disclosure?" Bugcrowd. Retrieved from https://www.bugcrowd.com/resource/what-is-responsible-disclosure/
9. Janita. (2016, November 7). "DDoS Attack Halts Heating in Finland Amidst Winter." Metropolitan.fi. Retrieved from http://metropolitan.fi/entry/ddos-attack-halts-heating-in-finland-amidst-winter
10. Morris, E. (2018). "Medieval Castle Defence: Defending a Castle." Exploring Castles. Retrieved from https://www.exploring-castles.com/castle_designs/medieval_castle_defence/
11. Schoenfield, B. (2015). *Securing Systems: Applied Security Architecture and Threat Models.* Boca Raton (FL): CRC Press.
12. Horowitz, M. (2014, December 9). "Securing the Internet of Things." Retrieved from https://www.slideshare.net/Stanford_Engineering/mark-horowitz-stanford-engineering-securing-the-internet-of-things

Chapter 3

The IoT Security Economy

I rob banks because that's where the money is.

— Sutton's Law (Apocryphal bank robber's maxim)

3.1 A Toy Is Not a Plaything, It's a Tool for Cybercrime

What's the first toy you remember playing with as a child? Perhaps it was a teddy bear, a car, or if you are a memory savant, you might even remember your teether. Whichever it is, you probably remember its color, texture, and how it made you feel.

If you are a parent of young children, or intend to be, you have probably wondered about what toys you would like to get for your kids. Toys that would spark their imaginations, engage their senses, and help them learn how to interact with others. In fact, some of us are already the Santa Claus for our kids, grandkids, nieces, or nephews. What if we told you that it is possible to buy toys that can be remotely hacked and controlled by cybercriminals who desire to deploy malware on those toys as money-making machines? Yes, that is alarming, but it is also our very present reality.

Nowadays, consumer stores are increasingly filled with an array of "smart" toys. A smart toy uses onboard electronics to enable it to behave according to programmed patterns or alter its actions, depending upon environmental stimuli. Modern smart toys may support communications linking them with other smart toys, game consoles, or computers over a network to enhance the play experience for children.

Why is this a concern? Well, as we revealed in earlier chapters, any device that is connected to a network is at risk of a remote hack. In February 2018, ZDNet reported the use of a worm to mine or digitally create the Monero cryptocurrency via a network of hacked smart devices such as smartphones and TVs.[1] Later on, in December 2018, Anomali® Labs reported the discovery of a strain of malware called Linux® Rabbit that was deployed to mine Monero, a type of cryptocurrency, on hacked Internet of Things (IoT) devices such as edge gateways and routers.[2] Similarly, a hacked smart toy is at the mercy of cybercriminals, and such a toy can be used to mine cryptocurrency or steal confidential information. (We discuss cryptocurrency in more detail in Section 3.3.1.) The dangers here go beyond just cybercriminal activity on your child's smart toy. There is also the added concern of this activity possibly causing the toy to overheat as its processor crunches through complex algorithms required for cryptocurrency mining. This places any children playing with the toy at risk of physical harm.

Still feeling this might be more fiction than fact? An example of a smart toy that sparked national concern is "My Friend Cayla." In February 2017, the German government banned the use of this seemingly innocuous smart toy.[3] The blond, childlike doll connects to the Internet via Bluetooth and responds to children's queries by using a concealed microphone. According to complaints filed by consumer groups, Cayla recorded conversations, translated them to text, and shared that data with third parties.[4] If consumer groups were right, Cayla could very well be an example of a potent tool for nation state espionage. But even if that were not the case, a hacked Cayla could be used to spy on families, steal the identity of children,* and mine cryptocurrencies. In Chapter 8, we will explore more about personal privacy and how seemingly innocuous small leaks can turn into large privacy concerns.

As the saying goes, crime goes wherever the money is. To help us understand and appreciate the need for IoT security, it's helpful to learn more about the motivation behind most IoT attacks. In this chapter, we will explore the IoT economy and the impact the criminal pursuit of money has on the current and future state of IoT security. We shall learn more about the actors who seek to profit from poor IoT security—their capabilities and their modus operandi. We will also explore the challenges facing the greenest defenders—builders of consumer IoT products such as smart toys—whose inexperience with technology makes them a prime target for the bad guys.

3.2 Understanding the IoT Economy

What exactly do we mean by the *IoT economy*? An economy refers to the resources of a community (or country), including all producers, distributors, and consumers in that community who are working toward the generation of some goal, which usually implies the accumulation of wealth or acquisition of necessities for self-sustainment. The IoT economy encompasses the production, distribution, and consumption of data and compute capabilities with the goal being the fulfillment of an IoT system's mission. The resources of an IoT system that we ascribe to the IoT economy are:

- The computing resources—sensors, gateways, and cloud computers, including both hardware and software
- The communication channels and bandwidth necessary to interconnect the different computing resources
- The data collected, stored, and utilized within an IoT system by those computing resources

---

* Child identity theft: This usually happens when the child's Social Security number is stolen and used to commit fraud. Many school forms require personal and, sometimes, sensitive information such as social security information. If this information is accessible to the child, a hacked smart toy could potentially be used to request the information from the child.

For the IoT economy to thrive, its resources must be available to its producers and consumers so that they may use them to benefit their community. Just as in any economy, if the resources or products of the economy are disrupted, the economy fails. It doesn't take much to disrupt a single IoT node, as we will discuss further in Chapter 4. And cybercriminals won't focus on only a single node—if they can take over one node, they will likely do so to many nodes.

Cybercriminals are not necessarily interested in destroying your IoT economy, they are more interested in making their dark economy thrive. They do this by stealing your resources, supplanting your production, and taking over your distribution. In other words, they replace the activities of your IoT nodes with their own activities. It is helpful to understand where cybercriminals see value in the IoT economy, and what resources they target and for what purpose. Understanding this requires that we understand a bit about the cybercriminal's economy.

3.3 The Cybercriminal Economy

Broadly speaking, cybercrime is the recent evolution of criminal activity wherein computers and networks of computers are used to carry out activities that are against the law. The cybercriminal is no different from the normal criminal. They follow the money. But rather than allow themselves to be bogged down in the regulations and rules placed on legitimate businesses by society, cybercriminals choose to create businesses that steal resources from others and build money-making engines that leverage the vices and weaknesses of society in order to make money for themselves. And usually these "businesses" steal, exploit, and cheat in order to make themselves more comfortable and increase their profits, all at the expense of the weakest and most vulnerable in society. The line between a criminal and a legitimate business is the activities that they undertake. Some of the activities of cybercriminals include:

- Offering the sale and distribution of illegal contraband in violation of import restrictions or to escape tariffs
- The distribution and sale of illegal drugs using hidden websites (see the dark web later in this chapter)
- The distribution and sale of counterfeit prescription drugs targeting the elderly or less-fortunate in society who cannot afford adequate healthcare
- The collection, storage, and distribution of pornographic materials, especially the most illegal of such materials
- Human and sex trafficking, including the identification, tracking, and abduction of vulnerable persons, including children, to trap them into such activities

A different class of cybercriminal seeks the destabilization of another entity via the destruction of their assets. In these cases, indirect profit is often the motive (profit because the competitor is no longer in business), but at times profit isn't a motive at all. Nation state actors, cyberwarfare, and cyberterrorism fall into this category. The methods used by criminals often change, but their goals don't. Table 3.1 outlines the major types of cybercrime activity, the motives of the criminals, and their relevance to IoT systems.

As we will see when we explore privacy in Chapter 8, the IoT, with its far-reaching surveillance capabilities and big data analytics power, can be a potent force in society, and even more so if taken over by cybercriminals and turned to evil. But many of these activities are not beneficial to the cybercriminal unless they can be converted into some form of currency that is highly liquid, portable, and can easily be hidden. Cash has always been that preferred currency in the real world. Criminals have even spent considerable effort to counterfeit different currencies. But in the cyber world, cash is being quickly replaced by cryptocurrency. And as we will see in the next section, cybercriminals are also interested in creating or "mining" cryptocurrency from nothing merely by stealing resources from IoT systems.

Table 3.1 Types of Cybercrime and Their Relevance to IoT Systems

Cybercrime Activity	Motive	Relevance to IoT Systems
Illicit Sales: This involves the sale of goods (drugs, export-controlled items, pornography) and services (hacking, cryptocurrency mining, sex, child labor, slave trading) via hidden dark websites	• Profit	• IoT resources are typically not well protected and may be used as free storage devices to hold illegal data, child pornography, or other data that cybercriminals don't want easily traced back to them. • Cybercriminals may also use the compute resources on IoT devices to run websites to perform these services.
Internet Fraud: This involves misrepresentation of fact or unauthorized use of information. Examples are credit card fraud,[5] identity theft,[6] and the theft and sale of classified information.	• Profit via theft • Profit via extortion • Destabilization of an entity	• IoT places many more computers in and around us that contain sensitive data about us, such as our health, credit cards, email addresses, phone numbers, and user accounts. All the information a criminal needs to perform identity theft. • Some IoT devices, such as the Cayla doll, can be used to perform surveillance on us, acquiring data that we never placed in the IoT systems, but was overheard or visually captured from the environment.
Cyberextortion involves the use of computers or the Internet to demand money or other goods or services (such as sex) by threatening to inflict harm on a person or entity, or their property. Examples of these are: 1. Distributed denial-of-service (DDoS)[7] attacks where multiple computers are directed to send traffic to another computer or server, rendering it unable to process legitimate traffic 2. Ransomware,[8] which involves threatening to publish a user's data or perpetually block access to it using encryption.	• Profit via extortion	• In many situations, IoT devices are privy to sensitive and valuable information. For instance, web cameras, home automation, and home security systems could have access to sensitive video and audio feeds concerning people in the home. Likewise, IoT devices used to assist with parking or toll gates have access to payment information and control the usage of parking garages and roadways, respectively. • Whenever sensitive or valuable information is accessible, there is a risk that the wrong individuals could access that information for the purpose of extortion and blackmail. • In addition, since IoT devices are so numerous, hacked IoT devices can be (and have been) used to perpetrate DDoS attacks (see Chapter 2, Section 2.1).

(Continued on following page)

Table 3.1 Types of Cybercrime and Their Relevance to IoT Systems (*Cont'd*)

Cybercrime Activity	Motive	Relevance to IoT Systems
Cyberwarfare is the use of computers and computer networks by nation states in order to attack other nation states.	• Destabilization of an entity	• IoT devices are used to augment and improve the operational efficiency and reliability of critical infrastructure such as the power grid, road networks, and waste-disposal systems. • In an IoT-driven economy where critical infrastructure makes use of IoT, a successful IoT-based attack on the critical infrastructure of such a nation or any of its cities directly impacts the economy, stability, and well-being of the nation and its inhabitants. • This ability to harm people through insecure IoT devices makes IoT a potent tool for cyberwarfare. • As recently as 2017, the security firm Symantec™ discovered that hackers had successfully obtained operational access to the interfaces at certain US power companies. Those are the interfaces that engineers use to send actual commands to equipment such as circuit breakers, giving the hackers the ability to stop the flow of electricity into US homes and businesses.[9]
Cyberterrorism involves an individual or entity threatening or coercing a government or organization to advance a political or social agenda via launching cyberattacks on the victim's computers or data on those computers. Cyberterrorism is usually cyberextortion tied to political or social agendas, with the victims as organizations or governments, rather than individuals.	• Indirect profit via extortion • Destabilization of an entity	• Same as cyberextortion and cyberwarfare. Hacked IoT devices can be used to arm-twist individuals or governments that depend on those device and the information they contain.

3.4 Cryptocurrency 01100101[*]

In the introduction, we discussed the possibility of cybercriminals hacking IoT devices to mine cryptocurrency. Cryptocurrencies are fast becoming a major financial medium for cybercriminals. They are used as a mode of payment, a mechanism to create money, and a mechanism to launder money. But what is a cryptocurrency?

A cryptocurrency is a digital or virtual asset designed as a medium of exchange that uses strong cryptographic algorithms to secure financial transactions.[10] A defining feature of a cryptocurrency, and arguably its biggest allure, is its organic nature; it is not issued by any central authority or banking system, rendering it theoretically immune to government interference or manipulation. Since cryptocurrencies can be spent on the Internet without the use of a bank

---

[*] 01100101 is the binary representation of the decimal number 101.

account, they offer a convenient system for anonymous purchases, making money laundering possible. As such, they have become the primary mode of payment or value exchange for cyber-criminals and other cyberattackers.

Cryptocurrencies are decentralized systems based on blockchain technology, a distributed ledger that is managed by an unstructured network of computers.[11] The ledger is essentially a database of transactions, managed by a set of computers (nodes) on the Internet, with each node in the network maintaining a duplicate copy of the database. It is important for the nodes to be owned by many different people or entities to prevent one person from being able to manipulate the currency. Each network node also plays a part in verifying transactions processed by the blockchain.

The first blockchain-based cryptocurrency was Bitcoin®, and it remains the most popular and most valuable. Today, there are many other alternate cryptocurrencies with various functions and specialities. Some of these are clones of Bitcoin, whereas others are forks or new cryptocurrencies that split off from an already existing one.

3.4.1 Mining, Minting, and Verifying Transactions

Transactions in cryptocurrency are verified by nodes in the cryptocurrency's network that run an open-source* software program to check the transactions submitted by coin owners and then attach them to the blockchain ledger.† Attaching the transactions to the blockchain requires the nodes to perform a certain amount of work to solve a cryptographic hash problem. This cryptographic hash problem links the new transaction into the chain of transactions already in the blockchain. The people (nodes) that perform this service are called "miners," because they are also the people that can "mine" or create new coins in the cryptocurrency. Miners get paid a small transaction fee for clearing a transaction and adding it to the blockchain, but the real value for miners is in creating new coin.

The way that cryptocurrencies create the money or coins within their system is different than with traditional currencies. In traditional banking systems, a central organization is responsible for printing or minting money, which is then placed into circulation. With cryptocurrencies, a certain amount of currency is initially programmed into the distributed software that runs the crypto network. Then, new coins are released at a fixed rate with a "proof-of-work" problem attached; the first miner to solve the "proof-of-work" problem associated with the new coin gets to keep the coin.‡ This proof of work problem is a cryptographic hash problem just like the one used to verify transactions. This is why the verifiers are well positioned to also mint new coin, and why they are called miners. And the problems that need to be solved to create new coins are very compute intensive—they require a lot of computer resources and a lot of electricity.

People normally acquire cryptocurrencies by purchasing them directly using traditional currencies like the US dollar, British pound, Chinese yen, or any other fiat currency in common use. And cybercriminals often do this to launder money. But minting coin using stolen compute resources is becoming big cybercriminal business.

---

* Retrieved from https://opensource.com/resources/what-open-source
† Retrieved from https://bitcoin.org/en/faq#economy
‡ Retrieved from https://www.investopedia.com/tech/how-does-bitcoin-mining-work/

Cryptocurrencies have fluctuated wildly, especially in the last couple of years. This is common with new currencies, even when they are national currencies. Nations have many ways to stabilize their currencies, but such controls are not available for cryptocurrencies. However, there are some built-in controls. Cryptocurrencies are usually designed to gradually decrease production of the currency, with a known public cap on the amount of that currency that will ever be in circulation. This fact is thought to create a general stabilization toward an increase in the currency's value over time. However, there are many factors that create destabilizing influences in cryptocurrencies, such as attacks on the verification exchanges, limitations in market acceptance, defects found in the open-source software, fear of crypto weaknesses, the potential for government regulation, and no insurance against loss of being hacked, just to name a few of the major reasons. But just like IoT, the draw of a new technology and prospects for opening new business ventures abound, creating counter-stabilizing pressures.

3.4.2 The Draw of Crypto Mining

One can see the draw to become a cryptocurrency miner. It is like free money, or finding gold. However, it isn't really free. The "proof-of-work" required to verify transactions and to mint new coin requires a lot of computer-processing power. And a lot of computer processing power requires some powerful computers and a lot of electricity. And both of those cost real money.

Some miners pool resources, sharing their processing power over a network and then split the reward equally, according to the amount of work they contributed to the validation of a transaction or part of a transaction. But this still requires computers, electricity, and network bandwidth—none of which are free.

Additionally, crypto mining usually requires expensive, specialized hardware that consumes large amounts of electricity, which can offset the gains in cryptocurrency rewards for the average miner. But this is where the cybercriminal steps in. If they can steal computer resources and electricity from someone else, using essentially free resources to mine for cryptocurrency, then the coin they collect actually is free.

3.4.3 The Monero Cryptocurrency

In 2014, a cryptocurrency surfaced, which has since become a darling of the netherworld of the Internet. It's called Monero.

Monero uses an obfuscated public ledger, meaning anybody can broadcast or send transactions, but no outside observer can tell the source, amount, or destination.[*] Additionally, "stealth addresses" can be generated for each transaction, which makes it impossible to discover the actual destination address of a transaction by anyone else other than the sender and the receiver.

Where Bitcoin once reigned supreme, Monero has become the goto currency for purchases such as stolen credit cards, online account credentials, guns, and drugs. As mentioned earlier, the Linux Rabbit malware, which was found in IoT devices in December 2018, was designed

---

[*] Retrieved from https://www.getmonero.org/

to mine Monero. This gave hackers the computing power of thousands of devices without the costs associated with buying hardware, power, or Internet access.

Last but not least, Monero is designed to support efficient mining on consumer-grade hardware, which is usually found in IoT devices such as x86, x86-64, Arm®, and GPUs,[12] unlike most other cryptocurrencies, which require specialized ASICs. This makes Monero mining particularly desirable to the average miner, since expensive hardware is not required. It also makes IoT devices, which contain consumer-grade hardware, attractive targets for Monero-mining motivated cyberattacks.

3.5 Where Cybercriminals Go to Hide

As mentioned earlier, cybercriminals break the law to achieve their own ends, using computers and related tools. Unsurprisingly, most cybercriminals prefer to remain undetected by the legal entities that enforce the law. Cybercriminals who are caught by policing authorities and convicted by the judiciary can expect to be fined huge amounts of money, spend years in prison, and in some rarer cases in some countries, the death penalty is an option.

For as long as crime has existed, darkness has attracted criminals. In the dark, it's easier to remain anonymous and thus avoid detection. It is no different with cybercriminals. They also seek the cover of darkness to increase their chances of perpetrating crime without detection.

Table 3.2 Three Views of the Web: Surface Web, Deep Web, Dark Web

Surface Web	The surface web is the portion of the web that standard search engines and browsers can access, without the need for special configurations or tools. Most conventional usage of the web involves the surface web. Interestingly, the surface web, which most of us use, only includes 10% of the entire web.[14]

Deep Web	This refers to any web content that is not indexed by standard web search engines—i.e., not searchable by the public. It includes dynamic web pages, restricted websites (such as those that require a password or captcha verification to go beyond the public page), unlinked sites, private sites (such as those that require login credentials), non-HTML/-contextual/-scripted content, and limited-access networks. Many such pages are used for legitimate purposes and businesses.
	
	    Examples of legitimate content on the deep web include email accounts, bank accounts, databases that are private to any organization or entity, medical documents, and legal records.
	
	    Generally, content of the deep web can be located and accessed by a direct URL or IP address and may require a password or other security access to get past the public website page.

Dark Web	The dark web is a small portion of the deep web. **This is where cybercriminals go to hide.** It is a portion of the web that relies on darknets, which are portions of the Internet (network of computers supporting the World Wide web protocols) that are deliberately hidden from the public.
	
	    Darknets include friend-to-friend or peer-to-peer networks, as well as large, popular networks like Tor®, Freenet, and I2P™. These cover all those resources and services that wouldn't be normally accessible with a standard network configuration and so offer interesting possibilities for malicious actors to act partially or totally undetected by law enforcers.
	
	    The dark web operates with a high degree of anonymity. It hosts harmless activities and content, as well as criminal ones.
	
	    Access to the dark web requires specific software, configurations, or authorizations.

An example from the dark underground of crime, which occurred frequently in the not too distant past and still occurs today, is travelers using phone numbers to request illicit sex and drugs in cities or locations along their route. Those phone numbers, and sometimes residential addresses, are usually exchanged via hearsay among trusted confidants. To access the services that are being offered, a caller or visitor would have to supply a password, which is likely to be changed periodically.

With modern technology and the Internet, the philosophies of crime haven't changed, but the tools and methods have. The World Wide Web (or just web) actually has more layers than at first meets the eye. The web includes dark recesses where cybercriminals can hide.[13] Table 3.2 describes the three layers of the web.

3.6 Accessing the Dark Web with Tor

At the time of this writing, the Dark Web is primarily accessed through two networks—Tor (The Onion Routing project) and I2P (Invisible Internet Project). These networks are favored because they provide a much higher degree of anonymity for web surfers, in comparison with more conventional means of accessing the web. We will detail the workings of Tor here, while leaving the exploration of I2P* to the curious reader. Both work in a very similar fashion.

Tor consists of a virtual network and software client that enable its users to browse the web anonymously. It is an implementation of the onion routing protocol, which was first developed in the mid-1990s at the U.S. Naval Research Laboratory by employees Paul Syverson, Michael G. Reed, and David Goldschlag to protect U.S. intelligence communications online. It was further developed by the Defense Advanced Research Projects Agency (DARPA) and patented by the Navy in 1998.[15] This isn't the first time and almost certainly won't be the last time that a military project has made such an impact on the general public. Other famous examples of military research projects that have become major consumer tools include the Internet itself, and Global Positioning System (GPS) used for navigation.

To understand the premise for onion routing, it's necessary to understand the inherent privacy weaknesses of the Internet. Data packets that are used to transmit data over the Internet consist of two main parts: the data payload, which contains the information being sent, and the header information, used for routing the packet from the source to the appropriate destination. Even when the data payload of your communications is encrypted, skilled attackers using a technique called traffic analysis are still able to reveal a great deal about what you're doing and, possibly, what you're saying. That's because traffic analysis focuses on the header, which discloses source, destination, data size, timing, and so on. Encrypted communication connections such as that provided by the Transport Layer Security (TLS) protocol protect the data payload but not the routing information contained in the header.[16]

As a result, the basic problem for anyone who is privacy minded is that the recipient of your communications can deduce who sent it by looking at headers. So can authorized intermediaries such as Internet service providers, and sometimes unauthorized intermediaries as well. This is because the Internet Protocol (IP) address can often be traced to an organization, an individual, or Internet service providers who lease IP addresses to organizations and individuals.

---

* Retrieved from https://geti2p.net/en/about/intro

A very simple form of traffic analysis might involve sitting somewhere between the sender and recipient on the network and just looking at headers. Instead of taking a direct and predictable path from source to destination, onion routing reduces the risks of traffic analysis by distributing traffic over a randomly chosen path of relay points on the Internet, before delivering the traffic to its final destination.

In an onion-based network, messages are encrypted in layers, which are analogous to the layers of an onion. The encrypted data is transmitted through a series of network nodes or relays called onion routers, each of which "peels" away a single layer, uncovering the data's next destination. To get a message from point A to point B, the user's software or client (such as Tor) incrementally builds a circuit of encrypted connections through relays on the network. The circuit is extended one hop at a time, and each relay along the way knows only which relay gave it data and which relay it is giving data to. No individual relay ever knows the complete path that a data packet has taken. The client negotiates a separate set of encryption keys for each hop along the circuit to ensure that each hop can't trace these connections as they pass through.

The sender remains anonymous because each intermediary only knows the location of the immediately preceding and following nodes.

The Tor implementation of onion routing only works for Transport Control Protocol (TCP) streams and can be used by any application with SOCKS support. I2P supports both TCP and UDP protocols. For efficiency, the Tor software uses the same circuit for connections that happen within the same ten minutes or so. Later requests are given a new circuit, to keep people from linking your earlier actions to new ones.[16]

Figures 3.1, 3.2, and 3.3. depict the functioning of Tor.

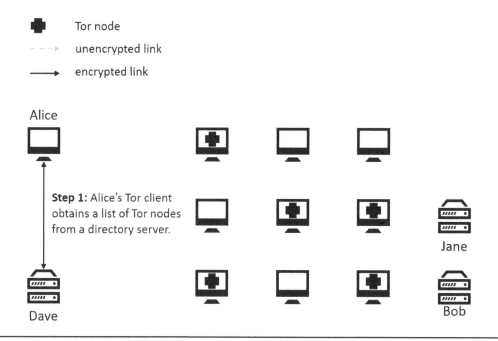

Figure 3.1 How Tor Works—Part 1 (*Source:* Adapted from https://www.torproject.org/about/overview.html.en [CC BY 3.0].)

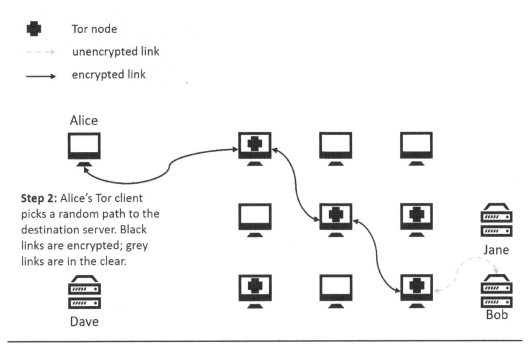

Figure 3.2   How Tor Works—Part 2 (*Source:* Adapted from https://www.torproject.org/about/
overview.html.en [CC BY 3.0].)

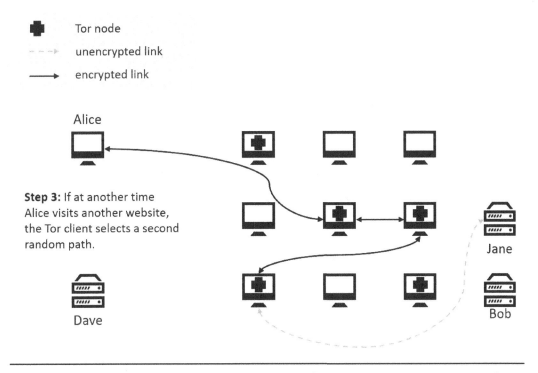

Figure 3.3   How Tor Works—Part 3 (*Source:* Adapted from https://www.torproject.org/about/
overview.html.en [CC BY 3.0].)

Tor does not provide protection against end-to-end timing attacks. If an attacker can watch the traffic coming out of the source computer, and also the traffic arriving at the chosen destination, statistical analysis can be used to discover that they are part of the same circuit. This is an inherent weakness of onion routing. You can check the listed references for more details.[17,18]

3.7 Money Money Money . . . Making Bank on the Dark Web

Not only do tools such as Tor allow users to access the web anonymously; they can also enable users to host web services anonymously. Once again, one of the most popular tools used to host web application services anonymously is provided by Tor, and it is called the Tor Onion Services Protocol.* This technology makes it possible to anonymously host web services via the

Table 3.3 Illegal Trading and Web-Based Onion Services in February 2016[20]

Category	Description	% of Total	% of Active
Violence	Murder for hire and instructions for violent attacks.	0.3	0.6
Arms	Weapon trades.	0.8	1.5
Illegal social	Forums, boards, and social networks sharing illegal information.	1.2	2.4
Illegal hacking	**Hacking for hire, such as the distribution and operation of security exploits that take advantage of zero day (newly discovered) security vulnerabilities, distributed denial-of-service (DDoS) tools and capabilities, botnets as a service, etc.**	**1.8**	**3.5**
Illegal links	Links to illegal content on the dark web.	2.3	4.3
Illegal pornography	**Pornography involving children, violence, animals, or sexual material that was obtained without consent.** Hacked IoT devices, such as cameras and microphones for home security or home automation systems, can be used to obtain images, videos, and audio recordings of people without their consent.	**2.3**	**4.5**
Extremism	Extremist ideologies including text, support for violence, how-to guides for terrorist attacks, and forums.	2.7	5.1
Ilegal other	Miscellaneous illegal goods such as fake or stolen identities.	3.8	7.3
Illegal finance	**Stolen user accounts (usernames or passwords), stolen credit card information, money laundering, counterfeit physical money, etc.**	**6.3**	**12**
Illegal drugs	Illegal drugs, including prescription medicine obtained illegally.	8.1	15.5
Non-illegal + unknown	Includes political or ideological content, secure information repositories, other legitimate services, and unknown/uncategorized content.	22.6	43.2
Totals			
Illegal total	Total number of illlegal sites.	**29.7**	**56.8**
Inactive	Inaccessible or no visible content, placeholder content.	47.7	
Active	Total number of active, reachable sites.	**52.3**	

* Retrieved from https://www.torproject.org/docs/onion-services.html.en

Tor network, thereby hiding the identity of the individuals or entities who author or deploy such services. Their locations are kept secret, while still enabling them to offer various kinds of Internet services, such as websites, webmail, or an instant messaging server.

Anonymous web-hosting capabilities have led to the proliferation of dark web marketplaces that specialize in the buying and selling of illegal merchandise. In such ecosystems, the online marketplace is hosted anonymously, accessed anonymously, and the mode of payment is an anonymized cryptocurrency, such as Bitcoin or Monero. One of the foremost dark web marketplaces was Silk Road.[19] It was hosted via Tor and made use of Bitcoins for transactions.

Table 3.3 outlines some illegal items that are routinely traded on the dark web, based on research from 2016. The items that are most relevant to IoT systems are emboldened.

3.8 Challenges in the Regular IoT Economy: Out of the Dark, and into Naïvety

Let's step of the dark for now. Yes, you are probably sighing in relief. It is important to know what goes on in the dark web, and it can be quite intriguing. But after a while, reading and thinking about all those illegal motives and desires can get pretty heavy.

So we'll step back into the light, or at least the legal world, since legal does not always mean bright light. For you see, naivety and silliness often dwell in the light. I'll give you an example. Consider a bus stop on a poorly lit road which is in a poorer part of town. On such a road, at any time in the night, very few people will leave their bags unattended, even if they wish to throw a banana peel into the trash can that is just a few feet away.

On the other hand, let's consider a green lush park in an affluent suburb. It's Saturday around 10:00 a.m., and the kids are running amok, squealing in laughter. Their parents and chaperones watch while sitting or taking leisurely strolls around the kids. It's a fun, relaxed morning as everyone enjoys the warmth of the sun. In such an environment, it will be unsurprising to see signs such as, "Do not litter." Hence, we can imagine an upstanding member of society who finds himself or herself seated a few hundred feet away from the nearest trash can. As our friend is diligently determined not to litter, she or he strides over to the nearest bin to dispose of a granola bar wrapper. Unfortunately, our diligent friend's bag and laptop computer are left unattended, just for about two minutes. Many times nothing happens, sometimes something does happen. The culprit? Naivety in the bright light of day.

This is no different in the relatively new and rapidly evolving world of IoT systems manufacturing and development. We began this chapter by examining the present danger in connected toys that are insecurely designed. Such insecurely designed IoT devices have been used to steal confidential information, spy on households, and even commit blackmail. An evident problem with smart toys is that toy companies are not technology companies. Toy companies do not have the background or expertise in designing and building secure technology. It's quite rare for such a company to have a cybersecurity department or the budget allocation that allows them to sink cash into cybersecurity. As a result, the company lacks an understanding of the security threats posed by their connected devices and certainly is not versed in the mitigation of those threats.

Adding to this problem of lack of expertise, toy companies have tried to keep the costs of children's toys as low as possible in order to aid sales. Customers have grown accustomed to the concept of affordable toys for their children. A parent can accept the extra cost for a smart, network-connected doll, but how much extra? Secure Internet devices can cost from $80 to $100. The new features enabled through an Internet-connected toy might be worth some extra cost, but few parents are willing to pay $100 or more for a doll. This creates a conundrum for the manufacturer who finds themselves trying to create a tech upgrade on the cheap. A well-designed, secure IoT system or device requires sound investment, in people and technology.

As you have probably guessed, this situation is not restricted to toy companies. The increasing ubiquity of Internet-connected devices or appliances along with aggressive consumerization of such IoT technology have created an attractive market that is too alluring for many. From fridges to thermostats, door locks, speakers, coffee makers, and more, manufacturers keep finding ways to connect the unconnected for "the greater good." In this chapter, we will not get bogged down in the quagmire of debating whether connecting the unconnected is for the greater good or not, although we will examine IoT connectivity and the difficulty of securing connections in greater detail in Chapter 7, and then explore some interesting concerns and ethical conflicts in Chapter 8, as we discuss privacy. For now though, it is safe to say that many IoT systems manufacturers do not have the appropriate expertise and investment in IoT security.

Perhaps it is wrong to lay all the blame on system designers and manufacturers. As the saying goes: "You've gotta give the people what they want." Sadly, many customers are yet to grasp the security threats inherent in connected devices. Hence, we are yet to have the majority of IoT-device consumers demanding good security as a feature of their smart toys, fridges, garage doors, and landscape-watering systems. For the manufacturer who is trying to create a product on a tight budget, if the customer doesn't seem to care about security, good security becomes a nice-to-have feature. They do the basic security features (if that is even anything at all), and move on. Thus, the doll saying "Hello!" in six different English accents becomes more important than the doll automatically destroying sensitive information that the manufacturer should not retain, and securely storing or transmitting only the smallest bits of information that is required for the business case and customer-usage model.

Lastly, as we alluded to in Table 3.1, we are entering into the age of the IoT-driven economy, in which critical infrastructure such as the power grid, road networks, waste disposal, and water-distribution systems are powered with IoT. Many technologically advanced nations are taking major strides in that direction, and as was also mentioned in Table 3.1, even the United States has experienced cyberattacks on their IoT-driven critical infrastructure. In a world or nation where IoT drives critical infrastructure, the security of IoT directly influences the security of the broad economy. An attack on such critical systems is a direct attack on all people who depend on the systems.

3.9 Why You Should Care

In this chapter, we have taken a necessary walk on the dark side, exploring the motives of cybercriminals. We have also explored how those criminal desires and motives are brought to bear on all of us. The objective? An understanding and appreciation of the magnitude of what is stacked against us as we seek to build and use IoT systems for good. As we have discovered, cybercriminals aren't driven by technical flights of fancy, so our problems cannot be solved by simple awareness programs or "one-size-fits-all" security solutions. Instead, cybercriminality is fuelled by the same human ills of greed and selfishness that fuel all crime. It is also fair to add that cybercriminality is influenced by worldly ills of poverty and hunger. For instance, there are countries with great universities, but poor or unstable economies. In such an environment, there's a greater chance that a capable graduate of Computer Science who has a hard time finding work, finds employment or an entrepreneurial outlet in the development and distribution of malware or other criminal endeavors.

Since cybercrime is driven by human motives, it is dynamic and evolving. This means that defenders who build and deploy IoT systems must be just as nimble and shrewd in their methods of defense.

The rest of this book is written to equip you with the skills and knowledge required to design and build secure IoT systems that are resilient against cybercriminals.

References

1. Osbourne, S. (2018, February 6). "ADB.Miner Worm Is Rapidly Spreading Across Android Devices." Retrieved from https://www.zdnet.com/article/adb-miner-worm-is-rapidly-spreading-across-android-devices/
2. Anomali Labs. (2018, December 6). "Pulling Linux Rabbit/Rabbot Malware Out of a Hat." Retrieved from https://www.anomali.com/blog/pulling-linux-rabbit-rabbot-malware-out-of-a-hat
3. Solomon, F. (2017, February). "Germany Is Telling Parents to Destroy Dolls that Might Be Spying on Their Children." Retrieved from http://fortune.com/2017/02/20/germany-cayla-doll-privacy-surveillance/
4. Criss, D. (2016, December 8). "These Dolls Are Spying on Your Kids, Consumer Groups Say." Retrieved from https://www.cnn.com/2016/12/08/health/cayla-ique-ftc-complaint-trnd/index.html
5. Consumer Action. "Credit Card Fraud—Consumer Action" (PDF). Consumer Action. Retrieved 2017-11-28.
6. Hoofnagle, C. (2007, March 13). "Identity Theft: Making the Known Unknowns Known." Social Science Research Network.
7. Zargar, S., Joshi, J., and Tipper, D. (2013, November). "A Survey of Defense Mechanisms Against Distributed Denial of Service (DDoS) Flooding Attacks." IEEE Communications Surveys and Tutorials, pp. 2046–2069. Retrieved from http://d-scholarship.pitt.edu/19225/1/FinalVersion.pdf
8. Young, A. and Yung, M. (1996). "Cryptovirology: Extortion-Based Security Threats and Countermeasures." IEEE Symposium on Security and Privacy, pp. 129–140. Retrieved from https://ieeexplore.ieee.org/document/502676
9. Greenberg, A. (20017, September 6)."Hackers Gain Direct Access to US Power Grid Controls." Retrieved from https://www.wired.com/story/hackers-gain-switch-flipping-access-to-us-power-systems/
10. Greenberg, A. (2011, April 20). "Crypto Currency." Retrieved from https://www.forbes.com/forbes/2011/0509/technology-psilocybin-bitcoins-gavin-andresen-crypto-currency.html
11. The Economist. (2016, October 31)."The Great Chain of Being Sure About Things." Retrieved from https://www.economist.com/briefing/2015/10/31/the-great-chain-of-being-sure-about-things
12. Courtois, N. and Mercer, R. (2017). "Stealth Address and Key Management Techniques in Blockchain Systems." In *Proceedings of the 3rd International Conference on Information Systems Security and Privacy—Volume 1: ICISSP.*
13. "What Is the Difference between the Web and the Internet?" Retrieved from: https://www.w3.org/Help/#webinternet
14. Pinkhattech. (2017, December 4). "What Is the Difference between the Surface Web, the Deep Web and the Dark Web?" Retrieved from https://www.pinkhattech.com/2017/12/04/what-is-the-difference-between-the-surface-web-the-deep-web-and-the-dark-web/
15. Syverson, P., Goldschlag, D., and Reed, M. (1997). "Anonymous Connections and Onion Routing." Retrieved from https://ieeexplore.ieee.org/abstract/document/601314
16. Tor. "Tor: Overview." Retrieved from https://www.torproject.org/about/overview
17. Soltani, R., Goeckel, D., Towsley, D., and Houmansadr, A. (2017, November 27). "Towards Provably Invisible Network Flow Fingerprints." Retrieved from https://arxiv.org/abs/1711.10079

18. Bangeman, E. (2007, August 30). "Security Researcher Stumbles Across Embassy E-Mail Log-Ins." Retrieved from https://arstechnica.com/information-technology/2007/08/security-researcher-stumbles-across-embassy-e-mail-log-ins/

19. Gayathri, A. (2011, June 11). "From Marijuana to LSD, Now Illegal Drugs Delivered on Your Doorstep." Retrieved from https://www.ibtimes.com/marijuana-lsd-now-illegal-drugs-delivered-your-doorstep-290021

20. Moore, D. (2016, February 1). "Cryptopolitik and the Darknet." Retrieved from https://www.tandfonline.com/doi/full/10.1080/00396338.2016.1142085

Part Two

Chapter 4

Architecting IoT Systems
That Scale Securely

*Which of you, wishing to build a tower, does not first sit down
and count the cost to see if he has the resources to complete it?*

— Luke 14:28

4.1 The IoT System Architecture

System architecture is the organization of elements in a system to address the needs of the business, the environment, operational constraints, and system users. It is important to outline a system architecture for IoT so that as we explore different elements of an IoT system and how they operate, we can refer to a common picture and common understanding of the system itself. However, as identified by the IEEE,[1] there is no widely agreed upon IoT architecture, and, in fact, the terminology used by different groups and companies across the IoT field can be diverse, leading to confusion. An interesting survey of thirteen different academic papers on IoT reference models identified six different types of architectural models, with each model having between three and seven layers, and using more than nine different naming schemes for these layers.[2] Although this appears bewildering at first, when one remembers that divergence in models and terminology is not new in the realm of evolving information and communication technology, we can look past this array of opinions to the crux of what these IoT models are communicating to us.

Various industries and standards groups have developed and proliferated system architectures in order to better communicate the issues and requirements relating to systems for which they are concerned. Anyone working in international standards for telecommunications has likely been exposed to the seven-layer Open Systems Interconnection (OSI) Reference Model for a communications stack.[3] However, different industries and different groups often adjust terms and modify the architecture to suit their particular audience and needs.

Figure 4.1 shows some stacks side by side to highlight these differences in terminology. The Internet Engineering Task Force (IETF) has popularized the TCP/IP Reference Model,

45

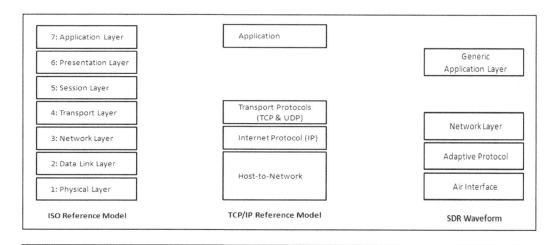

Figure 4.1   The Seven-Layer OSI Reference Model, the IETF TCP/IP Reference Model, and an Example Software-Defined Radio (SDR) Waveform Stack

which uses an abbreviated stack model, having only five of the seven layers identified in the OSI model. In an altogether different environment—that is, the software-defined radio (SDR) community—a completely different set of terminology developed to describe elements of a communication stack, which is referred to as a "waveform" not a stack. A waveform refers to the software implementation of a radio system, which includes the Air Interface (the physical layer from the OSI Reference Model). The term waveform denotes the actions occurring in a software-defined radio's physical layer, which modulates and demodulates the electromagnetic energy to create specific radio waves—it "forms" the radio waves. But the term "waveform" encompasses more than just the generation of electromagnetic waves of the lowest layers of the communications stack; it includes the data link and the network layer as well, which blurs the commonly understood lines between communication layers and confuses those not familiar with the technology-specific terminology.[4,5] While at first blush, the different terminologies create confusion, in reality, they are not very different at all. As one studies the different communication stack architectures in Figure 4.1, the architectures themselves communicate the elements and ideas that are most important to that architectural community.

The Internet of Things (IoT) community is no different from any other architectural community. The primary difference is that IoT spans many different industries and ecosystems, and has a diverse set of concerns. These different industries already have architectures and vocabularies that communicate their concerns, so it is not surprising that as these communities adopt IoT systems, they morph their existing terminology into IoT. For example, industrial manufacturing has a complete architecture and terminology around the Purdue Reference Model, which incorporates Operational Technology (OT) for control of physical manufacturing systems. As these industries adopt IoT architectures, they must evolve their existing systems into the IoT model, and many of the terms and structures remain in the new hybrid descriptive model.

Cisco Systems, in an attempt to create a unifying IoT architectural model, created a detailed seven-layer model (shown in Figure 4.2) and published their interpretation of what an architectural model for IoT should look like.[6] Their perspective was based on data communications, which matches their market perspective as a communications and Internet router company.

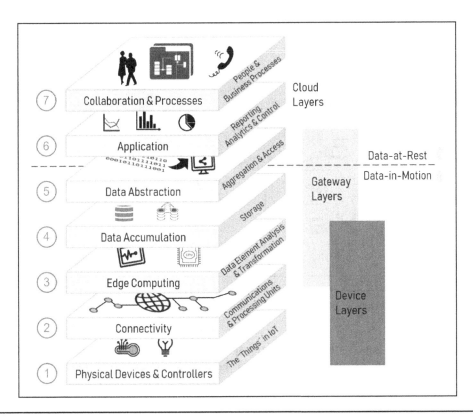

Figure 4.2 Cisco Systems IoT Reference Model (*Source:* Adapted from http://cdn.iotwf.com/resources/72/IoT_Reference_Model_04_June_2014.pdf)

Their model defines layers based upon where data is generated, transmitted, processed, and stored. They make no claims on the physical positioning of these layers and functions, and instead specifically identify that there can be "many, many exceptions."[6] However, almost no one else in the industry has adopted as deep a model with so many layers. Perhaps due to Cisco's communications background, they were too heavily influenced by the seven-layer OSI Reference Model and felt that any IoT architectural model should also have seven layers.

The trouble with many of the published architectural models is the attempt to combine both a physical rendition of the architecture and a data model of the architecture. The physical rendition of the architecture deals with where the end nodes, routers, gateways, and servers are positioned; the data model of the architecture deals with where the data is captured, formatted, combined, analyzed, and stored. Combining these two models creates confusing terms and awkward layers.

As researchers and companies strive to find the best descriptive words and the right number of layers to describe IoT architectures, most of the industry has settled on a three-layer model.[7] We introduced you to this as the MGC architecture in Chapters 1 and 2, which neatly breaks down into the devices layer (also referred to as the edge layer, perception layer, or physical layer), the gateway layer (also the fog, or network layer), and the cloud layer (also called the application & business layer, server & storage layer, or application layer). In Figure 4.2, we have added the three MGC layers next to Cisco's seven-layer model to show how they compare. The three-layer model leans heavily toward a pure physical rendition model. And although the

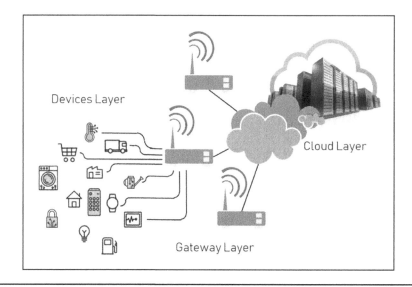

Figure 4.3   The IoT Device-Cloud-Gateway System Architecture

data model is loosely intertwined by assigning specific data and communications services to a specific layer, it is acknowledged that these data services may float between layers, depending on the industry and the specific installation needs.

So for the remainder of this book, an IoT system architecture is considered to be composed of three layers, as shown in Figure 4.3: a devices layer made up of sensors, actuators, and data collection devices; the gateway layer, which is responsible for data consolidation and transformation; and a cloud layer for performing data analysis, storage, and business value extraction.[8] This architecture is designed to maximize the benefits from IoT systems at the lowest possible cost. And the primary mechanism used to extract value from IoT systems is the use of large-scale analysis of the consolidated data from IoT devices. This data analysis enables improved business decisions, better logistics, trend recognition, and creation of new services. Let's examine these layers in more detail and provide specific definitions to the services and functions they supply to an IoT architecture.

4.1.1 The Cloud Layer

The cloud is the not-so-new paradigm of massive server farms accessible globally over the Internet, with multiple "As-A-Service" models for access[9]—such as Platform-as-a-Service (PaaS), Infrastructure-as-a-Service (IaaS), Software-as-a-Service (SaaS), and Function-as-a-Service (FaaS).* We cover the cloud in detail in Chapter 6, so this section will just briefly review

---

* Function-as-a-Service (FaaS) is a newer service model being explored by cloud providers and is easiest to explain by contrasting it with the other types of service models. With Platform-as-a-Service (PaaS), you get an entire hosting environment, including development tools; you build your application in the provider's environment and deploy it on their infrastructure. With Infrastructure-as-a-Service (IaaS), the provider gives you a platform that includes an operating system, and you deploy whatever applications you like; the environment is usually a virtualized platform, but you get an entire virtual

the high points essential for our current discussion. In IoT, the cloud refers to either a private data center instance; a public cloud offering such as Amazon®, Google Cloud Platform™, or Microsoft® Azure®, or a combination of these. From our IoT model in Figure 4.3, the IoT cloud layer represents the back-end services required to set up, manage, operate, and extract business value from an IoT system. In the cloud layer, sophisticated business analytic engines, artificial intelligence (AI), and web applications are used to process and massage the data collected by an IoT system and use that processed data to present sophisticated models of the system to operators and users. These models allow the IoT system to be managed or controlled with a greater degree of precision, thus extracting more business value and/or reducing operating costs. The models may even open up completely new business areas by exposing data that was previously hidden or too diverse and complex to collate into an understandable whole. With AI, these back-end cloud services may even run the IoT system almost or entirely autonomously.

The power of the cloud is the ability to, nearly seamlessly, reorganize massive computing resources to provide data services (e.g., storage, compute, and communication services) to multiple parties, and effectively share those resources on demand to customers. This model reduces the overall number of resources the cloud must actually contain, since every customer, application, or process does not need resources every minute of every day. This sharing of resources becomes more efficient as the number of independent usages grows, since the opportunity for downtime (e.g., lulls in the demand for services) increases as the number of services grows. This is, in fact, the push behind FaaS and is creating the need for ever-decreasing slices of a computing system that can be bound to a single customer.

The cloud is essential to an IoT system because it houses centralized applications and contains the data model of the IoT system, making those applications accessible to its users wherever they are, and is wholly independent of the actual location of the IoT devices themselves.

4.1.2 The Gateway Layer

The gateway layer is an intermediate layer in the IoT architecture and is concerned with data handling, communications, and control. Gateways are communication and processing hubs that directly service the devices layer. Gateways provide data relay services back to the cloud, offloading the cloud of some of the time-critical processing required to properly manage control loops and provide network and data security to devices. These responsibilities help form what a gateway will look like. Figure 4.4 shows a picture of an actual IoT gateway.

Because of the varied services provided by IoT gateways, and the fact that not all gateways need to provide all of these services, gateways will look different in different IoT systems. But the common factor across all gateways is communication to devices and communication up to the cloud. Therefore, all gateways support a large number of narrow-bandwidth, usually wireless, communications channels to talk with devices, and usually one wired connection to communicate up to the cloud. There are many instances, however, where a wired connection

---

server dedicated to your use. With FaaS, you share a virtual server with other customers, and you deploy a single function, or application, usually built into a container. With each service model, the customer receives a smaller and smaller chunk of computing resources, and the cloud provider has greater and greater opportunities to economize the use of actual physical resources. The different types of cloud services are discussed in Chapter 6.

Figure 4.4   Advantech UTX-3115 Fanless IoT Gateway Supporting Dual Ethernet, WiFi/3G/LTE, USB, and Serial Connections for Devices and Cloud Communications (*Source:* https://www. advantech.com/products/bda911fe-28bc-4171-aed3-67f76f6a12c8/utx-3115/mod_fa00d5cd-7d2b-430b-8983-c232bfb9f315)

to the cloud is not possible: large-scale agriculture deployments, forestry monitoring stations, and remote oceanographic stations, to name just a few of the extremes. In these cases, the gateway will necessarily include a long-haul communication system to reach the back-end systems and the cloud, which could be a 3G connection or even a satellite data service for very remote stations.

Gateways also have higher processing power and, therefore, higher power requirements. When these power requirements cannot be met by line power, for example, by tying into the power supplied to a street light that shares the same pole as the gateway in a smart city, then the power must be met by some other means—for example, a solar array and battery backup.

Gateways are critical to most large-scale IoT systems, but it is easy to imagine that, in some smaller IoT deployments, gateways may not be needed at all or may disappear into the communication systems that are connecting the devices to the cloud. Later in this chapter, we will discuss why gateways are so necessary and some of the IoT requirements that drive their inclusion in most IoT systems.

4.1.3 The Devices Layer

The devices layer of an IoT system is composed of the edge nodes that interact with the physical environment, either by mechanically or electromagnetically changing physical systems or by collecting data. The edge nodes that collect data are called sensors, and the edge nodes that interact with physical systems are called actuators. The devices layer is sometimes referred to as the *perception layer,* since it perceives or sees the environment through sensors and can manipulate the environment through actuators.

When referring to edge devices, by implication, we include the device itself, a microcontroller unit (MCU), and accompanying software to run the device; these parts together create a small system that can perform some rudimentary functions and communicate with other parts of the IoT system. This, of course, is an elementary view of devices. IoT devices range from such "dumb" devices that perform a single action, to more complex devices that are composed of multiple sensors and actuators. There is an effort in many areas of IoT to create "smart devices" that can directly use their own sensor data, as well as sensor data received from other nearby devices, to determine how the device itself will interact with the environment without any (or minimal) outside control or oversight. Many of these devices are too small to even contain a battery and, thus, are powered through energy harvesting.

What Is Energy Harvesting?

Energy harvesting, or energy scavenging, is the process by which a system or device collects energy from the environment around it. Solar and wind energy collection are familiar to many of us and are becoming popular for large-scale energy harvesting projects. And these are viable harvesting techniques for many IoT devices. However, there are also other harvesting techniques that operate on a smaller scale that are often used for IoT devices. These include vibration and kinetic energy conversion, thermal energy, and wireless energy extraction.

Vibration and kinetic energy harvesting translate some type of movement into electrical energy. Piezoelectric devices are used to accomplish this and can usually generate a few microvolts of electricity. These types of devices are popular in IoT wearables.

Thermal energy collection uses heat sinks attached to a thermoelectric generator (TEG) to turn heat into electricity. TEGs are made of a semiconductor material that generates electricity as heat moves through it from a hotter side to a cooler side.[10] TEGs do not produce much electricity and are a bit expensive but are useful in areas where heat generation (e.g., motors, industrial equipment, etc.) is expected.

Wireless energy extraction generators (WEG) uses RF radio waves to generate electricity. RFID tags are a common usage of this technology; this is an example of purposely directed RF energy to power a device. But it is also possible to use antennas to capture stray radio waves and use that energy to power a device. Some claim RF energy WEGs can provide almost a 70% power conversion capability,[11] making them very efficient for IoT devices, but other researchers report a maximum of only 30%. And in areas where lots of stray radio waves are expected, for instance, where cellular communication is prevalent, this can be a good source of energy for devices.

All of these forms of energy harvesting can be used to collect energy in capacitors or super capacitors, which can then charge batteries or be used directly to power microcontroller devices.

With all these differences, it is quite impossible to do justice to all the possible edge devices in all IoT systems due to the sheer magnitude of possibilities and combinations of functions. It would take an entire book to catalogue and characterize them all. However, edge devices can be easily classified into three basic categories, according to an IEEE report[1]:

- **Tags,** which are devices that contain an identifier and transmit that identifier when queried
- **Sensors,** * or edge data collection devices, which translate physical properties of the environment into a digital representation for reporting, sharing and consumption
- **Actuators,** or electro-physical devices, which can interact with a physical system

---

* The IEEE[1] adds in a device category called a *Reader,* which reads tags; in our treatment, we lump readers into the same bucket as sensors.

Tags are often overlooked edge devices because their operation is so simple. A tag merely reports back some identifier. This identifier may be unique within an IoT system, allowing the tag to report the unique identity of some element, for instance, a particular pumping station or a specific position on a robotics assembly line. A tag could instead report general information, such as the model number for a particular motor. Alternatively, a tag could be programmed to return different information depending on the query, and it could report the date of last maintenance or the identity of the last person that passed by this station. A tag is merely a storage device that reports a preprogrammed value when activated.

Tags have other uses as well. Tags can provide additional and more accurate data for navigation and may be used to augment GPS to perform finer-grained course plotting and physical routing operations, or provide information along a specific route. As an example, tags can be set up on the perimeter of a farm field, signaling information on an autonomous robot's specific location. Or they can be set up along virtual segments of the field itself, identifying what crops are planted in this part of the field, thus allowing the robot to perform specific actions, such as increasing or decreasing water delivery or administering specific types of fertilizers.

A sensor, on the other hand, collects data from the physical environment and translates that data into a value or set of values. Common sensors include temperature, humidity, barometric pressure, and light-level sensors; if the environment for a sensor is a pump or motor, the sensor might detect vibration, revolutions per minute (RPM), or current draw.

In many systems, sensors are combined to create unique tools; this is called *sensor fusion*. There are numerous apps for smartphones that provide stellar examples of sensor fusion. For instance, there are health apps that use the flash and camera plus data from the accelerometer to detect reflected light and measure movement to calculate the heart rate of an individual; tilt and accelerometer sensors are used to measure sleep patterns, or even level a table or picture.[12] The sensors by themselves just provide data—it is the application behind the sensor, and the way the sensor's data is being interpreted and used, and perhaps combined with other sensor data, which makes a system unique and valuable.

Actuators convert electrical energy into physical motion; actuators are electrical devices that open or close a contact or physically interact with the environment by moving a shaft, switch, or valve. Motors and servos in a remote-controlled toy car are excellent examples of actuators. There are really only two different types of actuators in IoT systems—one being electrical and one being physical:

- A relay is an electrical actuator that opens or closes an electrical circuit depending on an electrical control input; relays are an electronic on-off switch.
- A servo-mechanism is a physical actuator that turns a shaft a specified number of times or a certain degree of rotation based on an electrical input, usually a pulse wave.

From these two types of actuators, many different controls can be created, such as valves, switches, and motors.

A developing new field in actuators is Micro-Electro-Mechanical Systems (MEMS) technology.[13] MEMS are related to nanotechnology but typically involve a mechanical system that is larger than one micron, up to several millimeters. MEMS elements typically combine some mechanical system (an actuator) with one or more sensors and a microcontroller to provide processing and communications.[14] Because MEMS elements are so small, and so cheap, proliferation of MEMS to provide data and control for IoT systems is a promising growth area. Whereas MEMS units are small, and thus their mechanical operations must also be limited to the very miniature, the strength of MEMS comes in the power of many hundreds or thousands of MEMS units working together. Consider a window screen that is outfitted with MEMS units that communicate with the thermostat inside a building. When the weather is warmer, the MEMS units can close micro-shades in the screen to allow less light into the building during the summer, but they remain open during the winter, thus reducing heating and cooling costs. Though small, MEMS units have been used to superior effect controlling large-scale processes and are even being used on airplane wings to control the flight of an aircraft without the aid of other aircraft control surfaces.[13]

4.2 IoT Must Be a Low-Cost System

As we previously described, IoT architecture is designed to maximize the benefits derived from the IoT system components while simultaneously keeping costs at the lowest possible levels. This makes cost a primary driver in IoT systems. The whole reason for IoT is the promise to increase monetization of services, decrease costs of operation, reduce manpower expenses, and/or speed operations. All of these reasons come down to costs. Therefore, if the cost to install or operate an IoT system is high, or above a certain threshold, then the return on investment (ROI) of the whole IoT system decreases, or even disappears altogether. The entire IoT system architecture is built to drive cost down—low-cost devices, low-cost communications, no-cost maintenance, and low operations costs.

But if low cost is a primary goal, then why create three layers in the architecture? Can't we just dispense with all the gateways, and have only two layers? After all, large-scale data analysis of the consolidated information collected from IoT devices seems to only require the devices layer and the cloud layer.

Gateways seem like an unnecessary expense. After all, client–server architecture has been around for a long time and has been very successful. Cloud systems seem to effectively use client–server technology to service many millions of clients simultaneously, so why does IoT require a different architecture from Facebook® or the Amazon Webstore™?

Do not be mistaken, it is quite possible to create an IoT system using only devices and the cloud. Many such proof-of-concepts exist, including Amazon's IoT Core™ and the Arm® Mbed™ Cloud. And it is entirely possible to use those systems for some IoT deployments. But there are several reasons why it is not possible to manage a full-fledged, large-scale IoT system using only a cloud infrastructure. These reasons include client volume, energy, cost of communications, security, and scaling.

4.2.1 IoT Gateway Layer: Reason 1—Client Volume

In client–server architectures, lots of independent clients communicate with a single-server instance. Clients submit requests to the server asking for a particular service to be performed. The server completes the request and provides the response to the client. The client–server system can grow and scale by adding additional servers behind a load balancer or adding additional processors to existing servers.[3] These servers and load balancers run in the cloud. And as we mentioned, many large-scale services today rely on just this type of architecture. So why does IoT need an additional architectural layer?

The reason for the evolution of these tried and true systems is the sheer volume of devices involved in an IoT deployment. Single client–server systems are architected to deal with thousands of system users, and multi-server cloud-based systems are architected to deal with millions of clients. But IoT systems are intended to service from tens to hundreds of millions of devices. This intended magnitude requires a new architecture just to allow the servers to handle the number of potential devices that are feeding data into the system. IoT systems can be viewed as a necessary evolution of the client–server architecture and cloud computing. This architecture leverages the global scale of cloud computing coupled with the successful client–server architecture to deal with a massive number of end-user devices.

It is, however, entirely possible to spin up enough servers to handle all those devices, but those servers will be mobbed with communications traffic, and their processors won't be busy enough to warrant the cost of keeping all those servers active. And even if the servers could be balanced so that they were sufficiently busy, this two-layer architecture places a drain on the devices layer.

4.2.2 IoT Gateway Layer: Reason 2—Energy Costs

In many IoT systems, the devices are small and spread out, perhaps over a large agricultural field, or over an entire town or city. Many IoT devices are low powered and don't have the ability to plug-in to a primary power source. In these cases, the sensors and devices may be running from harvested energy or batteries. The IoT system must be very careful to balance how those devices consume energy; otherwise the power source will drain, and the device will stop operating.

The energy cost to transmit data is high, and the longer a device is required to be transmitting and monitoring for a response, the larger the drain on energy stores. If an IoT device is not directly connected to a power source and must use energy harvesting to power itself and its communication subsystem, there may not be enough power for the device to use traditional communication protocols. When using energy harvesting, communications must be limited in frequency and kept to a short duration, in order to allow the harvesting technology to build up more stored power.

If a device were communicating to the cloud, the time lag to communicate from the device to a cloud instance can be large, especially when considering that the time spent communicating directly drains the stored energy.

Have you ever hopped on your web browser to buy that really cool thing from Amazon and experienced excruciating delays? There are many reasons for this, including congested communication links, time to spin up a new server in the cloud, infrastructure delays in resolving network addresses, as well as software delays in constructing messages and getting them submitted into the communication channel itself. There might even be a Mirai botnet flooding traffic in some segment of the Internet causing delays. If an IoT device experienced these delays, it would have to stay connected to the Internet, burning precious battery power, just waiting for the server to respond.

And exactly because such delays to connect to a cloud instance are likely, IoT systems generally insert a gateway layer to prevent the devices from wasting power during communications. The gateway is much like the server in a client–server model but is located geographically close to the devices it serves, allowing it to respond quickly to those devices, saving the device's energy costs in communication and reducing the likelihood of congestion. In some cases, the gateway can even use directed RF beams to power the device through WEG technology. The gateway, which is generally line powered (directly connected to an always-available energy source, e.g., "plugged-in"), uploads the data collected from devices to a cloud instance for data analytics and further processing. Therefore, the gateway layer in IoT systems with low-powered devices actually plays a very crucial role in balancing the power–communications costs of the system.

4.2.3 IoT Gateway Layer: Reason 3—Long-Haul Communications Costs

Another reason IoT systems require the gateway layer is the cost of long-haul communications. Many IoT devices may be far away from any wired communications infrastructure. Supporting such devices with cabling would necessitate laying copper lines to reach all devices at an enormous cost; it may not even be possible to directly connect devices to cabled communications—think large-scale agricultural installations, or forest service monitoring for wildlife. Leased lines or even 3G data communications can be expensive when considering long-term access and operations of such networks, which cuts into the ROI for IoT systems. If every device required a 3G account, this could amount to millions of 3G subscribers for a single IoT installation—think of the recurring overhead cost just to subscribe all those devices! A gateway placed in proximity to the devices can enable a WiFi (or ZigBee®, SigFox®, LoRa®, IEEE 802.15.4®, or Bluetooth® Low Energy) network to capture data from the devices and transmit that data up through a wired or a single long-haul wireless connection. Suddenly, a single 3G account can service a few hundred devices, depending on the range and spread of the devices themselves. The gateway also has the processing power to perform data compression and data consolidation to further reduce the communications costs, something most small devices cannot afford due to limited processing power or energy restrictions.

There is another reason the Gateway layer may be required for communications in IoT systems, and that reason has to do with control loops. Control loops may be found in many IoT systems, but they are of highest priority in Industrial IoT (IIoT). IIoT is the term used to refer to the use of IoT in chemical plants, oil refineries, water and waste treatment systems, or any other industrial manufacturing operations. In IIoT systems, the primary concern is the control of valves, pumps, and sensors that make up the manufacturing system. The proper operation of these controls is of critical importance. Consider an oil refinery, which makes different types of gasoline products. At the gas station, you can choose between regular gas, unleaded and premium, or even diesel fuels. And different gasoline brands have different additives to clean your engine during use. Each of these products requires a slightly different mixture of chemicals, different temperature treatments of the raw oil, and a thousand other different variables. All of these variables are controlled by a series of valves, pumps, and liquid-flow control systems within the refinery. For each of these control points in the industrial system, there are a series of sensors to read temperature, chemical composition, viscosity, etc. If just a few of these control points are wrong, instead of getting premium gasoline out the back end of the refinery, you get sludge—chemical waste that you cannot sell but instead must pay to dispose of properly. You can imagine how the cost to run such a system can skyrocket, especially if it takes too long to get the whole industrial system into balance and keep it there! And if you have ever seen an oil refinery up close, it isn't difficult to imagine the hundreds of thousands of valves, pumps, and sensors that are contained in one of those systems!

Each one of those valves and pumps are part of what is called a control loop. The valve (for instance) is put into a particular position (e.g., fully closed, 3% open, or 12% open). Several sensors then read the effect of this valve on the industrial process. Is the pressure too high? Is there too much of a particular chemical at a particular point in the flow? Is the viscosity too low? Whatever the sensor reads, it returns that information back to the industrial operations center. All the sensor values are then analyzed, and adjustments are then sent out to the actuators that are controlling one or more valves and pumps. This loop—from valve, to sensor, to control center, to valve—is called a control loop. The faster an IIoT system can execute these control loops, the faster the entire industrial operations center can bring the whole system into balance to produce the intended product. This means smaller waste output from the factory and lower cost to produce the end product. An IIoT system can do this faster and more efficiently than humans can, as long as the control loops are running efficiently.

In a real industrial system, there are thousands of control loops. There are also control loops within control loops. Some control loops are used in the day-to-day operation and maintenance of the refinery, whereas others are necessary to prevent a catastrophe—for instance, the whole refinery blowing up and spreading a poisonous cloud of chlorine gas over a nearby population center! Those are human-safety control loops, and the system cannot wait for a cloud operations center to process the sensor data that a particular valve is open too much, creating a dangerous pressure condition in the system; that valve needs to be shut down immediately, or perhaps a different valve needs to be opened to relieve the pressure. The short-cycle control loops can be as small as a few milliseconds. This is where gateways become important in an IIoT system. Because the gateway is closer to the valves and sensors, it is possible for the gateway to operate within the tolerances of these critical control loops. The gateway also has the processing power and energy required to run the analysis programs, something the end devices may not be capable of performing. So instead of just blindly relieving the pressure in the system to prevent an explosion, the gateway has the processing power, and perhaps even the AI, to determine the optimum way to avoid a disaster without completely messing up the product that the refinery is currently producing. A gateway can provide a critical role in reducing waste and balancing costs while also protecting human life and property. Therefore, the gateway is by no means a useless third wheel.

4.2.4 IoT Gateway Layer: Reason 4—Security

All communications to and from the cloud must be secured, and the primary mechanism to secure this communication flow is the Transport Layer Protocol (TLS); TLS is the protocol that replaced the Secure Sockets Layer (SSL) protocol as an industry standard in IETF RFC 8446.* Much like SSL, TLS is defined by a series of Internet standards that specify how data is to be encrypted and integrity protected to prevent disclosure and tampering during

---

* This newest TLS version 1.3 was just recently approved in August 2018.

transmission. TLS also specifies a series of key exchange steps that allow the two communicating parties to securely generate shared keys for encryption and integrity operations. The trouble with these operations, especially the key exchange, is that they take a lot of processing power to compute these values—power that we already said the small devices lack.

In fact, to give you an idea of just how expensive these calculations are, when SSL first came out in the mid-1990s, and servers were set up to communicate with web browsers using SSL, the servers were enhanced with special "rainbow cards" to perform the SSL calculation because they were such a drain on the servers at that time. Today, servers include special instructions and special hardware to help speed up these calculations. But the inexpensive devices of IoT don't have these performance enhancements, mostly because they eat up too much power.

Some small MCU processors in IoT devices do have limited cryptographic operations (i.e., encryption and integrity), which are used by the devices for secure boot or even encryption of messages. But even if the IoT devices can perform basic cryptographic operations, they cannot afford to perform the expensive key exchange operations on every data upload, and this is required at the beginning of every TLS exchange.

Because most small devices cannot perform the expensive operations required by SSL/TLS, the gateway provides this essential security service. The gateway enables customized secure channels to low-power, low-energy devices, and then performs full SSL/TLS to the cloud.

Additionally, beyond securing the end-to-end communications path, the gateway can perform the network security for the small IoT devices. This is critical, because we all understand how dangerous it is to be on the public Internet. We use firewalls and NAT translation even in our home networks to separate our computers and devices from hackers on public networks. Our laptops are all running host intrusion-prevention software, and virus scanners and other software that drag down the performance of our machines. But all these protections are necessary to prevent the malware, worms, and all other forms of evil that hackers throw at us from taking over our devices.

Imagine the IoT system without a gateway, the under-powered sensor and actuator devices, directly connected to the Internet with a public IP address necessary to communicate with the cloud. All of a sudden, this small device needs to be running a firewall to protect it from crafted packets used by hackers to needle their way into industrial computer systems. The device should run some form of virus scanning to ensure no memory-resident malware is able to squeeze past the firewall, and the device also has to do all these heavy-weight TLS cryptographic calculations to communicate with the cloud. And all this has to be done while running from a small coin cell battery recharged by some milliwatt energy harvesting subsystem. OK, point well made.

The gateway acts as the firewall for its connected devices, handling the network security, as well as carrying out the heavy-weight communications tasks to the cloud. This is just another example of an essential service the gateway layer provides for real-life production IoT systems.

4.2.5 IoT Gateway Layer: Reason 5—Scaling

The final reason that real-life production IoT systems require a gateway layer is scaling. Most cloud instances are intended to scale to the level needed to service only so many devices simultaneously, and additional servers are spun-up on an as-needed basis. Even if the IoT devices had access to the long-haul communications channels and could afford to discharge the energy and processing power to perform a TLS key exchange with the cloud, most cloud installations would not have enough servers to immediately handle a glut of incoming requests. To the cloud, a million IoT devices simultaneously uploading their current temperature readings might look the same as the Mirai botnet attack we described in Chapter 2. Additionally, the fewer the number of servers that are running and waiting to service a request, and the shorter time those servers are active, the lower the cost to run the cloud part of the IoT instance. The cost is based on what the cloud operator charges, which is usually based on how many server instances are operating and how much traffic is being ingested by the cloud.

In addition, it is important to remember that use of the cloud is throttled by the bandwidth available to the actual data center where the cloud instance is housed. Remember, the cloud is made up of larger server farms in data centers. And these data centers are geographically spread all over the world. Each of these data centers taps into a large-fiber backbone of the Internet and gets all its traffic from that pipe. So no matter how many servers are in a cloud instance, the amount of traffic that a particular data center can consume is based upon the bandwidth of that incoming network connection. It is like filling up a pool with water—the speed at which you can get water into the pool depends on how big your hose is. So add another hose . . . er . . . fiber cable to get more data into the data center. That only works to a certain point. Just like in your home, you can add more hoses to fill up your pool—one from the front yard, one from the back yard, one from your kitchen . . . you get the idea. But at some point, the water main that feeds water into your house is going to be tapped out, and the water coming from each hose is going to be reduced. The same is true for the Internet. Geographically, there is only so much data that can be fed into a geographic region. The main backbone from the Internet gets saturated, the routers in that area become overloaded, and other Internet services begin to get affected. This is the reason that public cloud services, and huge Internet services such as Netflix®, Facebook, and Amazon, have data centers in different regions of the country and the world—it is in order to distribute the network load (when necessary) and also avoid long delays for network packets to run halfway across the world.

The same thing happens with IoT systems. The cloud could redirect your traffic to another server halfway around the world and spin up a server to service your request. Technically, that works. But now your devices have all that latency and the additional initial connect time and are required to maintain power to the communications subsystem to listen for responses. This type of scaling is fine for a cloud provider, but it works against an IoT system with limited energy. Again, the gateway layer solves this problem.

It certainly isn't practical for the costs savings generated by an IoT system to be swallowed up in a high cloud service rental cost. And the intended scale of an IoT system can easily swamp a single cloud data center, causing extended communication latency and increased energy costs for devices. The gateway layer removes this encumbrance from the cloud, allowing cloud instances to be more prudently used, resolving congestion, keeping costs low, and allowing IoT systems to scale up to billions of devices.

4.3 Details of the IoT Architecture Layers

Now that the baseline IoT architecture is laid out, and the reasoning for the primary requirements of the layers of the IoT architecture are understood—for example, volume, energy, communications, and scale—it is time to look at each layer of the IoT architecture individually in a bit more detail. A closer look at the hardware and software in each of the layers will provide an essential backdrop for our next chapter, which analyzes the security of IoT systems.

4.3.1 Basic IoT Edge Device Architecture

The most basic IoT edge devices are simple sensors and actuators, usually powered by a battery and using an MCU to perform the logic and run the basic software for the device. The most primitive sensors "run on bare metal," which means that the software performing the sensor functions is written to run right on the processor without any operating system or software frameworks, to make writing the sensor software easier. However, it is more common today for sensors and actuators to run on a Real-Time Operating System (RTOS). As its name implies, an RTOS is designed to allow a program to run in real time and eliminate as much overhead as possible, while still providing some basic "conveniences" for the software developer. These conveniences include file reading and

> A microcontroller unit (MCU) is a small processor with a basic central processing unit (CPU) and all the I/O and memory components on a single piece of Silicon that allow it to perform its functions. In the past, MCUs commonly used an 8-bit or 16-bit processing word, and were referred to as an 8-bit or 16-bit processor. However, today it is common to have a 32-bit processor in an MCU. The definition of an MCU has become even more blurred since the advent of systems-on-a-chip (SoC) processors used in smartphones and tablets. The primary differentiation is the amount of energy the MCU consumes in order to operate, which directly relates to how fast the CPU runs (e.g., the clock speed).

writing, allowing the programmer to read configuration information and write temporary files containing the sensor data readings; input/output drivers for serial and USB ports allowing the programmer to more easily interface to a modem or Ethernet card; and process and/or thread management allowing the developer to split their program into different parts and schedule each part in an appropriate way. An RTOS can have many other service functions, of course, and the ones listed above are only a few of the most rudimentary examples.

Figure 4.5 shows the hardware and software architecture of an IoT edge basic device, indicating the split between hardware and software. The software is typically built as one

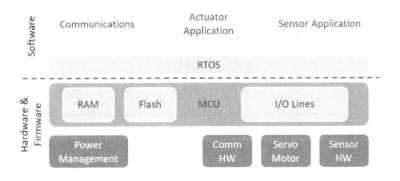

Figure 4.5   Basic IoT Edge Device Software Architecture

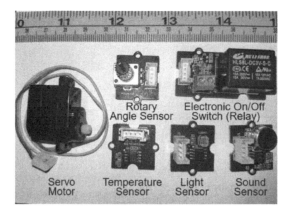

Figure 4.6   Different Types of Sensors and Actuators from the Grove Kit

monolithic application, meaning the application is compiled into a single program. But as shown in the diagram, it is possible to split different operations into different threads of execution, representing a specific function. In this case, the RTOS might include the main thread or interrupt handlers that call the specific functions, as needed. For example, when an interrupt occurs that indicates there is data on a communications port, the RTOS will cause the communications application to run; the RTOS might schedule the sensor application to run every five minutes, or some other appropriate delay, so that the sensor can take a reading from the environment. The actuator application would then be signaled from the communications application that a particular command was received, requiring it to perform some function.

The specific hardware in a device depends on its particular intended functionality and communications capability. In Figure 4.6, several sensor and actuator components are shown. These are elements from Seeed®'s Grove Kit,* a commercial hardware kit that can be purchased to interface with a Raspberry Pi® or Arduino® boards. Therefore, these components only show

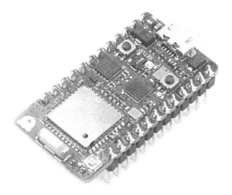

Figure 4.7   An MCU Board with USB and Bluetooth (DUO Board) (*Source:* http://digitalmeans.co.uk/shop/image/cache/catalog/seeedstudio/model/10299/102990523/redbear_duo_wi-fi_ble_iot_board-001-500x500.jpg)

---

* Retrieved from https://www.seeedstudio.com/iot-kits-c-921.html

the actual sensor hardware and do not include the MCU and the communications hardware. This figure gives some perspective of the size of each type of sensor. However, careful inspection of the photograph reveals that the largest components on the small sensor boards are the white connectors used to connect the sensors to an MCU. The MCU chip itself would be about the size of your pinky finger's fingernail. A complete sensor would require the MCU, some type of communications chip, a power source, and the sensor hardware shown in the photograph. The white connector would disappear and be replaced by printed circuit board connections between the MCU and the sensor. All combined, this would make a complete sensor board only a few centimeters in size. This is primarily due to the size and spacing requirements for physical connectors, the physical layer implementation for any communications devices, as well as memory and flash, if these are external to the MCU package. All in all, a fully functional sensor or actuator can be easily built on a rectangular board that is one centimeter wide by two or three centimeters long! An example of such a board is shown in Figure 4.7. Obviously, for some of the larger devices, like the servo motor and electronic on-off switch, the size would be a few centimeters larger. With continued miniaturization, MEMS, and nanotechnology, sensor and actuator devices can be nearly invisible.

4.3.2 Simple IoT Gateway Architecture

In contrast to devices, an IoT gateway is quite large and more complex, and is more akin to a common desktop or laptop computer. However, there are some significant differences, which can make an IoT gateway a bit more expensive.

The first difference is the need for additional communications capability. Remember that one of the gateway's primary responsibilities is to provide communications services to devices. This is typically more than just a point-to-point link between the device and the gateway, as it may involve enabling a mesh network for devices to talk among themselves. This machine-to-machine (M2M) capability is becoming more common and enables smart devices to utilize data collected by other devices around them and use that data to drive their own behavior. This requires the gateway to act as a router for these device messages. And, of course, the gateway must provide communications services to as many devices around them as possible in order to reduce the number of gateways required of the entire IoT system.

The second primary difference is the environment in which the gateway must operate. Typically, gateways are not running in the same type of climate-controlled environment in which your laptop normally operates, such as your living room or a Starbucks coffeehouse. The gateway must be able to operate continuously in extremely hot or extremely cold conditions. Consider a "smart city" gateway in Quebec City, Canada, which must survive the harsh winters of −30° C as well as hot summers reaching 32° C with 75% humidity. Regular semiconductor products are not designed to operate in these extreme conditions, and, therefore, IoT gateways require a special version of the processor and the other chips needed on a gateway computer board; these are referred to as industrial grade chips and are rated to operate from −40° C to 85° C. Additionally, because most IoT gateways are designed to operate without cooling fans in order to conserve power, the industrial grade chips are vital for the gateway to be able to operate in these unregulated, harsh environments.

Figure 4.8 shows a reference board for an IoT gateway. The board is designed to run off of only 5 volts and 7 Watts. It uses a dual-core Intel® Atom processor with 2 USB ports (on

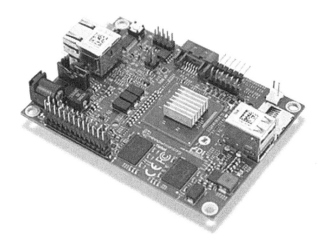

Figure 4.8 Turbot Minnow Board—An Example IoT Gateway Reference Board Design (*Source:* https://store.netgate.com/Turbot2.aspx)

the right-hand short side of the board) and an Ethernet port (on the left-hand short side of the board). It includes 2 GB of RAM and a microSD slot for external flash (next to the USB ports). This board does not supply Bluetooth or WiFi communications, but that can be added as a USB device. The bank of pins on the left-hand long side of the board is referred to as the "low speed expansion" (LSE) connector and includes two serial ports, an Inter-Integrated Circuit (I2C) connector, a SPI bus connector, and ten GPIO lines. All of these pins can be reconfigured through the CPU if desired, providing options to connect different devices to the gateway. The Grove sensors shown in the last section could be connected to the I2C bus on the LSE. Of course, because this is a reference design board, not an actual IoT gateway product, the number and type of I/O pins are for example purposes only. A real IoT gateway board would be configured specifically for the type of devices it was intended to connect to and control.

Gateways usually run a standard operating system. Linux® is a very common choice due to its support for many different CPUs and many different types of devices. The Windows® operating system is also used, both the enterprise versions (Windows 7, Windows 8, and Windows 10), as well as Microsoft's own IoT-specific operating system—Windows IoT Core. Because Windows IoT Core is headless (i.e., it does not provide a display or keyboard to directly interface with the device), more designs tend to use the standard operating systems with a full display and keyboard support. This preference is due in large part to the difficulty developers and system integrators experience using devices without a directly connected user interface.

Figure 4.9 shows an IoT gateway software and hardware stack based on the features and responsibilities assigned to gateways in the previous section. The first difference one notices is the hardware and firmware added to support long- and short-range communications. The long-range communications might just be a wired Ethernet port, as we saw in Figure 4.8, or it could be a 3G or 4G/LTE modem with a SIM card enabling cellular communications. These communication subsystems are purposely separated from the short-range communications hardware and software to avoid accidental (or maliciously deliberate) cross-over in communications. The short-range communications system is most likely some type of wired

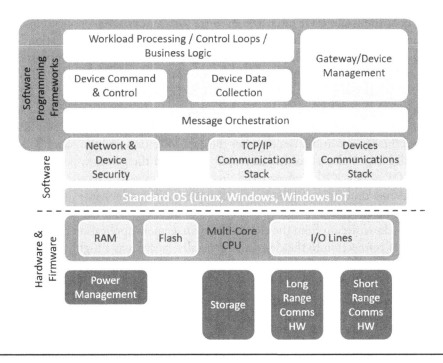

Figure 4.9 IoT Gateway Reference Software Architecture

bus for existing systems, such as industrial IoT. These communications links would likely be I2C, serial ports (RS-232 or RS-485), MODBUS®, or 4-20mA loops and could use the LSE connector from Figure 4.8. We will discuss these and other common IoT protocols in more detail in Chapter 7. More common in new IoT deployments are wireless connections to the devices, including WiFi, Bluetooth® Low Energy (BLE), ZigBee®, LoRa®, or 6LoWPAN. On top of these hardware/firmware communications devices, the gateway maintains two communications stacks. The gateway always uses a TCP/IP communication stack to talk with the cloud, but it normally has a custom communications stack to talk with the devices. Although it is conceivable that both the cloud and the devices could communicate using TCP/IP, it is usually not the case. Running a full TCP/IP stack on the devices requires a lot of power and memory, and requires many more protocols than just TCP and IP. TCP/IP also requires protocols for link connections, address resolution, routing, configuration, and other services. For this reason, a more simplistic stack on the device side is chosen, and the gateway provides the translation between the device's simplistic protocols and the TCP/IP communication required for the messages to reach the cloud.

To make programming for the gateway easier, developers usually create a message orchestration layer that hides the complexity of writing messages to the cloud or to the devices. The top-level application programs just write a simple message to the message orchestration layer, and that layer performs the necessary formatting and conversion required to deliver it to the correct place. In addition, the message orchestration layer is able to accept messages from the two communications stacks and deliver them to the appropriate application software service. This is essentially a pub–sub (Publish–Subscribe) software design pattern.

At the top of the stack, there are generally three different types of applications: command & control applications, data collection applications, and management applications. The command & control and the data collection applications perform the basic operations of listening to devices, collecting their data, and sending them commands. Typically, this software is split into two layers—the lower layer understands the specifics of the devices, and the top layer performs higher-level workload functions. The top layer contains the business logic and control loops used to command and query the devices. It is the top layer that might contain some AI to better control the devices, and it ties together the data collection and the command & control. This layer will likely make use of the increased storage on the gateway, perhaps even using a full-fledged SQL or No-SQL database to retain copies of all the data received from the devices, and perhaps even a time-stamped log of the commands sent to devices.

The management application provides a type of control plane for the gateway as well as the devices that are downstream from the cloud. The management application retains a completely different type of connection to the cloud and generally executes as a highly privileged application on the device, allowing it to change configuration files, update software and firmware, start and stop applications, clear error events, purge log files, and even shutdown or reboot the gateway or the downstream devices. All of these functions are things that a local system administrator would normally do, but in an IoT deployment, there are no human beings downstream to perform these functions, so they must be done remotely. The management application is normally separated from the rest of the applications on the gateway, although the command & control application typically has the ability to send certain commands to devices that only the management application should be able to send (e.g., shutdown, restart, and change frequency of reporting). The command and control application is the software that actually sends commands—such as restart—to downstream devices, but it is the management application that receives the command from the cloud, authenticates the command, and then instructs the command and control application to send the command to the device. These highly sensitive administrator commands are controlled by the management application.

Figure 4.9 also shows a network and device security function. The IoT gateway is responsible for protecting its downstream devices and, therefore, must protect the network from attackers. This includes both the TCP/IP network connected to the cloud, as well as the devices network, whatever that may be. The TCP/IP network security is fairly straightforward. Many host intrusion protection (HIP) and network intrusion protection (NID) programs exist for enterprise systems, and the gateway network security for the TCP/IP stack would fundamentally not be any different. On Linux systems, services such as tcpdump and iptables, for packet examination, and snort or Suricata®* for network threat detection, are common open source examples of network security applicable for gateways.

However, the device's network is extremely different, and there are not many solutions specifically designed to protect such networks. The gateway typically must cobble something together for the specific device network it is supporting, or it may not provide anything to protect the device communications network at all.

The final block in Figure 4.9 is the power management block. Obviously for devices, power management is important, but it is equally important for gateways. The power management unit controls sleep periods and powers down communications hardware when it is not in use.

---

* Retrieved from https://suricata-ids.org/

This helps the gateway conserve energy, which is important if the gateway partially depends on harvested energy. But even if the gateway is line powered, it reduces the energy consumption and, therefore, the actual cost to run the gateway. These low-power periods may be for scheduled communication blackouts because the gateway has told all devices to sleep for a particular timeframe, or the hardware may be able to maintain a type of fast wake-up and quickly turn on when it detects an incoming communication message. The power management unit is normally controlled by the operating system, but on IoT gateways the standard operating system power management code has some customizations to make the gateway operate in an optimal fashion for the particular mission for which it was designed.

As a final note on the IoT gateways, it is important to mention the use of software programming frameworks. Software developers are commonly using different types of software frameworks that enable them to easily interface with IoT infrastructure like the cloud and different devices, allowing them to share and reuse code and also speed up the development of IoT applications. The framework may be infrastructure specific or may be language specific. For example, there are frameworks that allow developers to easily interface with Amazon Web Services™ (AWS™) infrastructure, like storage, databases and server lambda functions. Other frameworks are language dependent—for example, Node.js®, which provides a JavaScript® environment for developing server-type applications and which can interface with many different functional libraries in JavaScript, off-loading the developer from creating much of the required infrastructure code themselves. These frameworks are extremely useful and also widely used. As seen in Figure 4.9, a software programming framework is a kind of supporting software behind the gateway's applications, allowing the developer to use functions from the framework instead of developing the entire application functionality from scratch. The framework may even provide all of the functionality required for one or more of the application software components themselves, such as the orchestration layer. This make the applications easier to write and speeds the development of IoT services.

4.4 Fundamental IoT Cloud Architecture

The fundamental design of IoT devices and gateways are simplistic compared with cloud infrastructure. In Figure 4.10, we show the generalized architecture of an OpenStack design. Other cloud services have similar infrastructures, and include (from left to right, in Figure 4.10), large-object unstructured storage (Object Store), disk image storage and management (Image Service), compute service for the actual server applications (Compute), standard storage volumes for saving files and streams (Block Storage), software defined-networking service (Networking), and authentication and authorization capability (Identity Service). These, of course, are logical services, and have servers, software, databases, hypervisors, and virtual machines behind them.

The full details of a cloud operating system (COS) or cloud computing platform (CCP) are far beyond the scope of this book, but the complexity and number of interfaces of a cloud service is important to comprehend when designing an IoT system. Typically, only the compute application—the actual server-side application running in the cloud—is considered when looking at the cloud piece in an IoT system. However, to get the server-side application running, a whole lot of configuration of the cloud infrastructure must be created, including

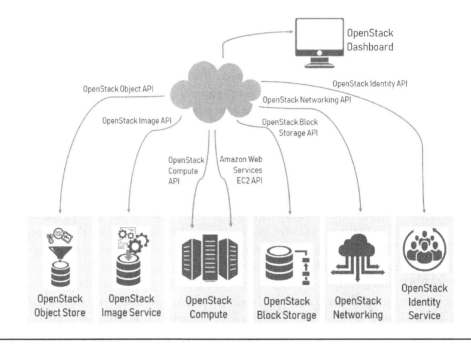

Figure 4.10   OpenStack Cloud Infrastructure Generalized Architecture (*Source:* Adapted from https://docs.openstack.org/arch-design/design.html)[15]

databases, network connections, and image repositories. Chapter 6 examines cloud and cloud architectures in greater detail.

4.5 Why Security Is Hard in IoT Systems

A common truth often repeated in security circles is that an attacker has to find only a single defect in a system in order to be able to break in, but the security professional has to find every single weakness in the system and fix them all. When this truth is considered, and the breadth of the hardware, firmware, and software that makes up an IoT system is considered, it is not difficult to imagine why protecting an IoT system is so difficult.

The security of an IoT system is extremely complex and crosses many security disciplines. It is very unlikely for a single person to be able to span all the different subject areas adequately. IoT system security takes a team of experts in network security, computer security, hardware and software security, cloud infrastructure security, secure protocols, and cryptography. It is also impossible for a single product to protect an entire IoT system by itself. Unlike in the early days of the Internet, where the installation of a single network security and anti-virus product adequately protected your computer, IoT does not have such a silver bullet.

In this chapter, we discussed the architecture of IoT systems in some depth, and we discussed the differences between the three layers. Those differences create an assorted set of security concerns that must be addressed by a variety of mechanisms and services.[16] Beyond the differences in the layers themselves, there is a huge diversity among the types of devices within each layer—including the hardware, platform software, and applications. This diversity makes it impossible to define a one-size-fits-all security solution. As an architect and IoT developer, you have to be able to analyze your system and uncover the security requirements for yourself. In the next chapter, we will review the process used to identify all the threat areas in an IoT system and map out the protections required to secure an IoT deployment.

References

1. IEEE. (2015, May 15). "Towards a Definition of the Internet of Things (IoT)." Retrieved from https://iot. ieee.org/images/files/pdf/IEEE_IoT_Towards_Definition_Internet_of_Things_Revision1_27MAY15.pdf

2. Bakhshi, Z., Balador A., and Mustafa, J. (2018). "Industrial IoT Security Threats and Concerns by Considering Cisco and Microsoft IoT Reference Models." 2018 IEEE Wireless Communications and Networking Conference Workshops (WCNCW), Barcelona, pp. 173–178.

3. Tanenbaum, A. (1996). *Computer Networks*, 3rd Edition. Upper Saddle River (NJ): Prentice Hall PTR.

4. Sundquist, T. (2006, April 25). "Waveform Development Using Software Defined Radio" (Thesis). Retrieved from http://www.diva-portal.org/smash/get/diva2:21831/FULLTEXT01.pdf

5. Dillinger, M., Madani, K., and Alonistioti, N. (2003). *Software Defined Radio: Architectures, Systems and Functions*. West Sussex (England): John Wiley & Sons, Inc.

6. Cisco Systems. (2014). "The Internet of Things Reference Model." Retrieved from http://cdn.iotwf.com/resources/71/IoT_Reference_Model_White_Paper_June_4_2014.pdf

7. Nagasai. (2017, February 18). "Classification of IoT Devices." Retrieved from http://www.cisoplatform.com/profiles/blogs/classification-of-iot-devices

8. Horowitz, M. (2014, December 9). "Securing the Internet of Things." Retrieved from https://www.slideshare.net/Stanford_Engineering/mark-horowitz-stanford-engineering-securing-the-internet-of-things

9. Reese, G. (2009). *Cloud Application Architectures: Building Applications and Infrastructure in the Cloud*. Sebastopol (CA): O'Reilly Media.

10. Allen, J. (2015, January 4). "What Is Thermal Energy Harvesting: An interview with Dr John Allen on Thermal Energy Harvesting." Retrieved from http://www.europeanthermodynamics.com/news/what-is-thermal-energy-harvesting

11. Kamalinejad, P., Mahapatra, C., Sheng, Z., Mirabbasi, S., Leung, V., and Guan, Y. (2015, June). "Wireless Energy Harvesting for the Internet of Things." *IEEE Communications Magazine*, vol. 53, no. 6, pp. 102–108. Retrieved from http://dx.doi.org/10.1109/MCOM.2015.7120024

12. Archer, J. (2015, November 27). "5 iPhone Apps That Use the Gyro/Accelerometer in an Interesting Way." Retrieved from https://medium.com/make-school/5-iphone-apps-that-use-the-gyro-accelerometer-in-an-interesting-way-33f5814ac84c

13. MEMS & Nanotechnology Exchange. (2018). "What is MEMS Technology?" Retrieved from https://www.mems-exchange.org/MEMS/what-is.html

14. Gubbi, J., Buyya, R., Marusic, S., and Palaniswamia, M. (2013, September). "Internet of Things (IoT): A Vision, Architectural Elements, and Future Directions." *Future Generation Computer Systems,* vol. 29, no. 7, pp. 1645–1660. Retrieved from https://doi.org/10.1016/j.future.2013.01.010

15. OpenStack. (2018, November 29). "OpenStack Logical Architecture Diagram." Retrieved from https://docs.openstack.org/arch-design/design.html

16. Vashi, S., Ram, J., Modi, J., Verma, S., and Prakash, C. (2017, February). "Internet of Things (IoT): A Vision, Architectural Elements, and Security Issues." *2017 International Conference on I-SMAC (IoT in Social, Mobile, Analytics and Cloud) (I-SMAC)*, Palladam, pp. 492–496. Retrieved from https://doi.org/10.1109/I-SMAC.2017.8058399

Chapter 5

Security Architecture
for Real IoT Systems

A good seaman weathers the storm he cannot avoid,
but wisely avoids the storm he cannot weather.

— Author Unknown

5.1 Preparation for the Coming Storm

As discussed in Chapter 4, an IoT system is composed of many different types of devices, combined with gateways and cloud systems. The security engineer is responsible for protecting this diverse system and for ensuring that it can weather any storm some hacker might dream up. How does a security engineer figure out which architectures present security storms they can weather and which would create a security storm they should avoid?

A comprehensive understanding of how an attacker can enter your system, overcome its defenses, and plunder at will, requires design-with-security in mind—as opposed to being an afterthought. This is the concept of treating your IoT system as a castle and reviewing its defenses, as we discussed in Chapter 2. This is easier said than done, and the attacker has the advantage. Where the security architect has to think of all possible vulnerabilities across the system, the hacker need only find one that he or she can exploit and leverage. This has been true for as long as we've had networks, and the security architect has risen to the challenge. However, complexity in networked systems is growing—heterogeneity of platforms and operating systems, multifunction protocols, ubiquity of network access, all increase the complexity of systems. And now the explosive growth of the IoT ecosystem has even expanded the diversity in systems and software, creating the perfect storm for the security professional. Every growth-savvy technology firm on the planet has created some IoT offering, many just cobbling together various protocols and open source solutions, slapped onto one operating system or another, without much, if any, thought to security. Complicating things further, IoT systems are composed of a conglomeration of such devices with limited compute power and restricted energy budgets, often resulting in whatever security capabilities that might be available on the devices being turned off to conserve power or to meet performance requirements. No longer can the security engineer use *ad hoc* methods, scribbling attacks on the back of a napkin—the complexity is too high. And management cannot budget for a long security analysis process. Time-to-market is king. "Damn all those torpedoes heading straight for us! I don't care! Full speed ahead!"

The IoT world's equivalent of the perfect storm can be navigated successfully if we are strategic and disciplined in dedicating expert resources and analytical engineering early on in the process. And that process can be bounded and controlled, with clear risk trade-offs allowing us to meet time-to-market deadlines. But securing any system requires careful analysis of the system, as well as an understanding of the threats to the system and its value to both the organization and an attacker. In fact, your attackers will be performing a similar analysis on your system, attempting to find that one weakness that will allow them to penetrate the system and steal resources or wreak havoc.

This chapter is split into three equal parts. The first part is a methodical security architecture process that includes measureable criteria for completeness; the second part discusses security principles and the security engineer mindset to guide the security architecture process in making trade-offs and prioritizing decisions. The chapter concludes with some security analysis examples of IoT use cases that explore IoT unique security challenges and solutions.

5.2 What Is Security Architecture?

In the last chapter we talked about system architecture and defined it as the organization of elements in a system to address the needs of the business, the environment, operational constraints, and system users. Security architecture is similar but focuses on technological solutions to ensure that the goals or mission of the system are maintained even in the presence of an adversary attempting to subvert that mission.

Security architecture extends a system architecture with controls and countermeasures enabling the system to be run and managed in a way to minimize the risk of loss of information and misuse of system resources and prevent the corruption or unauthorized disclosure of information.[1] Security architecture leverages the work of system architecture by reusing its many different views of the system. The security architect must role-play through those different views to determine how the system is going to be used and misused. Oftentimes, the security architect becomes overly enamored with the hacker role and how the system is going to be attacked. The attacker perspective is vital, but other views are often missed, including the security management view.

A common misperception is to ignore security management altogether, because it is viewed as just policies. Security management includes policies and standards, but it also requires procedures that ensure the confidentiality, integrity, and availability of information, applications, and systems which also preserve the network for authorized users only.[1] Planning for security management is a critical and often overlooked part of good security architecture. If the security architect does not use the perspective of security management, they will fail to ask themselves critical questions. What features are going to be built into the system to inform an administrator that there is something wrong with the system, or that an attack is in progress? How will those features be used by the administrator so that they are not spending hours trying to determine exactly what is wrong? What mechanisms are included in the system to allow for speedy recovery of the system after an attack? How will features in the system be controlled to limit or prevent the system administrator from becoming an adversary themselves?

Time and again Dave has sat on a security review board and asked the presenter how their security architecture aids the system administrator in detecting and isolating attacks. The common response, other than silence, usually is, "We provide logging." This answer does not take into account the system administrator's viewpoint. Try this experiment on your project. Grab a random log file from your security validation team and then try to separate the log entries that represent normal behaviors from entries that represent one of the test attack vectors. Finding those entries in the mass of extra log statements the developers never removed from the system is a herculean task. Just logging is not the answer.

Security management needs more than just logging or encryption of a particular piece of data, or integrity protection for a database record; security architecture must analyze system users and administrators, operational use cases, likely adversary actions, possible accidental user misuse, as well as utilize secure channels and regulatory-mandated cryptography to properly protect a system.

5.3 The Security Architecture Process

This section presents the security architecture process that Dave developed. Dave has used this process on dozens of projects, from small single-engineer projects to complex multiyear projects involving teams in multiple geographies around the world. Figure 5.1, and the following

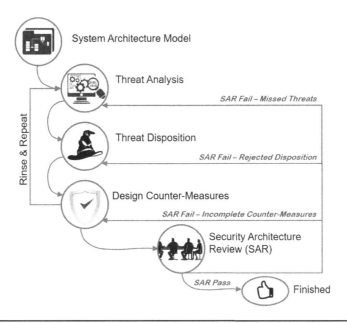

Figure 5.1 The Security Architecture Design Process

list, introduces our comprehensive security architecture process. We use this list as an outline for the remainder of this section. The steps we use for security architecture are as follows:

1. **Analyze the System Architectural Views:** Create an architecture that addresses the system goals and create a set of logical views of the architecture for analysis.
2. **Perform Threat Analysis:** Use the selected architectural diagrams to model and analyze threats to the system.
3. **Disposition Threats:** Like the Sorting Hat in the Harry Potter series,[*] the threat list produced from the threat analysis must be sorted into categories defining how those threats will be addressed. Each threat is classified as follows[†]:
   * Accepted threat
   * Transferred threat
   * Mitigated threat
   * Detected threat
4. **Incorporate Threat Mitigations:** Refine the system architecture to incorporate the technical features required to mitigate and/or detect the threats that were not accepted or transferred in Step 3; if no new threats were added in Step 3, then move to Step 6.
5. **Rinse and Repeat:** Clear your mind, and then repeat the process from Step 2 on the updated architecture. This requires an exceptional amount of discipline.
6. **Security Architecture Review:** Review the completed security architecture with a security review board to ensure that all the threats have been identified and all the mitigations are appropriate. If the security review board identifies new threats or additional mitigations, add these to the threat model or architecture, respectively, and repeat the process at the appropriate step.
7. **Approval:** When the architecture is approved by the security review board, the security architecture is complete.

The process above is designed to achieve a complete analysis of the system and add the appropriate level of countermeasures[‡] and controls[§] necessary to address the identified security threats as well as the needs of security management. The process involves a large amount of good judgment and experience on the part of the security engineer in order to properly characterize threats and identify effective countermeasures. However, a security review board is included in Step 6 to provide feedback and guidance for the less-experienced engineer. It also provides checks and balances to a senior engineer performing the analysis to ensure that they have not been overly confident and accidentally missed something.

---

[*] In the Harry Potter series, the Sorting Hat was placed on the heads of students attending the Hogwarts School of Witchcraft and Wizardry, and the hat spoke, assigning the student to one of four houses (like sororities) in the school. The houses included Gryffindor, Slytherin, Ravenclaw, and Hufflepuff. In much the same way as the Sorting Hat, the security engineer must discern details about the threats and assign each threat to a particular category.

[†] We will define the disposition categories later in the chapter. And, no, there is no Gryffindor category.

[‡] A countermeasure is a technical feature added to a system that mitigates (prevents) one or more attacks.

[§] A control is a security management procedure that delays or stops an attack, or diminishes the effectiveness or damage of an attack, removing or reducing the benefit an adversary gains from that attack.

The next sections will review each of these steps in some detail and provide pointers to background material necessary to execute the process effectively.

5.3.1 Analyze the System Architectural Views

Architecture is all about perspective. Good systems architecture creates different diagrams that communicate different perspectives of the system. So the first step in security architecture is to collect the system architectural views and study them to gain a comprehensive perspective of the system. But you should never settle for just a single architectural view. Clements' book[2] on software architecture describes the complexity of architecture and the need for multiple views this way:

> *A software architecture is a complex entity that cannot be described in a simple one-dimensional fashion. Our analogy with the bird wing proves illuminating. There is no single rendition of a bird's wing. Instead there are many: feathers, skeleton, circulation, muscular views, and many others. Which of these views is the "architecture" of the wing? None of them. Which views convey the architecture? All of them.* (p. 13)

Security architecture is primarily an analysis discipline. We use different views of the system to analyze its security properties. The importance of well-defined and complete architectural views of the system cannot be overstated. And Clements clearly guides us to use multiple views.

First, notice that we use the term *views*. That term is preferable to *diagram* because the word diagram connotes a specific type of picture with a specific methodology behind it. This often devolves into a secular battle regarding what type of methodology produces better diagrams. Object-Oriented. Data-Definition Language. Procedural. The list is practically endless.

The specific diagrams do not matter. It is the view or perspective that the diagrams capture and convey that is important. Additionally, the diagram must be understandable by the engineers doing the security analysis. This may seem really basic, but countless hours have been wasted on arguments over system diagrams. The best diagram in the universe is of no use if the person needing to receive the information cannot decipher the diagram. Release yourself and your peers from any useless devotion to a particular diagram or methodology and, instead, seek to convey the important views of the system.

What are the important views that should be consumed in security architectural analysis? I'm glad you asked. In *Applied Software Architecture*,[3] Hofmeister, Nord, and Soni (from Siemens Corporation) describe architectural views from their industrial practice that seem valuable to IoT. Merging their ideas with Clements and colleagues[2] and adding a specific view for security assets, we recommend the following list of basic views:

1. **The Conceptual View**—Presents the primary elements of the system and their interrelationships.
2. **The Module Interconnection View**—Decomposes the primary conceptual elements in logical functions and layers, and shows the flow of data between modules (source-sink relationships).
3. **The Execution View**—Displays the actual hardware–software deployment with functionality ascribed to libraries, processes, and services, and shows how the system starts

up with particular attention to the system's dynamic structure, including how users, administrators, and operators interact with the system.

4. **The Security Asset Allocation View**—Using the views from the prior three models, shows the data and functionality (or features) that represent critical or valuable assets to the system. A data or functional asset is critical if the system security is negatively impacted when an attacker seizes control of, destroys, or disrupts the asset.

Without these clearly defined architectural views, it is impossible to determine all the avenues an adversary could use to attack the system. Not only are these views necessary for the threat modeling to analyze how an adversary could attack the system, but they will be used in the security architecture review to defend the comprehensiveness of the security analysis.

5.3.2 Perform Threat Analysis

Once the system views are collected or created, the process of analyzing the system for threats can begin. Threat modeling, or threat analysis, is a technical exercise used to identify all the potential weaknesses (or attack points) in a system. Threat modeling does not attempt to correct the problems but instead only lists the potential problems that exist. In many instances, it is not even important to describe exactly how an attack could be perpetrated, as much as it is important to identify potential areas of attack. After all, hackers are coming up with new ways to attack systems all the time. Just because the security engineer cannot describe every nuance of an attack, it does not mean that the threat of such an attack can be discarded. If that reasoning were used, only the attacks a security engineer could actually perform would ever be considered. That leaves the system vulnerable to a lot of actual attacks.

Just like security architecture, threat modeling is performed by iterating through a series of steps to create a list of potential attacks and then repeating those steps until no further attacks can be identified. When learning to perform threat analysis, people are often confronted with several different methodologies. The list below describes the most common threat analysis methodologies.

- **Asset-Based Threat Analysis**—Focuses on the assets and the ways those assets can be misused.[4]
- **Attack-Based Threat Analysis**—Focuses on the attacks and the ways the system is susceptible to certain attacks. Example: Attack Trees.[5]
- **Adversary-Based Threat Analysis**—Focuses on the adversaries along with the motives and capabilities of the attacker.
- **Architecture-Based Threat Analysis**—Focuses on an architectural model of the system and the exposed interfaces and potential attack points.[6]

It is important not to be intimidated by different methodologies and to understand that threat analysis (or threat modeling, depending on your preferred terminology) is an iterative process that takes into consideration all these pieces. As shown in Figure 5.2 and listed below, the iterative steps of threat modeling are:

Figure 5.2   Iterative Threat Analysis Methodology

- Identify the assets the system contains.
- Identify the system adversaries that could be attackers of the system.
- List the architectural entry points and exit points where an adversary could access the system.
- Enumerate all the ways an adversary could use a particular entry/exit point to observe, capture, manipulate, steal, or destroy an asset.

Where you start in this process depends on where you are most comfortable and what information you know best about the system under analysis. Do not accept the lie that there is only one right way to do threat analysis. If you are comfortable listing the assets first, lead with an asset-based threat analysis. If you know the system design really well, then an architecture-based threat model may be easiest for you to get started. Some engineers may find it easier to begin by enumerating attacks, effectively creating attack trees using hacking methods they have read about or practiced during pen testing. The important thing to understand is that the process is iterative, and you will eventually have to think through and define all the parts of the threat model—assets, adversaries, entry/exit points, and attacks. If you start with attacks, you can then go back and figure out what type of attackers would be able to carry those attacks out. Are they software attackers or hardware attackers? Is this a network attack, or does it require physical access to the system? If you start with an architectural listing of attack points, list the assets that are accessible from those attack points and then list the attackers that can leverage those attack points to get at the assets. Can you see the iterative process? As you work through the process, it gains momentum and builds on itself.

Since threat modeling is an iterative process, how do you know when you are finished? Put in a more formal way—how is *completeness* defined in regard to threat modeling? Completeness is defined as the cross-product of the attributes of the threat model, as shown in the following equation.

$$\forall\ A \in Assets,\ V \in Adversaries,\ E \in Entry\text{--}Exit\ Points,\ T \in Attacks$$
$$A \bullet V \bullet E \bullet T = Complete\ List\ of\ Potential\ Threats$$

This equation shows that when all the system variables are known (assets, adversaries, entry–exit points, attacks), the complete potential threat list is the set of unique combinations created by the cross-product of the four system variables. The complete list of all possible threat combinations can be large, but what one finds through the analysis process is that huge sections of the list are easily discarded as not applicable. The reasoning will be obvious. For example, an asset is not accessible via a particular entry point so all attacks through that entry point for that particular asset can be discarded; or an adversary cannot perform a particular attack on an asset, so that asset–adversary combination is ignored. The actual list of threats that needs to be analyzed becomes very manageable. By approaching your threat analysis using the cross-product, you can show that your analysis is complete, assuming your list of system variables is complete. And if you document your threat analysis properly, it is easy for anyone to come back at a later time and verify completeness, or show that a variable is missing.

This methodology also enables you to respond to management's time-to-market pressures. It is easy to show the complete breadth of analysis that needs to take place, and the process can be prioritized to look at the high-valued assets first. Management can make a risk-based decision to release the product early and know exactly what assets are not covered or what mitigations are not implemented. The security analysis and implementation of mitigations can proceed following the initial release, and an update to the system can address any new issues uncovered during the completion of the analysis.

The Threat Analysis Worksheets

Walking through parts of an actual example can be instructive. A very practical and easy-to-use tool to document the threat analysis process is a spreadsheet. In our example, we create our spreadsheet (as shown in Figure 5.3), with a worksheet for each system variable found in the cross-product equation above and one worksheet for a *heat map*. The following paragraphs explain each worksheet.

The cover worksheet contains metadata about the threat model, including the date the model was created, the product and product version—or versions—if there are more than one, the people involved in the threat modeling exercise, and any other relevant information to help readers associate the model with the correct product and release.

| Cover | Assets | Entry&Exit Points | Adversaries | Threat Matrix | Threat List | Arch References |

Figure 5.3   A Threat Analysis Spreadsheet—The Worksheet Tabs

The assets worksheet contains the list of assets. Assets are discovered by reviewing different views of the system architecture and asking pertinent questions. A configuration file shown in the Module Interconnection View could be an asset if that configuration data were corrupted and caused the system to operate in an open fashion. Trusted certificates used to verify servers over a transport layer security (TLS) connection would be considered another asset. The assets are specific to the architecture, but they should be something that either an adversary might want (e.g., a database of names and credit card numbers or passwords) or might want to disrupt in order to harm the system (e.g., a configuration file listing the trusted servers). In the asset worksheet, each asset should be assigned a unique number to be used for cross-referencing to that asset in other parts of the spreadsheet. In many cases, it is acceptable to create an *asset class* instead of listing every individual asset. For example, there may be no real difference between the configuration file for TLS and the configuration file for the database, other than their use in the system—the attackers are likely the same, the attacks that would be mounted are similar, and the entry points for the attack are the same. Reducing the total number of assets, while not compromising the details of the threat model, reduces the overall size of the threat matrix. This speeds the process and also allows us to think at a higher level of abstraction.

The entry and exit point worksheet contains a list of every way into or out of the system. The list should contain both physical and logical entry/exit points. Physical access points include USB ports, printers, terminals with screens and keyboards, removable hard drives, backup devices, hot-swap sites for business continuity, and the physical computers and CPUs themselves. In IoT systems, it is important to understand where all the physical components actually reside so that physical theft of the devices can be considered. Logical access points include listening network ports, databases, files, REST APIs, system services, network services such as DNS and trusted time servers, user input, consolidated log repositories, backup files, and anything else you find in the Conceptual View of the system. Each entry/exit point needs to be listed as a potential avenue for an attacker to get into the system or retrieve data from the system.

The adversaries list contains all the persons or roles that could attack the system. This includes different classes of users, the network attacker, the system installer, and other insiders. Because insider attacks are often overlooked, the adversaries list should include roles such as system administrator and backup administrator. If you choose to ignore them as adversaries, merely mark them as trusted in the adversary list. This serves to document the assumption that although these roles interface to the system, their special access is being marked as an accepted risk. As an example, the following adversary list has been used for an actual enterprise application and includes the following adversaries:

- An IT administrator that manages the servers in the enterprise, including creating and managing backups (IT Admin)
- A system administrator that manages the application running on the application servers (S. Admin)
- The users that use the enterprise application (User)
- The security officer that sets up the roles and permissions of users (Sec Officer)
- The auditor who has access to read all the logs, database entries, and system event files (Audit)
- The SPID manager that is basically the database administrator (SPID Mgr)
- The developers that created the cloud application (Develop)
- An outsider who is anyone not listed above, and could include hackers, or someone in the organization not authorized on the system (Outsider)

Figure 5.4 Threat Analysis Worksheet—The Threat Matrix Page

2200	Entry/Exit Point (Access Id)													
Asset ID	**1**		**2**		**3**		**4**		**5**		**6**		**7**	
17	IT Admin	S. Admin	34	34	2,15	2,15	IT Admin	S. Admin	IT Admin	S. Admin	IT Admin	S. Admin	2	2
	User	Sec Off	34	Sec Off	5	2	User	Sec Off	User	Sec Off	User	Sec Off	5	2
	Audit	SPID Mgr	Audit	SPID Mgr	Audit	5	Audit	SPID Mgr	Audit	SPID Mgr	Audit	SPID Mgr	5	5
	Develop	Outsider	34	34	9	5	Develop	Outsider	Develop	Outsider	Develop	Outsider	5	5
18	IT Admin	S. Admin	25, 28	S. Admin	25	S. Admin	25	S. Admin	25	S. Admin	25	S. Admin	25	S. Admin
	User	Sec Off	User	Sec Off	User	Sec Off	User	Sec Off	User	Sec Off	User	Sec Off	User	Sec Off
	Audit	SPID Mgr	Audit	SPID Mgr	Audit	SPID Mgr	Audit	SPID Mgr	Audit	SPID Mgr	Audit	SPID Mgr	Audit	SPID Mgr
	26	27	26	27	26	27	26	27	26	27	26	27	Develop	Outsider
19	IT Admin	S. Admin	IT Admin	S. Admin	IT Admin	S. Admin	IT Admin	S. Admin	IT Admin	S. Admin	IT Admin	S. Admin	IT Admin	S. Admin
	User	Sec Off	User	Sec Off	User	Sec Off	10	Sec Off	User	Sec Off	User	Sec Off	User	Sec Off
	Audit	SPID Mgr	Audit	SPID Mgr	Audit	SPID Mgr	Audit	SPID Mgr	Audit	SPID Mgr	Audit	SPID Mgr	Audit	SPID Mgr
	Develop	Outsider	Develop	Outsider	Develop	Outsider	Develop	Outsider	Develop	Outsider	Develop	Outsider	Develop	Outsider
20	IT Admin	S. Admin	IT Admin	S. Admin	IT Admin	S. Admin	IT Admin	S. Admin	IT Admin	S. Admin	IT Admin	S. Admin	IT Admin	S. Admin
	User	Sec Off	User	Sec Off	User	Sec Off	User	Sec Off	User	Sec Off	User	Sec Off	User	Sec Off
	Audit	SPID Mgr	Audit	SPID Mgr	Audit	SPID Mgr	Audit	SPID Mgr	Audit	SPID Mgr	Audit	SPID Mgr	Audit	SPID Mgr
	Develop	Outsider	Develop	Outsider	Develop	Outsider	Develop	Outsider	Develop	Outsider	Develop	Outsider	Develop	Outsider
21	IT Admin	S. Admin	IT Admin	S. Admin	49	49	IT Admin	S. Admin	IT Admin	S. Admin	IT Admin	S. Admin	IT Admin	S. Admin
	User	Sec Off	User	Sec Off	User	49	User	Sec Off	User	Sec Off	User	Sec Off	User	Sec Off
	Audit	SPID Mgr	Audit	SPID Mgr	Audit	SPID Mgr	Audit	SPID Mgr	Audit	SPID Mgr	Audit	SPID Mgr	Audit	SPID Mgr
	Develop	Outsider	Develop	Outsider	Develop	Outsider	Develop	Outsider	Develop	Outsider	Develop	Outsider	Develop	Outsider
22	IT Admin	S. Admin	11	11	55	Sec Off	15, 37	15, 37	IT Admin	S. Admin	IT Admin	S. Admin	1	S. Admin
	User	Sec Off	11	11	User	Sec Off	User	Sec Off	User	Sec Off	User	Sec Off	User	Sec Off
	Audit	SPID Mgr	11	11	Audit	SPID Mgr	Audit	SPID Mgr	Audit	SPID Mgr	Audit	SPID Mgr	Audit	SPID Mgr
	Develop	Outsider	11	11	56	Outsider	37	15	Develop	Outsider	Develop	Outsider	Develop	Outsider
	IT Admin	S. Admin	25, 28	S. Admin	56	Outsider	37	15	57	52	25	Outsider	5	S. Admin

Many of these people would not be considered adversaries, but they could be. Listing them and then describing which ones we consider trusted informs management regarding the risks that they are taking.

As shown in Figure 5.4, the threat matrix worksheet contains a map of the threats of the system. It looks complex at first, but it is really quite simple. The rows are the list of the assets, the columns are the list of the entry/exit points, and the individual cells in the worksheet that are defined by the intersection of an asset (row) with an entry point (column) contains a list of adversaries and the threat or threats that such an adversary poses. The threats in the cell represent the cross-product of that cell's specific asset (the row) with the cell's specific entry/exit point (the column) against one of the adversaries listed in the cell. For example, in cell H-68, the adversary is the user and the attack is against asset 17 (server software) using entry point 4 (ports & connectors); but no attack was found for this particular combination. The combination in cell F-68 using entry point 3 (the console) results in an attack listed as threat 5, where the user can update or load new software on the server. Looking carefully at Figure 5.4, three different types of cells are shown that require some explanation: a cell that contains just adversary names, a cell that contains all numbers referring to threats, and a cell with a mixture of numbers and adversary names.

In Figure 5.4, you will notice that the larger blocks encompassing multiple cells represent an intersection of an asset and entry-point combination and are regularly split into separate adversaries from our adversary list (look back at our previous bulleted list). In our specific example, the IT Admin adversary is always in the upper-left corner of the cell, and the outsider/hacker adversary is always in the lower-right corner. Every intersection of a larger block representing an asset and entry/exit point combination has the same adversary list in the exact same order. The cells that still show a particular adversary name are cells that have not been evaluated, or have been evaluated and no threat has been found for that adversary–asset–entry/exit point. In the worksheets that Dave uses to differentiate these two conditions, he initializes all the cells at the beginning as colored gray, meaning they have not been evaluated; this makes it easy to track where you are in the process. When you evaluate a cell and find no threats, change the cell color to green. Just by glancing at the worksheet, you can easily see what cells still need to be evaluated. When you do find a threat, replace the name of the adversary with a number corresponding to the threat you identified, as we did in F-68.

These actual threat numbers are an index to a threat list with details in another worksheet, the threat worksheet. Each unique threat in the threat worksheet has a unique number. To keep the threat matrix clean and readable, every cell entry that contains a threat only contains the threat's unique number from the threat worksheet. It is best to color code the threats in the threat matrix using a coloring scheme to help identify the threat's severity. That way, the threat matrix acts like a heat map and easily shows the assets and entry points that are most vulnerable. For example, entries that are green, with the adversary name still showing, would represent a reviewed asset/entry-point combination with no threats found, whereas entries that are gray have not been evaluated yet. If you look carefully, some of the entries (E71 and F91) show a little triangle in the upper right corner of the cell indicating that a worksheet comment has been added to the cell. Comments are added to document assumptions when classifying the threat or lack thereof. One possible comment might be that an attacker is considered trusted, and, therefore, a particular threat is not listed in the threat table.

Figure 5.5 shows an excerpt from a threat list worksheet. The important elements to describe the attack are listed again in this worksheet. Although it may seem redundant to write

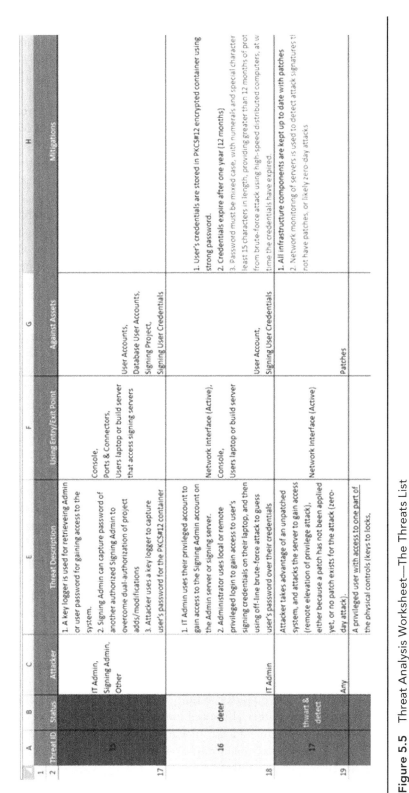

Threat ID	Status	Attacker	Threat Description	Using Entry/Exit Point	Against Assets	Mitigations
15		IT Admin, Signing Admin, Other	1. A key logger is used for retrieving Admin or user password for gaining access to the system. 2. Signing Admin can capture password of another authorized Signing Admin to overcome dual-authorization of project adds/modifications 3. Attacker uses a key logger to capture user's password for the PKCS#12 container	Console, Ports & Connectors, Users laptop or build server that access signing servers	User Accounts, Database User Accounts, Signing Project, Signing User Credentials	
16	deter	IT Admin	1. IT Admin uses their privileged account to gain access to the Signing Admin account on the Admin server or signing server. 2. Administrator uses local or remote privileged login to gain access to user's signing credentials on their laptop, and then using off-line brute-force attack to guess user's password over their credentials	Network Interface (Active), Console, Users laptop or build server	User Account, Signing User Credentials	1. User's credentials are stored in PKCS#12 encrypted container using strong password. 2. Credentials expire after one year (12 months) 3. Password must be mixed case, with numerals and special character least 15 characters in length, providing greater than 12 months of prot from brute-force attack using high-speed distributed computers, at w time the credentials have expired.
17	thwart & detect	Any	Attacker takes advantage of an unpatched system, and attacks the server to gain access (remote elevation of privilege attack), either because a patch has not been applied yet, or no patch exists for the attack (zero-day attack). A privileged user with access to one part of the physical controls (keys to locks,	Network Interface (Active)	Patches	1. All infrastructure components are kept up to date with patches 2. Network monitoring of servers is used to detect attack signatures t not have patches, or likely zero-day attacks

Figure 5.5 Threat Analysis Worksheet—The Threats List

out this much detail a second time, numerous times Dave has gone back to a threat list and could not remember exactly what threat he was trying to describe. More detail is always better. You will thank yourself later.

One thing that should jump out at you from the figure is the *status* column. It shows the process of dispositioning threats, which is discussed in the next section. The snapshot depicts three different dispositions of the threats, color coded so that it is easy to understand the current status of the threat. Thwart is the same as mitigate. Deter is a milder form of mitigation Dave uses to indicate that the mitigations may not be as effective against all attacks as he would like. A red-colored threat indicates that there are no mitigations in place, and the threat will likely become an accepted threat, unless some acceptable mitigation is found.

5.3.3 Threat Disposition

Once the threats have been enumerated and documented, they need to be dispositioned. Dispositioning the threats means classifying them into one of four buckets that define how the threats will be handled. Earlier in the chapter, we introduced the four classifications without any definitions. The following list provides definitions for each of the dispositioning buckets:

- An *accepted threat*—Accept the risk and do nothing to counter the threat; these become weaknesses in the system that should be disclosed to the end customer. Typically, accepted threats are situations that are extremely unlikely to occur or attacks that are expensive to perform and to mount the attack on the system would be more costly than the benefit achieved by the attacker.
- A *transferred threat*—Document and transfer the threat to some other party in the system, for example, the system integrator, the customer, or the end user. An example of a transferred threat could be a power loss that shuts down the system. This threat is transferred to the end customer, with a recommendation to add a battery backup to prevent sudden power loss.
- A *mitigated threat*—Add a technical feature or features to the system that completely prevents the threat from being executed by the attacker. A threat may be mitigated for some adversaries but not for others, or different technical features may be needed to mitigate the same threat posed by different adversaries.
- A *detected threat*—Add a technical feature or features to the system that responds to the threat but does not prevent the attack associated with the threat. This response is typically to identify and alert that the threat/attack occurred. An example of this would be a log message that alerts the network administrator to a failed login attempt, or signals that a corrupted database record was found. Detected threats should be linked to specific recommendations for security management controls, such as "monitoring of syslog events for illegal login should be performed on a daily basis." A better solution would track anomalous system events and link those events to attacks identified in the threat analysis. When a certain number of anomalous events linked to the same type of attack occur, a special report is constructed for the system administrator to warn of the potential attack, with details about the events that trigger the alarm. This report would show up as a warning condition on a system dashboard or sent through a message alert system. Alert systems

and email messages should be used sparingly because many of us ignore such messages when we receive a lot of them—it is the *cry wolf* syndrome in security management.

5.3.4 Incorporate Threat Mitigation into the System Architecture

After dispositioning the threats, the countermeasures and controls are designed into the system to handle the mitigated and detected threats, respectively. Later in this chapter, we will discuss some principles the security architect should use to accomplish this part of the process and provide some examples of IoT threats with countermeasures and controls.

5.3.5 Rinse and Repeat

Each time the system architecture is modified, you need to go back through the process to ensure that the system is still secure. This is very easy if you have documented your steps in the worksheet. If you don't have such a system, these types of updates are onerous and typically ignored, which creates a significant security risk to the system. Why? New countermeasures and controls often add new assets, for example, cryptographic keys or security protocols. A re-analysis of the system, even briefly, is required to ensure that the proper protections are in place for any new assets added to the system. These new assets should be added to the Asset Worksheet, and a new row, and potentially new columns for entry/exit points, should be analyzed in the Threat Matrix.

5.3.6 Security Architecture Review Board

One of the tenets of the secure development lifecycle (SDL), or any system of processes, is to ensure proper quality control. The security architecture review (SAR) board serves this purpose. Just like code reviews, the SAR process forces the security engineer to "dot all the i's and cross all the t's." Just the act of verbally walking through a design causes a person to see things they had missed before. It is not carelessness. When we walk through a presentation, explaining it to others, we use a different part of our brain, and that often ferrets out problems with our logic or helps us to see the problem in a different way. Sometimes, the presenter is the only one in the room that sees the problem. But that is OK, because the problem has been found, and identifying and correcting that problem further limits the way an adversary can break into your system.

It is very easy to set up a SAR. Just get several of the experienced security engineers in your organization to review the security architecture with you. A formal process set up by your organization works even better. But it is probably enlightening to point out some of the ways that SARs have gone wrong, or some pitfalls to watch out for.

The first pitfall is a SAR that is too adversarial or too amicable. If a SAR is only about making someone look foolish, or finding errors in every bit of design trivia, then the security engineer is going to protect themselves and their reputation, rather than use the experienced members of the SAR to help them create a better architecture. The SAR needs to be a place where engineers work together in a collaborative fashion. On the flip side, it is important that the SAR take a critical look at the architecture, and not become a place where buddies just waste time swapping stories about their latest project. If the security engineer is not held to a high level of competence, and reasonable questions about the architecture are not given due consideration, then the whole SAR process becomes a sham and a waste of everyone's time.

Another pitfall that occurs in the SAR is an overly shallow review of the security architecture. If the SAR is not thorough, and does not require a detailed explanation of the system, it is likely that real problems will be missed. One common way for a SAR to be too shallow is to limit it to a single system architecture view that does not represent the real complexity and operation of the system. A related pitfall is the SAR occurring too late in the development cycle, so that any recommended changes or additions to the security architecture cannot be implemented before the product release date. Dave sees this happen frequently. A security engineer is reluctant to come to the SAR because several of the architectural details are not yet worked out or are still in flux. Rather than come to the SAR and explain the situation— perhaps getting only a partial approval, with additional work to come back again later—the engineer delays the whole SAR process until all the details are known. But when they bring their work into the SAR, the board finds something they have missed, or disagrees with a fundamental assumption of the security architecture. The result is either a schedule slip or a product release with a known deficiency. Either way, an early review in the SAR is much better than a late one.

The last two pitfalls we discuss here relate to experience. In one case, the SAR board is composed of security engineers who do not have enough experience, or they have experience in the wrong field. A SAR composed of embedded device engineers who are reviewing a cloud system will most likely miss something. This does not mean that the SAR board should be composed only of engineers with direct experience in the subject matter, but it does mean that there should be a quorum of engineers appropriately experienced in the subject matter. A related pitfall occurs when a senior member of the SAR brings their own project in for review. Oftentimes, the senior person's perspective is not questioned, and junior members of the board feel intimidated or uncomfortable raising concerns. The best way to handle this situation is for the senior member to delegate the review to a more junior engineer on the team and to use the opportunity to mentor the junior engineer. Eventually, every security engineer has to bring their first project to a SAR. What better experience than with a senior engineer providing guidance all the way?

Be aware of these potential pitfalls and work to overcome or avoid them. Use the SAR as a way to refine and improve the good work security engineers are doing. Do not just go through the motions. Get every potential benefit you can out of the process.

5.3.7 After Security Architecture Approval

Once the security architecture is approved, it has to be used by the development team. This sounds silly, but it bears saying out loud. Dave has a mildly humorous story he encountered where, after shepherding a project through a particularly grueling and difficult security architecture process, he went away for an extended vacation. When he returned, he asked the development team how they were doing. They responded that they were fine, but were heads down on development, so Dave left them alone. He figured they would come find him and ask questions if they had any. However, at code completion for the alpha release, during the final implementation review, Dave asked to see the mapping from the architecture's security objectives to the implementation modules. After an awkwardly long bit of silence, the development lead said, "I thought the security architecture was your responsibility." Well, yes it was, but implementing the security architecture is everyone's responsibility. Dave has a bit of advice. After security architecture approval, check up with the development team and ensure that they understand what they are supposed do with the security architecture and that it gets implemented properly!

5.4　Design Principles for Security Architecture

In the last section, we casually glossed over the part of the process in which security mitigations and controls are developed in response to threats. The rest of this chapter is devoted to providing you with insights into that part of the process and to giving you the tools you need to be successful. Skill in this area of developing mitigations is gained only through experience, but we can speed our acquisition of skill by learning through others. This section starts that process by discussing the foundational security design principles all security engineers should use as a foundation for their designs.[7]

Security design principles have been around since the dawn of computing. Saltzer and Schroeder introduced them in the early 1970s, but noted that some of them had been discussed as early as the 1960s.[8] We include this discussion of design principles to arm you with the knowledge necessary to make trade-offs during security architecture and to ground your design of mitigations in time-tested design patterns. Using these principles should allow you to avoid common mistakes in security architecture and think through your design to ensure that you have not missed something critical.

How do we expect you to use these principles? During security architecture, you will be faced with a particular threat and need to evaluate a set of mitigations to determine which mitigation is most appropriate. The design principles allow you to evaluate a mitigation and compare aspects of its essential security qualities to other mitigations. Although we give you specific examples that reflect the design principles, it is the principles themselves that you should use to evaluate and guide your design. The design principles lead you back to the most rudimentary and important aspects of security that should be the bedrock of your designs.

5.4.1　Open Design Principle

The *open design principle* states that your security architecture should not depend on any secret constructions, and that even if your design is published, it should not weaken the security of your system.

An interesting use of the open design principle was the knock-knock protocol. Back in the analog modem days, this protocol was used to prevent hackers from war-dialing into computer systems. The knock-knock protocol worked by recording the series of TCP port connection requests that were sent immediately after a modem connected to the network switch of a computer system. The series of port connections were a kind of secret knock or secret key. If the combinations were 111, 722, 86, then the modem must connect to each of those ports in succession and immediately abort each connection, then finally connect to the port they actually wanted, say port 22. If the sequence was not followed, the network switch would drop the connection, supposedly preventing an unauthorized user from connecting to the system. The knock-knock protocol can be described completely in an open way, but that does not give enough information for an attacker to break into the system. There are 65,535 different ports, and the knock-knock protocol can have any number of ports as the required series. It could require a single port, or seven ports in succession; a requirement of eight ports would be equivalent to 128-bits of security. The knock-knock protocol was discouraged from use because it represented a global password to the whole computing system. The network switch could only easily be configured for one sequence, and everyone who needed access had to have the same combination. This global knowledge made the secret leak more frequently and was difficult to change when someone left the company or group.

In modern systems, including IoT systems, the open design principle still applies. It is the same principle used for cryptographic algorithms. When all the details of a cryptographic algorithm are published except for the exact value of the key, the algorithm should still be secure. Any algorithms used for confidentiality or integrity should stand up to the open design principle. This means that you should never use a *special* encryption algorithm that is known only to the designer. Another example of breaking the open design principle is modifying an open encryption algorithm in a special way. Once, I was told by a design engineer that he had changed the S-Boxes in the well-known encryption algorithm, DES; he surmised that because he was using a unique set of values in the S-Boxes, it would make breaking his encryption so much more difficult. His assumption was that the S-Boxes were just random, and that changing them would fluster any cryptanalysis. On the contrary, S-Box design is extremely complex, and random changes can significantly increase the ease of cryptanalysis.[9]

Your IoT security architecture should be crafted using the open design principle. Even if an attacker has all your design documents, it should not allow them to walk into your system. Proper use of cryptographic algorithms, authentication and authorization, and platform controls should deter any attacker. In fact, if you were to take one of the IoT systems we designed and coded and install it, changing passwords and keys and user accounts, our intricate knowledge of the system should not provide us with an advantage in attacking the system. Obviously, we all know that although this is the goal, having inside knowledge does aid an attacker. With inside knowledge, an attacker knows *where* to attack, and the most fruitful location for attack. But, technically, this should not make the system any weaker. The point is, we do not depend on any secret architectural elements.

5.4.2 Economy of Mechanism Principle

The *economy of mechanism principle* states that the most secure design is the simplest mechanism. Complex designs are more difficult to build correctly, and more difficult to test to ensure that they behave in the proper fashion. For example, creating containment can be accomplished with a complex set of rules that allow two different programs to share the same resources but not interfere with one another. A simpler design is to just provide separate resources for each program. The complex design is implemented in many operating systems but also creates many vulnerabilities. Most often, it is more secure to use the simpler design. However, there are reasons why complexity may be required. Beware of complexity, because complexity is the enemy of security.

5.4.3 Fail-Safe Default Principle

The *fail-safe default principle* is the default–deny condition. Security-sensitive operations in computer systems should be designed to allow access or give permissions only when certain conditions are met (e.g., presenting the right password or holding a proper key), and fail otherwise. The opposite of this principle is to check special conditions that would deny access to a resource (i.e., the requestor is over a WiFi network or dialing in over a VPN, so deny the connection). Default accept is harder to test and more likely to miss corner cases in the implementation. It is also more likely to create security holes when new features are added to the system in the future, because those corner cases in the default–accept situation are forgotten about or not tested appropriately. Default accept requires validation of every possible condition and scenario that might cause a denial.

IoT systems have many access control decisions: Should a particular sensor be allowed to upload data to a certain gateway? Can this gateway connect to this cloud system? Should this cloud system be allowed to perform device management on this device? How is your system architected to make these decisions, and on what basis is the decision made?

Access controls should always be implemented as fail safe defaults, preferably using cryptographic techniques as the underlying decision factor. This could be a cryptographic hash of a password, or better, an authenticated TLS connection using an asymmetric key pair. However, cryptographic keys should be limited in their accessibility only to authorized programs or users. This becomes harder when we consider the next design principle.

5.4.4 Separation of Privilege Principle

The *separation of privilege principle* implies that two independent mechanisms should be used to control highly sensitive data or functions. This means two independent failures are required to overcome the protection mechanism. Separation of privilege is used in safety deposit boxes where two keys are required to open the box. It is also used in two-factor authentication, where you need both a password and some access token sent via text message to your phone to log into your bank. This principle is similar to the *separation of duties* principle,[10] which requires two separate persons to perform a function, like signing a check. In computer systems, it is vitally important that separation of privilege and separation of duties be implemented using compartmentalized or separated software or hardware elements. This principle should be used to protect cryptographic keys. In IoT systems, this is harder than it may seem at first due recursion of the problem. How do you protect a bunch of keys? Let's put them in a file and encrypt them with a special key! OK, where are we going to put that special key? We'll protect the special key under a special account that we protect with a password! Where do we put the password? Do you get the recursion? Because there is no human being to login to an IoT device or unlock some file with something only that person knows, the device itself has to unlock its secrets, and how does it do that? Let us look more carefully at this problem.

Assume you have an IoT system with an encryption key stored in a secure element or trusted platform module (TPM). The encryption key should only be used by an authorized program, and we want to use separation of privilege with independent conditions guarding the encryption key resource. The first condition could be the user or task identifier that the program is running under. In some systems, this alone could be an acceptable access control factor. However, we know these IoT systems will come under attack, and if a network attacker finds a buffer overflow in a program, they can leverage that to run arbitrary code on the device. Therefore, a single, simple buffer overflow will overcome the task identifier protection. We need another independent condition. Since we do not have a user to present a password, we can save the password in a text file and have the program present the password, as well its task identifier, in order to gain access to the key. The task identifier satisfies the access control principle of "something you are," while the password satisfies the access control principle of "something you know." Voila, two factors, right? Wrong. The file system leverages the user identity to restrict access to files, so even though the password seems to be a second factor, it is not *independent* of the first factor. An attacker can leverage the user identity impersonated through the buffer overflow to access the password file. Separation of privilege turns out to be a really hard problem without a completely independent kernel-level service mediating the second factor.

One mechanism that is being used in container execution engines is integrity checks. The container image is signed and an integrity check (e.g., a hash digest) is used to verify that the container's code has not been changed. The container execution engine caches this measurement and can be queried later to determine if the container is executing the correct version of the code and can be trusted. This definitely improves security, but it may not fully satisfy the security requirement we are aiming for, as the next principle demonstrates.

5.4.5 Complete Mediation Principle

The *complete mediation principle* states there can be no caching of access control decisions and that each access control decision must be performed at the time of access to ensure that the access rights have not changed. This is directly related to a time-of-check-time-of-use (TOCTOU) attack. TOCTOU is the window of time between an executed security check and the actual use of the object that was checked, which is the amount of time an adversary has to execute an attack. In the case of integrity checks on containers, the check is performed at container launch time. This means that any time after launch, a buffer overflow attack can dynamically change the code inside the executing container, and the changes will not be detected and will not have an effect on access control decisions. One solution to this problem is microservices. The container itself is a very tiny service that is brought up only when an action needs to be performed and is then shutdown. This limits the time of attack only to the time the microservice is executing.

5.4.6 Least Privilege Principle

The *least privilege principle* states that a program should only have the privileges they require, and no other privileges. This principle is most often violated by running a process as the administrator or root. The reasons one might run a process as root vary, but they are excuses, not justifiable reasons. Running a process with administrator privileges because root was the process that installed the software is an excuse not a reason. Often it is easier for the developer to write the program using only administrator privilege and not having to worry about how to drop privileges. Of course, if an attacker finds a buffer overflow in a process that is running as root, then the attacker becomes root. And in that situation, you can pretty much kiss your system goodbye, because in order to recover control of your system, you will have to completely rebuild it from scratch, or replace it.

The least privilege principle requires the architecture to set up multiple users or task identifiers with different privileges, and each process runs under a different, or limited, number of user identities. This separates what files each process can access, as well as what other resources are accessible, including keys, devices, message queues, and network ports. This is one of the harder things to do on an IoT system, but it is one of the most effective platform hardening countermeasures.

5.4.7 Least Common Mechanism Principle

The *least common mechanism principle* states that any system action or activity should be performed as independently as possible from other users. Stated in a different way, sharing between processes and users should be minimized. Any functionality used by multiple processes should have the least amount of common resources among them.

Violation of the least common mechanism principle is at the heart of every side channel attack or covert channel. Shared resources always leak. Always. Leaks can be minimized or slowed down, but they cannot be wholly prevented. Whenever there is a choice between a common shared service, or implementing a function as library that is dedicated to the caller, choose the library.[7] Unless, of course, there is no concern over leakage of data.

Containment and separation of state are the primary tools used to create a secure shared resource or service. The state used by a shared service potentially has information that could leak between the service users. Purging the state between uses of the service is the safest approach, but in many cases, this can result in performance impacts. Containment between users of a service is the other protection mechanism. A secure service must ensure that users cannot see or access the service's state or that each user is buffered from the data used by other processes. This is more difficult than it may at first seem. Global variables include state. Hardware devices that are not re-initialized to a pristine condition may contain state. The location of data in memory buffers and caches may change access times and leak state information. It is really difficult to share resources and not leak state information. The recommendation is to carefully review your shared resources for state and identify how much leakage there is and whether that amount of leakage is acceptable. If someone tells you there is no leakage, they either aren't being honest with you, they haven't done the research to actually know how much leakage there is, they don't understand the system, or they don't understand this principle.

5.4.8 Defense-in-Depth Principle

Though not one of Salterz's original principles, the *defense-in-depth principle* has become a commonly touted mantra of security professionals and is related to the separation of privilege principle. This principle says that if a security mitigation is warranted in a system to protect an asset, there should be a layering of independent defenses so that an attacker cannot gain access to an asset by overcoming a single defensive measure. Layering of independent defenses makes a system significantly more robust against attack.

We will use a sensitive configuration file for an example of defense in depth. The configuration file is used to set up a communication protocol used to send and receive data, and it contains sensitive URLs that we do not want an attacker to know and perhaps some passwords used for authentication. The configuration file is stored on the file system protected with file system access rights; the user identifier associated with the file is only used by the communication protocol application that should have access to the configuration file. Great, one defense! Now, let's encrypt the file using a key that is bound to the application through a Trusted Execution Environment (TEE). The TEE program is tightly coupled to the communication protocol application, and the TEE binds the encryption key to the platform so that the key cannot be exfiltrated from the platform. Perfect, second defense! Now, let's partition the communication protocol so that the URLs are only used at startup to open the connection, and the TEE provides those URLs only once after decrypting the configuration file inside the TEE. The passwords are only used inside the TEE and are never passed outside the TEE.

Let's be clear. This design is not flawless. However, an attacker has to overcome several defenses in order to misuse the sensitive information, and the window and avenues of attack are limited. A single buffer overflow cannot extract the sensitive data, and misuse of the data by an attacker requires an active persistent malware on the device—something that other security controls and defenses should be able to detect.

5.4.9 Trust No One Principle

Trust no one is another principle we add to Saltzer's set. Inevitably, within a design, some element must be trusted. We typically define this as the root of trust. The root of trust must be as small as possible and is the first thing that executes out of reset of the device.[*] Following the execution of the root of trust, everything else that is executed must be verified. It should not be trusted outright. This is accomplished through digital signatures using cryptographic algorithms. When loading firmware and software, first verify that it is from an authentic source and has not been modified. When communicating with other entities, authentication of the other party against a trust anchor, a public key that is securely stored as part of the root of trust, is vital. As the design is developed, question where trust is placed and how trust is established. Keep implicit trust to the bare minimum, and use cryptographic techniques to build on the root of trust.

5.4.10 Secure the Weakest Link Principle

Within any design, there is going to be a weak link. There will be an easiest place for an adversary to succeed in attacking the system. What is the easiest way for an attacker to compromise your system? It might be a weak authentication protocol, or no protections against physical attacks. Determining the weakness in your design requires careful analysis to determine whether the best mitigations have been put in place. But oftentimes it is the simple things

---

[*] Reset includes both startup and soft boot or return from sleep. In all cases, we want to ensure that nothing in our system has been compromised, tampered with, or replaced.

that attackers go after, so don't overlook those. A recent blog post from the United Kingdom's National Cyber Security Centre shares some practical experience on securing the weakest link[*]:

> *Russian cyber actors do not need to leverage zero-day vulnerabilities, or install malware, to exploit these [network routers and] devices. Instead, cyber actors take advantage of the following vulnerabilities:*
>
> - *devices with legacy unencrypted protocols or unauthenticated services;*
> - *devices insufficiently hardened before installation; and*
> - *devices no longer supported with security patches by manufacturers or vendors (end-of-life devices).*

The weakest link in these cases was open telnet, with default passwords set by the manufacturer. Not some advanced attack by crack Russian hackers. Fixing the weakest link is vital to providing any real security in a system. Anything else you do is wasted effort because the adversary is going to circumvent all that wicked jazz you built into the system and attack your weakest link.

When you perform a SAR, ask the security architect to list their weakest links—the things that are accepted threats in the system. Ask them, "If you were going to attack this system, what attack would you perform, and what could you get out of the system as a result? And how much would it cost you?" If the security architect says there is no way to attack the system, or that they do not know, fail the SAR and mentor that security architect using these design principles.

5.5 Addressing the Security Concerns of an Industrial IoT System

Now that we've reviewed the security architecture methodology and have a process for comprehensive threat analysis, this section walks through actual use of that process. The example draws from our set of foundational security design principles to properly evaluate architectural issues and address some security concerns in an IoT architecture. For our example, we develop an Industrial Internet of Things (IIoT) architecture loosely following the Open Process Automation (OPA)[11] initiative spearheaded by Exxon Mobil. As we develop this architecture, we demonstrate our security architecture process to uncover threats, evaluate mitigations, and document a secure design for an IIoT system. Along the way, we focus on specific architectural and security topics that are widely relevant to IoT, including:

- Manageability
- Device Trust
- End-to-end security
- Longevity
- Intelligence
- Scale

Before jumping into our exercise, it is important to highlight that this is an *exercise*, and, by definition, it is incomplete. It is a struggle to decide what to put into this chapter and what

---

[*] Retrieved from https://www.ncsc.gov.uk/alerts/russian-state-sponsored-cyber-actors-targeting-network-infrastructure-devices

to leave out. In fact, every time we reread this chapter we find more information that we feel is vital for you to have at your fingertips. Alas, at some point we must stop and declare that this is enough. So if there is one piece of advice we must leave with you, do not stop your investigation and learning of IoT security with this chapter. For a more exhaustive review of IoT security recommendations, take a look at the 226-page recommendation from the UK's National Cyber Security Centre.[12] This guidance document is a treasure trove of helpful insights, organized around thirteen principles from the UK's cyber security code of practice and should guide development, installation, and management of your IoT systems. The thirteen principles bear repeating here, although we have reorganized them slightly to flow from design guidance to operational guidance.

- No default passwords, ever
- Minimize exposed attack surfaces
- Ensure software integrity
- Provide mechanisms for secure software updates
- Securely store credentials and security-sensitive data
- Communicate securely, including authenticating endpoints and implementing message protections on communicated data
- Validate input data
- Make installation and maintenance of IoT devices easy
- Ensure personal data is protected
- Implement a vulnerability disclosure policy to inform customers of security problems in your devices
- Make systems resilient to outages
- Monitor system telemetry data
- Make it easy for customers to delete personal data

This list includes some things that are clearly design oriented—for example, validate input data. And there is other guidance that is clearly operational, such as monitor system telemetry. But other guidance can be taken from both sides—for example, no default passwords. This clearly means that as a system designer, do not create a system that includes default passwords or worse, uses hard-coded default passwords. From an operational viewpoint, verify that all passwords on devices have been changed and do not include any default (or easy to guess) passwords.

For the rest of this chapter, as we develop our industry use case, keep these principles in mind. Use them to expand on the details in this chapter and ensure that this guidance becomes an integral part of the questions you ask yourself and your fellow security architects and system designers as you evaluate designs.

5.5.1 The Autonomous Factory

Industrial IoT (IIoT) covers many topics, from oil refineries, to mining, to food processing, to manufacturing. For our example, we use an autonomous factory setting because it provides a rich set of potential hazards of varying severity that does not require our readers to consume significant background information in order to understand the scenario. In the sections that follow, several use cases describing IoT architectural flows are described and then analyzed for

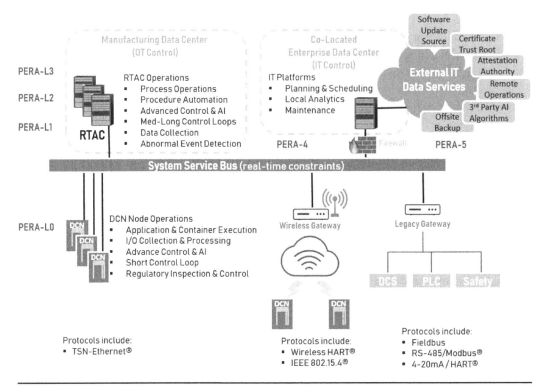

Figure 5.6   Exxon Mobil Industrial Vision (Adapted from Bartusiak, D. (2017, February 7). "Open Process Automation." Retrieved from https://arcadvisorygroup.sharepoint.com/sites/extranets/client-portal/arc-forums/_layouts/15/guestaccess.aspx?docid=0eecd23610e544dd7a7022dd097aad1ae&authkey=Abtp_UGVYT5OJg7VG5gGCm0.[11]

various security attributes. Due to space limitations, and because of the depth of architectural details that would be required for a complete architecture, our scenarios do not result in a fully developed autonomous factory. However, enough details are provided to demonstrate the security architecture process, allowing the reader to complete the process on their own actual architectures. For our scenarios, we chose elements that have broad appeal and applicability to many IoT ecosystems.

Today, industrial automation is morphing into what some call *Industry 4.0*,[13] loosely equating the IoT revolution in the industrial space to the fourth phase of the Industrial Revolution.* Two influential forces in this transformation are the Industrial Internet Consortium (IIC)[14,15] and the Exxon Mobil and Lockheed Martin partnership to develop the Open Process Automation (OPA) using standards published by The Open Group®. Exxon has publicly presented its vision for industrial automation, which is shown in Figure 5.6. From this vision and guidance from

---

* For the curious, the first Industrial Revolution was ignited by the steam engine in the 18th century; the second Industrial Revolution started with the use of electrical power just prior to World War I; the third Industrial Revolution began with the widespread use of digital electronics in the 1980s. IoT is said to have started the fourth Industrial Revolution.

the IIC, we develop a mock industrial system as a way to explore our security architecture process and practice developing security mitigations while addressing IoT requirements.

Our *autonomous factory architecture* (AFA) borrows from the baseline OPA vision developed by Exxon (see Figure 5.6). For some basic background, we review the underlying premise for the OPA vision. As Exxon points out, in the past, industrial systems have always been built in a hierarchical fashion. The Purdue Enterprise Reference Architecture (PERA) is a formal model describing this hierarchical organization of industrial systems into six different layers, with each layer being part of a particular zone. The physical implementation of PERA's hierarchical layers[16] is no longer preferred because it creates delays in information flow and increases costs to populate the different physical layers with equipment. In OPA, everything is connected like peers on a real-time bus, allowing for complete connectivity. There is no master hierarchy or series of layers that must be managed here, although the various PERA layers are logically created and annotated in Figure 5.6. The logical layers allow OPA to function in a similar fashion to PERA, although with reduced delays.

The operational controls for the system (PERA layers 1–3) reside in the real-time advanced computing (RTAC) platform—refer to the RTAC in Figure 5.6. The RTAC provides all the necessary Operational Technology (OT) controls, including configuration management, a historian to record the exact measurements from current processes, process simulation to verify configurations and commands before they are deployed, and performance monitoring and control of the downstream devices. Information Technology (IT) services (PERA layer 5) are provided by a private cloud co-resident with the RTAC, but are logically separate.

Analytics (PERA layer 4) are supported by a cloud instance that is a peer to the IT and RTAC; the analytics cloud may be a private cloud or a hybrid private and public cloud instance.

The downstream devices consist of a series of distributed control nodes (DCNs) that are attached to sensors and actuators in the actual factory. The specific devices and sensors connected to the DCNs are shown in Figure 5.7 and discussed shortly.

In Exxon's OPA, the DCNs are developed using specific types of design patterns following IEC-61499 and the OPC Foundation's OPC UA (see inset). A fair treatment of OPC UA and IEC-61499 is beyond our scope. In order to make our scenarios viable without going into the deep details of OPC UA, our architecture uses an open source Message Queue Telemetry Transport (MQTT) Publish–Subscribe protocol. This type of pub–sub model aligns well with the OPC UA, and we avoid the complexities of COM, DCOM, and OLE connectivity, with all the process automation enhancements created by the OPC.

As depicted in Figure 5.7, various physical machines are connected to the DCN nodes. A quick review of these machines may be beneficial. Automated storage and retrieval systems (ASRS) supply the factory line with the required parts and raw materials that are required by the autonomous factory's processes. These may be in the form of liquids delivered from tanks or electronic components, or plastic pellets delivered from bins. The level of raw materials must be monitored, and the flow must be regularly checked to ensure delivery remains consistent. Detecting when a material becomes low and replenishing the resource without taking the factory offline is important to maintain high-manufacturing flow and reduce waste from starts and stops. The "Tank Control & Liquid Flow" component in Figure 5.7 represents an ASRS.

The speed of the assembly line is controlled by the conveyer belt. The speed of the belt is verified by additional sensors, including a camera for artificial intelligence (AI) verification, as well as human visual checks.

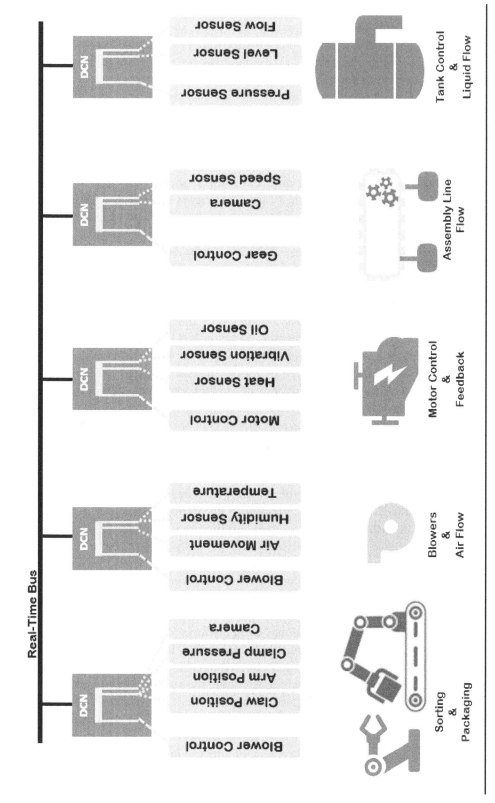

Figure 5.7 Example Open Process Automation Industrial IoT Installation

---

What Is OPC UA?

The OPC Foundation® was formed in 1994 to translate Microsoft Windows® Object Linking and Embedding (OLE) and Component Object Model (COM) software technologies to the world of process control. A first set of specifications was produced in 1996, which is now known as the OPC Classic model.

The Unified Architecture (OPC UA) is new set of standards that first appeared in 2006 and unifies the COM and DCOM models, providing considerable improvements, including security, binary and web translations, and multiple vendor information models for communication between devices.

OPC UA is a complex set of specifications (14 volumes and over 1000 pages). The specification is open only to members of the OPC.

OPC UA represents a significant attack surface for which we cannot do justice in this chapter. With this as the primary reason, we forego coverage of OPC UA to a more restricted protocol appropriate to the breadth of analysis that can be covered in a single chapter.[17]

Various machines on the factory floor are controlled by motors. The heat, vibration, and power usage of the motors can be monitored to detect maintenance requirements. Their operation can be controlled by loading programs from the RTAC onto the DCN that controls the exact operations of the machines.

Some factory processes are sensitive to temperature and humidity. Air handlers can be controlled to produce the right environment in the factory. If human beings are on the factory floor, monitoring for chemical contaminants or other environmental hazards can be integrated into the system, allowing immediate reaction to properly and safely vent dangerous fumes.

Robotic manufacturing can be controlled through programs on the DCN. Through the DCN, the RTAC can load the robot with the proper instructions and configuration based on the materials being processed in the factory. Automated sorting and packaging can be performed by robots and can be monitored for quality control.

In the following sections, we apply our list of IoT concerns to this general architecture and set of machines connected through the DCNs. We develop and explore specific architectural views of our AFA as needed for each section.

5.5.2 Architecting for IoT Manageability

The very first use case that comes to mind in our autonomous factory is *manageability*. The factory devices need to be configured, or reconfigured, for the current job running through our factory. The cost to manually configure all these devices is enormous. While working on an automation project in the late 1990s, Dave was told that the cost to switch an oil refinery from one product to another—for example, diesel fuel to unleaded gas—was more than one million dollars a day and took up to six days, and the whole time that the switch was going on, the refinery was producing hazardous waste that could not be sold as it was neither diesel nor gasoline. The benefit of automated configuration is that it not only increases profit, but it also reduces waste that poisons our planet.

The primary justification for automated manageability is to maintain a very low cost-of-ownership, allowing IoT systems to provide a reasonable return on investment (ROI). The other part of the reason is that IoT devices tend to be situated in places in which it is difficult for humans to perform management tasks. Therefore, nearly all administration must be done remotely and, whenever possible, automatically. Even in the factory setting, management tasks such as installing updates, rebooting the device, and reconfiguring connections to other systems should be done through virtual means, instead of sending a human being down to the DCN or devices.

Manageability is performed through a software agent on the device that is being managed. A management system in the cloud sends commands and receives status from the managed device. In the autonomous factory, there are two different management perspectives. The one most commonly understood is the IT management perspective. This perspective includes management of the operating system, software updates, and device health. The other management perspective is from the operational side. This includes installing certain applications and configuring the device to communicate with the appropriate sensors and actuators in order to accomplish the specific factory tasks assigned to the device.

From a security perspective, a management agent is a desirable target for an attacker, since very likely that agent must have administrative privileges somewhere in order to perform the update tasks and other management controls. And because the agent must communicate with the management system in the cloud, the management agent has open network connections that a remote network attacker might be able to leverage for an attack. For this reason, the management agent must be structured in the following way:

- **Separation of Privileges Principle:** The management agent must compartmentalize administrative-privileged actions into a separate process that does not have network accessibility.
- **Fail-Safe Default Principle:** The management agent must not have any listening ports; incoming connects are default deny. All network connections must connect out from the device to the management system, preferably using a whitelist of acceptable locations and certificates to verify the endpoint's authenticity.
- **Open Design Principle:** All management connections must be performed over a secure channel, using encryption, integrity, and message delivery protections on all messages. Message delivery protections include proof-of-source, message deletion detection, message replay protection, and message mis-ordering detection. The security of messages is dependent on open design cryptography, not obscurity or special formatted messages.
- **Complete Mediation Principle:** All management commands must be cryptographically verified as coming from a trusted source. Acting on a privileged management action should not be performed just because the management agent has such a command in its buffer. Each command carries with it a cryptographic authorization that is independently verifiable.
- **Least Common Mechanism:** The management agent should not share resources, memory, or services with other applications or services on the device. Such sharing represents an avenue for exploitation.
- **Defense-in-Depth:** The management agent should be protected by a host firewall on the device and utilize technologies that reduce the likelihood of buffer overflows (e.g., control-flow execution technologies, source code analysis during development, etc.)

The management agent in our AFA is built on the pub–sub model, with different parts of the management agent split into different processes, as shown in Figure 5.8 where each process

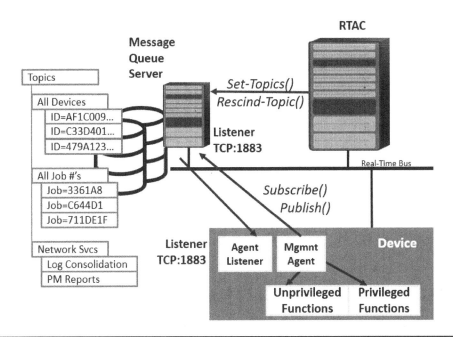

Figure 5.8   Autonomous Factory Manageability Architecture

is a separate box in the device. The management agent uses a separate listener process to connect to a message bus hosted by the Message Queue Server and subscribes to various topics, such as its unique device identifier, its model number or device type, and several other topics related to its current function, such as the job being processed, current network services the device may require, or special device services applicable to its connected sensors. The management system in the RTAC controls the message bus and sets or rescinds certain topics to the bus. The RTAC can then publish messages to the message bus, and these messages are sent to all the connected and subscribed devices. Through the bus and using only a single message, the RTAC can manage the entire system and notify all devices that are running a specific operating system, for example, that they should pull a new update or perform some action. The rough notional architecture for the message agent is shown in Figure 5.8. If we are to follow our process, our architectural diagrams should be more robust. The software architecture of the management agent should be provided in greater detail, as well as separate architectural diagrams for the message queue. Flow diagrams that show how messages are constructed and the order and content of messages between entities are needed. Finally, an asset diagram showing where critical assets are located and their importance is also required. Due to space restrictions, those diagrams do not appear here, but the outcome of their analysis is shown in the threat table for this design in Table 5.1.

5.5.3 Architecting IoT Device Trust

In our autonomous factory architecture, all devices are connected over the real-time bus and have access to every other device in the system. For all IoT systems, device interaction is essential to system operation. Devices report data up to cloud systems. Cloud management systems send commands down to devices. Interactions among devices allow sharing of environmental

Table 5.1   Autonomous Factory Manageability Threats

Status	Attacker	Threat	Assets	Mitigations
✔	Network attacker or software attacker on a device	An adversary injects unauthorized messages into the message queue, causing devices to perform management actions that were not intended, including rebooting devices, taking functions offline, or reconfiguring endpoints.	Commands to device and device availability	• Access to the message queue is restricted to authenticated devices. • Authentication is performed by device ID and password, presented during message queue channel establishment. • Topic management is restricted to the RTAC, and all operations to add or rescind a topic require a command digitally signed using the RTAC's keypair.
		An adversary rescinds topics in the message queue without authorization, causing topics that are required for system operation to disappear.	Channels to devices represented by topics	
✔	Network attacker	Data sent over the network to publish or subscribe can be modified by a network adversary; this could cause operations in the AFA to fail due to missing or incorrect data.	Data during transmission over message bus	• Connections to the message bus shall use TLS to protect the connection with encryption and integrity.
✔	Network attacker	Device is tricked into connecting to the adversary machine instead of the real message bus; adversary now has control of the device and can receive its published data.	Device data and device management control	• When establishing TLS connection to the message bus, devices perform authenticated TLS, verifying the public key of the message bus server against the trusted key in the device's local keystore.
✔	Device software attacker	Software adversary takes over a device and uses it to publish to control topics on the message bus, effectively performing management on other devices.	Device management control	• Publishing to topics is access controlled based upon the user identity associated with the channel to the message bus. • The message bus sets the channel's user identity based on the authentication information provided at channel establishment. • The RTAC assigns authorized topics writers to topics when the topics are created, and the message bus enforces those access controls.
Transfer	Software attacker on device	Data stored locally on the device from the message bus channel is accessible to other processes on the device, possibly malware; this could expose sensitive data in commands.	Information disclosure of data or commands to device	• Requires special configuration tests to verify device is properly configured—an operational not a development concern. • Local agent's message queue could be stored to local disk or database; configuration of device must ensure that this data is adequately protected from other processes and users on the device.
Partial	Network attacker	A network attacker could flood the message queue with many bogus messages causing denial of service on the queue.	Availability of the message queue	• Message queue detects overly chatty agents and disconnects agents from the message queue. • Repeated misbehaviors by agents are reported to the RTAC as potential devices that have been taken over by an adversary.

measurements and informing of device actions. All this interaction and connectivity is fine as long as everything is trusted and trustworthy. But how is this trust established? Is blind trust just a requirement of the system? I trust you because you are on my network?

Trust in our personal relationships usually happens in one of two ways: A previously unknown person can be introduced to us by someone we already trust, or we can have interactions with an unknown person, explore their background, and create an understanding in order to establish a *trust relationship* with them. Trust in devices is similar to the way trust is established in our own relationships, through introductions or interactions. However, an underlying assumption in this discussion is identity. Introductions require the identity of the person being introduced or the disclosure of the identity of the person or entity with whom we are interacting. Therefore, the first step of trust in devices requires identity.

Identity is not exactly a simple concept, and it requires some exploration. Your own identity, for example, is a complicated thing. At a particular point near the beginning of your life, you were given an identity in a name. This name was most likely unique within the context of your environment, which was your family or community. And you, or more likely your parents, received documentation to prove your identity, likely in the form of a birth certificate. As your environment expanded, some identity collisions occurred, which required you to add additional attributes to your identity to distinguish you from others (e.g., date of birth, place of residence). As your role changed, new identity credentials were issued to you to give you additional rights and capabilities, such as a driver's license and possibly a passport. These credentials allow you to quickly establish trust in certain venues. And you protect these identities to prevent others from impersonating you or misusing your identity.

Identity for devices is extremely similar. Processors today are manufactured with identifiers that are likely unique, such as a name. A processor (or chipset) unique number is built into the silicon itself, usually as a number burned into fuses, a value set into non-volatile locked memory, or perhaps a random value set by a series of Physical Unclonable Function (PUF) devices (perfectly unclonable functions). The identity is set by the manufacturer, often by a random process, and can include a manufacturer's signed certificate to attest that the value represents a silicon device the manufacturer in fact produced. When the processor or chip is incorporated into a device, the chip identity becomes the device identity. In fact, there may be several such identities in the device, and other identities may be injected into the firmware of the device as well. An important aspect of these identities is that they are immutable, and their usage is restricted. Additionally, the identity is normally tied into what is called a root-of-trust (RoT) and a RoT keypair.

The National Institute of Standards and Technology (NIST) defines a root of trust as:

> Roots of Trust (RoTs) are the foundation of assurance of the trustworthiness of a [device]. As such, RoTs are security primitives composed of hardware, firmware and/or software that provide a set of trusted, security-critical functions. They must always behave in an expected manner because their misbehavior cannot be detected. Hardware RoTs are preferred over software RoTs due to their immutability, smaller attack surfaces, and more reliable behavior. They can provide a higher degree of assurance that they can be relied upon to perform their trusted function or functions. (SP800-164 DRAFT,[18] p. 9)

There are different types of RoTs defined in SP800-164 and in common use in platforms today. The RoT for Measurement and RoT for Integrity are involved in the protected boot of a platform; the ROT for Verification performs verification of firmware of a platform to ensure that the firmware is not corrupted. Identity is part of the ROT for Reporting:

> Root of Trust for Reporting (RTR)—provides a protected environment and interface to manage identities and sign assertions for the purposes of generating device integrity reports. It has the capability to reliably cryptographically bind an entity to the information it provides. In the

case of device integrity reports, the RTR serves as the basis for the capabilities of integrity and non-repudiation of these reports. It necessarily leverages capabilities provided by other RoT in the system, including the RTM and RTI. (SP800-164 DRAFT,[18] pp. 9, 10)

The RTR is just a protected function on the device, implemented in firmware and hardware that contains the device's identity, and a cryptographic key, which is the RoT keypair. The RTR is used to present the device's identity in a trustworthy way along with other information that allows another entity to evaluate the information and possibly trust the device. The RTR's cryptographic key is normally an RSA key or an ECDSA key that it can use for signing. The information that the RTR signs is usually a set of claims it signs together with the identity. Let's give some concrete examples of RTRs in use today:

- A Trusted Platform Module (TPM) is a discrete chip or a firmware function on a processor that is provisioned with an Endorsement Key (usually an RSA key) and an EK Credential. The EK Credential is signed by the manufacturer proving that the TPM is genuine. The TPM also includes a RoT for Integrity, a RoT for Measurement, and a RoT for Storage. A TPM's signature on a claim or message can be verified by an independent trusted third party that the EK Credential is from a valid, unrevoked TPM.
- An Intel® processor with Software Guard Extensions (SGX), or Dynamic Application Loader (DAL), has access to a platform unique EPID* key that can be used to sign claims about software running in SGX or the DAL. The signature on the claims can be verified by an Intel service that the platform has a genuine Intel processor with an unrevoked and trusted EPID key.
- A key burned into fuses on an Arm® platform, where the fuses represent an RSA or ECDSA key, which is only accessible to the secure world applications in the Arm Trustzone® partition. Secure applications in Trustzone can use the key to present identity claims.

Given this background on the ways devices can contain and present an identity, let's explore how IoT devices use that identity to establish trust into an IoT system. There are really only three ways that devices establish this trust, each with increasing orders of complexity:

- During device manufacturing
- During device provisioning
- During information exchange

The easiest method of establishing trust between devices is during manufacturing. A specific key is generated or installed during device manufacturing, and then the public portion of the key is registered with the cloud management service. Microsoft® Azure Sphere® uses this method to register devices into their Azure® IoT Hub, Microsoft's device cloud. The drawback here is that the device must be allocated to a specific cloud service and customer at manufacturing time; this prevents (or at least significantly complicates) generic IoT device manufacturing.

---

* EPID—Enhanced Privacy Identifier—an elliptic curve type of key that can perform digital signatures which provide that the key is part of a group (e.g., a particular processor model number) without revealing the actual unique processor the signature came from. EPID is part of Intel offerings for advanced user privacy.

However, because manufacturing lines are very controlled and relatively secure environments, this method is secure. An additional drawback is speed of manufacturing; the devices must be powered up and communicate with the cloud back end in order to register their public key, which can significantly slow down a manufacturing line.

The next method of establishing trust is during provisioning. There are two general ways of performing provisioning: by hand with human action, or through an automated protocol. Provisioning by human interaction is common in IT today because most devices are issued to human beings. In IoT, the goal is to limit the human interaction because it is expensive. The technically skilled person capable of performing this type of provisioning usually demands a large salary; couple that with the trust this individual must be given to handle all the device keys, and the cost to perform human provisioning for IoT devices becomes a non-starter.

Automatic provisioning of IoT devices has been tackled by several companies and standards. Normally, this involves some type of minimal provisioning to a cloud service provider so that the cloud knows what devices to expect, and to which customer each device belongs, allowing the cloud to provision them. Intel's Secure Device Onboarding (SDO) is unique among these solutions as the cloud provider does not have to be chosen in advance, and no special credentials are required on the platform. SDO leverages Intel's CPU key pair (an EPID or ECDSA key) to authenticate a device with an Intel service, called the rendezvous service. The device owner registers their devices with the Intel rendezvous service and tells the service what cloud provider or private cloud system the device should be linked to. When the IoT device contacts the rendezvous service, the device is authenticated and redirected to the customer's preferred device management system. The SDO service can use a public rendezvous service or can be set inside a private network. The primary benefit to this solution is that any SDO device can be provisioned this way, and no special provisioning or tagging of devices is required during manufacturing.

Once a device has been provisioned into an IoT cloud management system, establishing trust between that device and other devices in the system is only a matter of having the cloud management system provide an introduction, so to speak. The cloud management system acts as a verifier of the identity of devices and provides the public key for one device's identity to other devices in the system.

This section has provided the essential background for device trust. Figure 5.9 depicts a RTR, with sole access to a set of hardware PUFs that define the device's identity. For our autonomous factory, registration of identities is performed through a proxy; this allows us to limit access to the IT cloud implementation of our trust services to a specific firewall filtering proxy. Specific attacks on an RTR and the device identity are very implementation dependent, as are the protocols for registering and using the device's identity. However, using the security design principles from this chapter, the following list identifies the security concerns when establishing trust between devices, and Table 5.2 provides a threat table outlining where things can generally go wrong.

- **Separation of Privileges Principle:** The device identity should be separated into a protected device function called the RTR. The RTR must produce consistent signed device attestation claims, with clear delineation between trusted device attributes and other device claims attached to the attestation.
- **Open Design Principle:** The security and immutability of the device RTR should rely on established hardware immutability principles (e.g., fuses and PUFs), not on a secret interface or unknown non-public API.

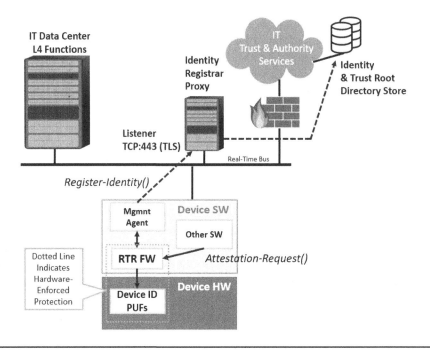

Figure 5.9   Device Identity Architecture

Table 5.2   Device Identity and Attestation Threats

Status	Attacker	Threat	Assets	Mitigations
✓	Physical or software attacker	Adversary lifts the device identity and clones other devices with the same identity, either during manufacturing or device operation.	Device identity	• Device uses PUFs technology, which is uncloneable. • The device's identity in the PUFs is readable only by the RTR firmware running on a special coprocessor on the device.
✓	Insider	Attacker records device identities during manufacturing and uses that information to clone devices.	Device identity	• PUFs are provisioned through random processes on the PUF silicon themselves and are not observable. • Registration of public identity certificate does not leak or reveal any information about the private identity in the PUF; the PUF value is used as part of a public key-generation function, such as ECDSA, where the public key is safe to reveal even to an adversary.
Partial ✓	Software attacker	Adversary on the Device uses the RTR to sign a false claim.	Attestation claim	• The RTR constructs a signed claim using information that the RTR knows to be true, including the true device identity and state of the machine, from a RoT for Integrity and RoT for Measurement. • The additional software claims added to the attestation may not be true and are the only thing that the SW attacker can manipulate—the receiver of the attestation must make security policy decisions based on the signed claim and any attached claims made by the software.

- **Complete Mediation Principle:** The RTR must be completely separate from other functions on the device so that the device identity key cannot leak or be misused.
- **Least Common Mechanism:** The RTR cryptographic functions, encryption algorithm, and/or hardware engines must be dedicated to the RTR function and may not be shared with other application usages on the device.

5.5.4 Architecting End-to-End Encryption

One of the buzzwords that travels around regularly is *end-to-end encryption*. It is a buzzword because in reality you can always claim your architecture meets it by merely redefining the word *end*. For instance, Does your browser connection from your laptop to the Amazon Webstore™ provide end-to-end encryption? The little lock in your browser URL bar tells you it is secure end-to-end. However, likely somewhere just after your data enters the Amazon® cloud, a load balancer decrypts your traffic and sends it in the clear to an Amazon web server that currently has a few spare cycles to process your request. Your data is not really encrypted end-to-end. But since you (mostly) trust Amazon, you consider that good enough. The same goes with IoT systems. End-to-end encryption of data may not reach each end entirely, but it is likely good enough. But let's look a bit closer at what should be good enough.

There are a couple of generic difficulties in creating end-to-end encryption, and one problem specific to industrial systems. The following list identifies these problems:

- **The Key Problem:** The first problem with encryption is getting the encryption key to both ends of the communication channel. For many years, key distribution was a major issue that limited the use of encryption. In 1976 when Whit Diffie and Martin Hellman published their ground-breaking paper on public key cryptography,[19] the problem of key distribution nearly vanished. However, the use of the Diffie–Hellman (DH) key exchange, while prolific in web and other standard protocols, is more difficult to perform on low-power microprocessors under restricted energy budgets. DH requires fairly large mathematical calculations, and these calculations take considerable processing power if dedicated hardware is not available. Many low-end microprocessors upon which small IoT sensor and actuator devices are based do not have the hardware acceleration for DH. This means that either the key exchange must be done in software, paying the price in energy and time, or manual key delivery would be required. However, even if the devices can perform the DH key exchange, if end-to-end encryption all the way to the cloud was supported, the cloud servers would need to maintain millions of keys, which is not practical. Alternatively, the end-to-end encryption is architected so as not to reach down to the device and instead terminates at the gateway. The cloud maintains a key with the gateway, and a different key is used between the gateway and the device, or there is, perhaps, no encryption from the gateway to the device, depending on the data sensitivity and the threats.
- **The Key Rollover Problem:** The second problem has to do with key rollover. An encryption key cannot be used forever; for security reasons, it must be changed because as more data is encrypted with a key, there is a higher probability that an attacker can successfully perform cryptanalysis and recover the encryption key. When we change a key, we call this procedure a key rollover, which must be coordinated between both ends of the encrypted channel. When the key distribution problem was easily solved using DH, key rollover became easy—both ends of the channel just negotiate a brand new key. But if

the key distribution reverts to a manually delivered key, then we must resurrect many of the old techniques that perform key derivation since key exchange is so expensive; but the rollover must also be coordinated to ensure both ends of the channel switch to the new key at the same time. Rolling over to a new key using these derivation techniques is not impossible. However, these techniques are not included in recent protocols and designs.

- **The Algorithm Upgrade Problem:** Another problem is the occasional replacement of an encryption or integrity algorithm. The recent upgrade from the SHA1 hash algorithm to the SHA2 family of algorithms caused numerous headaches for IT shops around the world. Although, these changes do not occur often, they do occur. This affects end-to-end encryption, especially when your cloud provider is using one set of algorithms and you want to use something different. And upgrades to IoT devices in the field are often more difficult than servers in the datacenter. One of the next sections, titled *Architecting for Longevity*, discusses some upcoming algorithm changes that will affect IoT and provides some recommendations to prepare for it.

- **The IIoT Operational Data Blackout Problem:** The final problem is rather IIoT specific. This problem occurs because end-to-end encryption effectively makes data in transit unreadable by intermediate nodes. In the industrial IoT setting, data is used for multiple purposes as it passes through nodes in the system, including short control loops, safety monitoring, and system-health analysis. If the data is encrypted, these intermediate nodes get blacked out and do not have the data that they need to operate. Some of the ways to solve the blackout problem include sharing the encryption key with all the intermediate parties (a difficult key management problem), transmitting the data separately to all parties (causing bandwidth waste), or not encrypting the data at all. The more likely solution involves encrypting the data hop-to-hop for privacy but providing an end-to-end integrity protection on the data to ensure that it cannot be modified. This last solution works best in IIoT.

Most sensors today do not have either the hardware capability or the software storage and processing power to perform encryption. Figure 5.10 shows the end-to-end security solution in our AFA. Data from the sensors is sent unencrypted to a control loop program running in a real-time virtual machine. That VM stores the data in a shared memory location as it processes the data. The control loop uses the data to update commands to the end device. A separate VM runs the integrity protection algorithms and tags all the data with meta-information before sending the data up to the RTAC. The data to the RTAC is not encrypted, but it is integrity protected. Because the industrial system should be a highly restricted network, confidentiality protection afforded by encryption is not a necessity. However, integrity protection using a message authentication code (MAC) or digital signature is warranted to prevent any accidental or malicious modification of the data in transit. The RTAC can then store this data in the historian function with the MAC value for long integrity protection and can also depend on the veracity of the data for other operational purposes. In designing an end-to-end encryption capability, the following list identifies the security design principles that should be observed, and Table 5.3 lists the analyzed threats:

- **Separation of Privileges Principle:** If encryption keys are stored instead of dynamically created using the DH key exchange algorithm, the privilege to access the stored keys should be separated into at least two independent privileges, as we previously discussed.

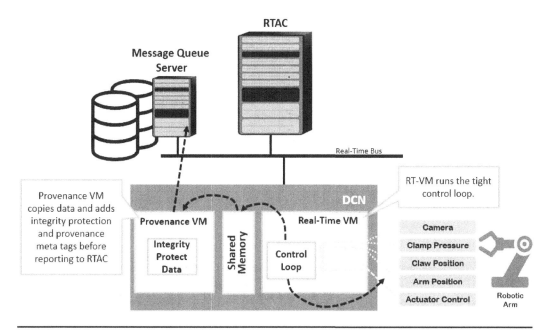

Figure 5.10   End-to-End Integrity Architecture for IIoT Control Loops

- **Fail-Safe Default Principle:** When data is deemed to be sensitive and requires protection, whether that is encryption or integrity protection, the architecture must ensure that such data does not transit without the requisite protections. Verification of encryption must be performed to ensure that a failure condition did not replace encryption with the null operation. Cleartext or unprotected data must be partitioned in the system to prevent accidental, unprotected release. In our AFA, the real-time control loop VM deals with unprotected data and does not have network connections onto the real-time network, so data cannot escape.
- **Open Design Principle:** All cryptographic algorithms and modes of operation for those algorithms should be from well-known and published standards with supporting academic cryptanalysis and security proofs. Standard algorithms such as AES, RSA, and ECDSA all fill this criteria. Modes of operation for symmetric ciphers are published by NIST.
- **Least Privilege Principle:** Software components in the design have the least privileges possible for them to complete their operations. Only a single software element should have access to encryption keys. Privileges include access to the network; therefore, system design must separate software elements that have network capability from other software elements, thereby minimizing the effect of network attacks.

5.5.5 Architecting for Longevity

Longevity is a critical aspect of IoT systems. IoT systems are generally expected to be in service for a decade or more. Although we do not specifically threat model anything in this section, guaranteeing security for long-lived systems is more difficult than protections we have

Table 5.3 End-to-End Encryption Threats

Status	Attacker	Threat	Assets	Mitigations
✔	Software attacker	Unauthorized software component accesses the cryptographic to access encrypted data or forge a MAC on unauthentic data.	Sensitive data	• Encryption keys are dynamically negotiated using DH with the remote party and only maintained in memory separated from other software elements (preferred). • Stored cryptographic keys are released only to a specific privileged software element of the VM by the hypervisor at VM-Start and are not available any other times. Refreshing the key requires VM restart.
✔	Network attacker	Network attacker modifies data in-transit or injects unauthentic data and sends it to the RTAC, or deletes traffic on the network.	Data in-transit	• All data is protected with an integrity MAC and metadata, including time-stamp and/or monotonic counter to prevent modification and detect missing data reports.
Partial ✔	Software attacker	Software attacker monitors shared memory to eavesdrop on data or modify data before it is integrity protected.	Data awaiting protections	• Hypervisor ensures shared memory is visible only to the two VMs; other VMs on the DCN do not have access. • VMs contain minimal software to perform their intended functions. • Misbehaving or compromised VM is still privileged to overcome protections and still a potential threat.
Partial ✔	Physical or wireless attacker	Attacker physically tampers with wired connections to sensors to disconnect or insert taps to record data, or attacker jams wireless sensor transmissions.	Data from sensors to DCN	• No mitigations against eavesdropping. • Nonresponsive sensors are reported to RTAC.

discussed for other design criteria. Updates for firmware and operating systems may not be available at all or be discontinued through normal distribution channels after only a few years, original hardware platforms may not be able to be purchased, and even programming languages and applications frameworks may be out of style and no longer widely used. As of this writing, systems running FORTRAN and COBAL are not unheard of, but they have become much more difficult to service and maintain. An IoT system must be architected for a long service life. But choosing the newest technology is not always the right answer either, as it might just be a passing fad. This often means that the newest languages and open-source projects should be passed over in place of packages that have been in service for a longer time period and are more likely to still be in service after a decade.

When considering longevity, choose software and hardware that are more likely to be available and in use even after a reasonable period of service—five to eight years or longer. Operating systems, databases, language frameworks, and programming languages, for example, should be sustainable. Hardware, and the firmware to run the hardware, must also be considered for longevity. If you cannot depend on others to provide those services for you, then you have to archive and maintain development systems to keep the IoT platforms operating, and integrate and test security and functional updates into the system yourself.

One aspect of longevity that is being discussed more frequently for IoT systems is the use of cryptography. The reason for this is the impending appearance of a large-scale quantum computer, and the threat such a computer would have on classical cryptography. Systems today are using cryptographic algorithms for encryption and authentication that are based on certain hard mathematical problems. The algorithms you use in your web browser on the Internet utilize these algorithms. The hard mathematical problems on which they are based are so difficult that a classical computer would need to spend decades to break them. Classical computers use a technique called *brute force*, attempting to try every possible combination of a key (the secret parameter to a cryptographic algorithm) to break the cryptographic algorithm. However, quantum computers operate differently from classical computers. Quantum computers utilize qubits to encode the state of a problem. Encoding of the problem creates entanglements between different qubits in the problem state. A solution to the encoded problem occurs as a kind of search across the qubits. It is theorized that if a large enough quantum computer could be built, the hard problems upon which our classical cryptographic algorithms depend could be solved very quickly. New cryptographic algorithms are needed that are based on even harder problems.

The debate is not *if* a quantum computer could break our classical cryptographic algorithms but *when* such a quantum computer will be built. As of 2018, quantum computers with 64 to 80 qubits have already been built. However, to break RSA or AES, one would need a quantum computer with thousands of qubits. There are significant engineering problems to be solved in order to build such a large quantum computer. Some speculate that it will be a matter of years, or perhaps a decade. Others claim scaling up the entanglement problem to thousands of qubits is more difficult than we realize, and it may take three or even five decades.

So the problem for IoT is its longevity—many IoT systems will still be operating in ten or twenty years. An IoT system built today that depends on classical cryptographic algorithms for security would be susceptible to quantum cryptographic attack if a quantum computer were built in ten years. If that IoT system were running a critical infrastructure, such as electrical generation or a chemical manufacturing plant, that vulnerability could result in a serious cybersecurity risk.

The solution should be to build quantum-resistant cryptographic algorithms into IoT systems today. But there are two problems with this. The first problem to address is algorithm selection, and the second problem is cost.

There are some cryptographic algorithms that we know how to harden against quantum computers. Symmetric encryption algorithms, such as AES, and hash algorithms, like SHA2, can be hardened by just increasing their key size. For AES, increasing the key size from 128 bits to 256 bits will increase the problem space enough to stump even a quantum computer. For SHA2, increasing the digest size from 256 bits to 384 bits will likewise protect it from quantum attack. However, public key encryption algorithms such as RSA and elliptic curve digital signature algorithm (ECDSA) will be completely broken by quantum computers; an increase in key size will not prevent a quantum computer from breaking the algorithm. Increasing the key size will prolong the algorithm's useful life, but we just do not know how fast quantum computers will grow in qubit size, so it is very difficult to know how large the keys should be. Complicating things further, there are very few algorithms available that can outright replace RSA and ECDSA. Hash-based signatures are one partial solution, covering digital signatures only. Although hash-based signatures have some drawbacks in key generation and key lifetimes, they are manageable. Other algorithms are being developed to perform key exchange, digital signature, and encryption, but they are still in the research phase under cryptanalysis and will not be production ready for five or more years.

The other problem is cost. There is impact in both development and in performance. Upgrading all the hardware to use algorithms with larger keys is going to cost more money. Larger keys in silicon will require additional fuses or other silicon structures that increase the die cost of processors. Additionally, as the key size used in algorithms such as RSA and ECDSA increase, the algorithms run more slowly. Slower performance results in a slowdown in device secure boot. Some devices would need to be redesigned to meet startup time budgets, with a greater dependence on hardware acceleration for all cryptographic usages. These are tweaks to existing hardware blocks and algorithms, so the development costs are not terribly high. However, this small cost per device adds up as you scale out to a large IoT system with millions of devices.

The prevailing wisdom from the National Institute of Standards and Technology (NIST) is to upgrade to larger key sizes for any system that will be in service beyond the year 2030. This means the following upgrades are recommended:

- Advanced Encryption System (AES) for symmetric encryption using 256-bit keys
- Standard Hash Algorithm 2 (SHA2) for hashing using 384-bit digest size or larger
- RSA Public Key Encryption using 3072-bit keys or, for additional security, 4096-bit keys

For long-life IoT systems, it is prudent to add a recovery system that can upgrade the firmware on the device to include new cryptographic algorithms. That recovery system should use a hash-based signature scheme that signs the firmware to verify the authenticity of the downloaded firmware. A hash-based signature scheme is secure even if attacked by a large qubit quantum computer.

5.5.6 Architecting IoT with Intelligence

Intelligence is a fundamental attribute of IoT systems. Intelligence can be used within the system to automate or improve a process, or to perform system management. Without intelligence, it is hard to distinguish an IoT system from merely a connected system. Intelligence causes the system to act or react autonomously based upon environmental and physical inputs. These inputs are typically not direct user or human inputs, such as activating a control, but may be indirect inputs based on human interaction with the system. For example, a traffic control system observes the volume of human drivers and may divert traffic to alleviate congestion. Or an autonomous robot may detect a human being is nearby and slow down its operation so a quick action does not take the human off guard.

Running AI in the cloud is relatively straightforward. The massive servers with multi-core processors and large amounts of memory and arrays of GPUs or specialized AI processors can easily churn through whatever is thrown at them. Increasingly, however, AI processes need to run closer to the devices, at the *edge* of the network. Edge computing with AI becomes rather difficult to perform on the constrained devices that make up the IoT devices on the edge. This is where the IoT gateway becomes indispensible. The IoT gateway can include the extra processors—GPUs and FPGAs—necessary to run AI algorithms, and their close proximity to edge devices allow them to react in short millisecond control loops. Let's perform our security analysis on such a control loop and discuss the protections and mitigations needed to run AI at the edge.

First, we need a more detailed architectural picture of our edge gateway and the resources needed to run AI at the edge. Figure 5.11 depicts our AFA, with a preventive maintenance control at the edge.

AI algorithms are used to predict when equipment needs service, or when the equipment is about to fail. Sensor readings for the temperature of the oil can determine if a motor is running with too much resistance, a vibration reading that is excessive can indicate when a bearing is about to fail, and heat readings can indicate when oil needs to be replaced. The placement of sensors is as important as the sensors themselves. Historical data provides input that the machine learning algorithms use to detect the failures.

The AI algorithms are stored in repositories and may be refreshed by the DCN as commanded by the RTAC. If the DCN is reconfigured to monitor other equipment, it can pull the appropriate code and algorithms from the repository. This allows the DCN function to be modified as conditions, functions, or loads change in the industrial factory. The DCNs do not have enough storage to maintain all the algorithms themselves, so they must depend on the repository to store them and download them to the DCN as needed.

Additionally, some algorithms may be leased third-party algorithms that must be downloaded from an external source. Third parties who have invested millions in research and development of AI algorithms are not going to release these directly to their customers, so some parts of these algorithms are likely going to execute in a trusted execution environment. This third-party upload is likely to happen only at specific times (at line startup) but may require upgrades to the algorithm at various intervals (e.g., scheduled maintenance or line change-up

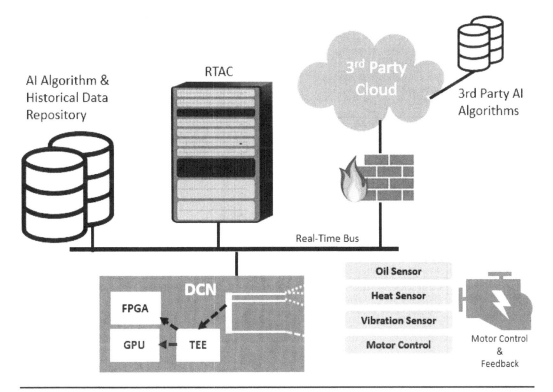

Figure 5.11   Artificially Intelligent Control Loop at the Edge

Table 5.4 Threats against IoT Artificial Intelligence

Status	Attacker	Threat	Assets	Mitigations
✔	Network attacker	AI algorithms are replaced or modified during download to the DCN. Entry: Network	AI algorithm	• AI algorithms are cryptographically signed to detect modification. • Signature verification key is stored in fuses, locked flash, or protected keystore in the DCN to prevent modification. • Verification of signature is performed in TEE or protected container to minimize threat of interference by malware.
✔	Network attacker	Replay of command to change the AI algorithm sent by attacker to the DCN, causing the AI algorithm to be changed. Disruption or destruction of the current operation is possible. Entry: Network	Control message	• Control messages are delivered over an encrypted and integrity-protected channel (TLS) to detect modification or spoofing. • Protected channel is set up only to authenticated RTC end points, verified with an authorization key in fuses, locked flash, or protected keystore. • Control messages include a sequence number to prevent replay attacks; sequence number is covered by encryption on the command packet.
Transfer	Network attacker Insider	Developer builds a back door into the predictive maintenance process, allowing the developer to send commands for the DCN to execute. Entry: Network	DCN control or destruction of attached equipment	• Not mitigated. • Factory operator must inspect or audit software, or inherently trust developers not to install back doors. • Secure development process with proper implementation reviews can be used to deter such attacks.
Partial ✔	Network attacker Insider	Attacker modifies historical data in repository causing AI algorithms to signal failures when there are none, or ignore warnings of failures. Entry: Network or physical attack on repository	Historical data Destruction of equipment on factory line	• Large data sets should be signed in a similar fashion to the containers and other code. • Verification of signature over the data may result in significant delay to startup time, if the data sets are large. Verification can be delayed and done after startup, with notification of failure to verify occurring after system is operational. Window of potential damage is minimized.
✔	Insider	Physical disconnection of sensors from DCN resulting in no change to read data, even though changes are actually occurring.	Destruction of equipment due to missed sensor readings	• Sensors must include a self-check operation that the predictive maintenance process can run at different intervals to verify proper connection and operation to the sensors.
✔	Network attacker	Radio jamming or interference with wireless network resulting in malfunctions during reading data from sensors.		
	Software attacker	Other software on DCN replaces GPU code, or overloads GPU with additional tasks that slows down AI algorithm.	AI algorithm	• No mitigation.
✔	Software attacker	Other software on DCN reloads FPGA, replacing AI algorithm with alternative implementation.	Destruction of equipment due to missed sensor readings	• FPGA should be locked after installing AI algorithm; changing the FPGA bitstream should require a soft reset of the device limited to privileged users. • For higher security, FPGA could be locked until soft reset of the DCN, which would require a privileged command from the RTAC.

times). Obviously, the network includes firewalls and intrusion-detection systems to protect the autonomous factory from attacks. For brevity, full details of network security will not be covered here as, in many ways, this network is no different from any enterprise network.

Whether the source is a third party or the local repository, the predictive maintenance process is loaded into a TEE or runs from a Linux Container. This process then loads the AI algorithm into the GPU or FPGA and controls the flow of data. As sensor data flows into the DCN from the motor, it is routed to the predictive maintenance process, which sends the data, as necessary, to the AI algorithm in the GPU/FPGA. The AI algorithm reports back out to the predictive maintenance process for appropriate action. At various intervals, the predictive maintenance process reports collected results data back to the RTAC. If a serious failure condition is detected, the predictive maintenance process can slow down or even halt the motor. This action might require coordination with other DCNs controlling other motors on the line so they can speed up and take up the slack, or it may start up backup equipment or slow down the operating line.

The edge sensors are connected to the DCN in device-dependent ways. Both wired and wireless connectivity are expected between the DCN and the sensors. Much of the time-budget in a control loop is spent in communications—preparing data to be sent, sending the data over a communications port, receiving that data at the destination, and converting the data into a usable command. Many shortcuts and optimizations can be architected into the system to reduce this time, such as preallocated buffers in shared memory and binary format commands with minimal translation between network representation and command consumption. The gateway resources are most important.

In our example, we will dispense with the preliminary threat modeling tables and list the adversaries and assets in bullet form. We will ignore the heat map table, and the entry and exit points will be incorporated into our threat table itself. The heat map is mostly to guide us through the completeness of the exercise and is not useful here; the entry and exit points will differ with your specific architecture and software stack, so concentration on those aspects is not as vital. Table 5.4 represents the threat list worksheet (Figure 5.5 from earlier in the chapter) and details the threat analysis of AI functions running in the IoT gateway, using Figure 5.11 as the architecture picture.

5.5.7 Architecting for Scale

Scale is an attribute of large systems that enables those systems to adjust to substantial, near simultaneous demand, or a rapid growth of system users. Scale is difficult to pin down because it is not a solitary element of the system, but many attributes spread across the system that sum together to produce robustness. Scale is not an architecture feature that can be addressed with our security design principles. But we are all familiar with the lack of scaling when the website crashes on Super Bowl Sunday, spurred by an instantaneous spike in visitors due to a halftime advertisement. Or the retail store's websites that go down on Black Friday or Cyber Monday due to extreme volumes of online shoppers. These types of crashes are caused by some element of the system that is overwhelmed by the volume of requests placed upon it, or by a sudden enormous burst of requests. There are many reasons why systems do not scale. Sometimes it is improper memory management, or other times it is poor control of sessions or threads. Other times, it is a failure to structure the database to respond to numerous queries in a reasonable amount of time. Eventually, any finite system can be consumed with too many requests. So how should we interpret scale in the context of IoT security? When we architect for scale we talk about *availability*, but in security we talk about *denial of service*. The primary difference is that the first is usually accidental, and the latter is malicious. In both cases, we must architect our systems to respond securely to a failure and then monitor the systems for those failures.

In IoT systems, we look at the three layers of the IoT architecture for different aspects of scale. The cloud systems must be architected to appropriately handle incoming traffic and data, while still being able to send out commands to gateways and devices. If necessary, the cloud should be capable of spinning up new instances to handle the influx of traffic; this is more easily configured when using a public cloud infrastructure such as Amazon, Azure, or Google Cloud™.

In gateways, incoming traffic needs to be properly prioritized so that messages and data are not lost, and commands from the IoT infrastructure are not delayed. This typically means segmenting the networks and ensuring that there is always sufficient processing power and network bandwidth to handle the gateway's tasks. For the gateway, one should also consider implementing a network firewall to ensure that connected devices are segmented from the rest of the network to protect them, and if they are attacked, they cannot be turned into zombies that fuel botnet attacks on your own network or others.

In devices, especially small devices, we do not usually think about scale of the device itself. Mostly we worry just that it has enough resources to do the simple job that it is intended to do with the meager set of resources we give it. However, we should consider what we enable on the device so that the device itself can participate in helping other parts of the system scale. For example, allowing the device to be configured with a time scheduler controlling when it will report back up to the gateway or cloud, or giving the device sufficient memory allowing it to buffer and consolidate data to minimize frequency of message transmissions.

Table 5.5 Threats to Scale for IoT Architectures

Status	Attacker	Threat	Assets	Mitigations
Partial ✔	Network attacker	Jamming wireless radio links between gateway and sensors.	Availability of sensor communications	• Non-responsive sensors are reported to RTAC by the DCN.
Partial ✔	Network attacker	Denial of service on a gateway upload link, severing its connection to the cloud.	Availability of gateway or DCN	• RTAC maintains a heartbeat to all DCNs; non-responsive DCN/ gateway will trigger an alert to IT team for investigation.
✔	Software attacker	Filling the message queue with bogus messages.	Availability of message queue	• Access to the message queue is restricted to authenticated devices.
	Network attacker	Transmitting bogus packets on the real-time bus.	Availability of RT bus	• No mitigation. • Access to the real-time bus is a physical attack that requires physical security.
	Network attacker	Network attacker that finds Software defect on DCN sends malformed packet that causes a DCN service to crash or entire device to reboot.	Availability of DCN service or DCN itself	• No mitigation. • Anomalous traffic should be detected by network SIEMS or IDS and identifiable from logs, so attacker or malicious software can be identified.

There are great books that are focused solely on building systems that scale. The architectural design patterns for scaling IoT architectures should cover databases, network stacks, applications, hardware, virtualization, as well as specific programming language techniques. Table 5.5 provides a limited discussion of scale, listing the primary security threats to scale for IoT.

5.6 Summarizing IoT Security Architecture

The process of security architecture is not complex but does require a disciplined approach. Completeness and predictable schedules in security architecture are achievable using simple techniques to chart the path through the systems assets, adversaries, entry and exit points, and · threats. Exposure to other's threat models and reading reports on how attacks are performed provide important data for the security architect to leverage in developing their own architectures. Experience and leveraging work from others, including reliance on established security design principles, are keys to success.

References

1. Killmeyer, J. (2001). *Information Security Architecture: An Integrated Approach to Security in the Organization.* Boca Raton (FL): CRC Press, Taylor & Francis Group.
2. Clements, P., Bachman, F., Bass, L., Garlan, D., Ivers, J., Little, R., Nord, R., and Stafford, J. (2003). *Documenting Software Architectures: Views and Beyond.* Boston (MA): Pearson Education, Inc.
3. Hofmeister, C., Nord, R., and Soni, D. (2000). *Applied Software Architecture.* Reading (MA): Addison-Wesley Longman.
4. Myagmar, S., Lee, A. J., and Yurcik, W. (2005, August). "Threat Modeling as a Basis for Security Requirements." Retrieved from http://people.cs.pitt.edu/~adamlee/pubs/2005/sreis-05.pdf
5. Amoroso, E. (1994). "Threat Trees." In *Fundamentals of Computer Security Technology* (Chapt. 2). Englewood Cliffs (NJ): Prentice Hall PTR.
6. Shostack, A. (2014). *Threat Modeling: Designing for Security.* Indianapolis (IN): John Wiley & Sons.
7. Bishop, M. (2003). "Design Principles." In *Computer Security: Art and Science* (Chapt. 13). Boston (MA): Pearson Education Inc.
8. Saltzer, J. and Schroeder, M. (1974, July). "The Protection of Information in Computer Systems." *Proceedings of the IEEE,* vol. 63, no. 9, pp. 1278–1308.
9. Gargiulo, J. (2002). "S-Box Modifications and Their Effect in DES-like Encryption Systems." Retrieved from https://www.sans.org/reading-room/whitepapers/vpns/s-box-modifications-effect-des-like-encryption-systems-768
10. Gegick, M. and Barnum, S. (2013, May 10). "US-CERT: Definition of Separation of Privilege." Retrieved from https://www.us-cert.gov/bsi/articles/knowledge/principles/separation-of-privilege
11. Bartusiak, D. (2017, February 7). "Open Process Automation." Retrieved from https://arcadvisorygroup.sharepoint.com/sites/extranets/client-portal/arc-forums/_layouts/15/guestaccess.aspx?docid=0eecd23610e544dd7a7022dd097aad1ae&authkey=Abtp_UGVYT5OJg7VG5gGCm0
12. UK Department for Digital, Culture, Media and Sport. (2018, October). "Mapping of IoT Security Recommendations, Guidance and Standards to the UK's Code of Practice for Consumer IoT Security." Retrieved from https://assets.publishing.service.gov.uk/government/uploads/system/uploads/attachment_data/file/747977/Mapping_of_IoT__Security_Recommendations_Guidance_and_Standards_to_CoP_Oct_2018.pdf
13. Schwab, K. (2017). *The Fourth Industrial Revolution.* New York (NY): Crown Publishing Group.
14. IIC. (2017, January 31). "The Industrial Internet of Things Volume G1: Reference Architecture." Retrieved from https://www.iiconsortium.org/IIC_PUB_G1_V1.80_2017-01-31.pdf

15. IIC. (2016, September 26). "The Industrial Internet of Things Volume G4: Security Framework." Retrieved from https://www.iiconsortium.org/pdf/IIC_PUB_G4_V1.00_PB-3.pdf

16. Obregon, L. (2015). "Secure Architecture for Industrial Control Systems." Retrieved from https://www.sans.org/reading-room/whitepapers/ICS/paper/36327

17. The OPC Foundation. (2018). "Unified Architecture." Retrieved from https://opcfoundation.org/about/opc-technologies/opc-ua/

18. National Institute of Standards and Technology. (2012, October). *Special Publication 800-164 (SP800-164)-Guidelines on Hardware Rooted Security in Mobile Devices (Draft)*. Retrieved from https://csrc.nist.gov/csrc/media/publications/sp/800-164/draft/documents/sp800_164_draft.pdf

19. Diffie, W. and Hellman, M. (1976, November). "New Directions in Cryptography." *IEEE Transactions on Information Theory,* vol. 22, no. 6, pp. 644–654. Retrieved from https://ee.stanford.edu/~hellman/publications/24.pdf

Chapter 6

Securing the IoT Cloud

Clouds come floating into my life from other days no longer to shed
rain or usher storm, but to give color to my sunset sky.

— Rabindranath Tagore[*]

6.1 The History of *The Cloud*

Take a moment, look outside the window—find a window, if you must, then look up! What do you see? Can you see the sky? If you can, you must see the clouds too. They are the fluffy-looking white or greyish objects floating around. A cloud is a visible mass of condensed water vapor floating in the atmosphere, typically high above the ground. Sometimes, one just wants to reach out and squeeze them. It's not impossible either, thanks to the thrill of skydiving— that all-time bucket list champion. However, your hand might just pass through the clouds, when you finally get up there.

You might wonder, as we did, about how the term cloud computing came to be used, although as you will see, the answer is fairly obvious. We will define "cloud computing" momentarily, but first it is worthy of note that the use of "cloud" symbols in fields of computing and technology dates back to the foundation of the Internet.

As we discussed in Chapter 1, the Internet started as a computer networking research project funded by the United States Department of Defense, called ARPANET. As early as 1977, cloud symbols were used to represent networks of computing equipment.[1] The cloud eventually became a metaphor for the Internet—the Internet connects us to information that's out there, information seemingly floating in the air, like clouds. Later on, in 1993, the term cloud was used to refer to platforms for distributed systems—systems whose components are located on different networked computers, which then communicate and coordinate their actions by

---

[*] Retrieved from https://en.wikiquote.org/wiki/Rabindranath_Tagore

passing messages to each other. Fast forward 13 years, Amazon® breathed new life into cloud computing with their 2006 announcement of the Elastic Compute Cloud™.

The Amazon Elastic Compute Cloud™ (Amazon EC2™) was and is a web service that provides resizable and configurable compute capacity in the cloud. EC2 is designed to make cloud computing easier for developers. It provides a web service interface that allows you to obtain and configure capacity with minimal friction. It reduces the time required to obtain and boot new server instances to minutes and allows you to quickly scale capacity, both up and down, to adapt to your changing computing requirements. Amazon EC2 changed the economics of computing by allowing users to only pay for the computing capacity that they actually use.[2]

Amazon's cloud compute offering, Amazon Web Services™ (AWS™), allowed customers to host and run applications seamlessly via Amazon-provided compute resources, and it quickly became the gold standard for cloud computing.

Today, there are many established cloud computing providers, besides Amazon, that prospective customers can choose from. Also, some organizations run private clouds, which allows them to have full control over the configuration and management of computing resources, as well as their data.

As described in Chapter 2, cloud computing represents a major attack surface for IoT solutions. Cloud-hosted services are usually used to administer, control, and configure myriad edge devices in an IoT deployment. For example, let's consider an immersive retail store with one thousand outlets. Digital cameras and infrared sensors are used to detect the presence and features of humans. That information is processed by machine learning models on edge gateway devices that are connected to digital kiosks next to clothing racks. The digital kiosks use the results of the machine learning inferences to make auto-suggestions on the right-fitting clothes for the customers. Afterwards, customers can use the kiosks to identify the location and stock of the recommended clothes.

In the scenario just described, let us assume that the IT department was to allow each outlet to manage its own technologies and processes for immersive shopping. Would you want to be the IT Director conscripted to ensure that immersive shopping works seamlessly across all one thousand outlets?

Instead, it is in the interest of the retail store to ensure that there is a centralized, adaptive, and automated means of deploying, handling, managing, and protecting software, hardware, and data across all one thousand outlets. That is precisely what cloud computing is able to offer IoT solutions vendors—that is, if well designed. As such, cloud-hosted services are a very powerful piece of any IoT system, making them an attractive target for attackers who wish to subvert or neutralize the system.

The objective of this chapter is to take you, the reader, on a hands-on journey into the security architecture and design of the cloud components of an IoT system. We will start by exploring what the cloud is; then, we will move on to explain why IoT needs the cloud. Afterwards, we will describe the major security concerns that are relevant to IoT systems. Finally, we shall define an IoT system, architect its cloud, and create a threat model that identifies the cloud-related threats to the IoT system, as well as the mitigation of those threats.

Rest assured, this chapter is not intended to be a conclusive volume on cloud computing. The field of cloud computing is a vast one, for which tons of books and blogs are already written, and continue to be written. Instead, this chapter focuses on the security concerns that cloud computing introduces to an IoT system. Some of those concerns are more "generic" cloud computing issues, which are well discussed in existing literature. We will mention such concerns in passing, while focusing closely on security concerns that are particular to the use of the cloud in an IoT system.

First though, let us get situated by gaining an understanding of what the cloud is and learning how the cloud is used to drive computing solutions.

6.2 So What Is the Cloud?

To design a secure system, we must understand that system, its functionalities, components, usages, and constraints. As such, a good place to begin our journey into the design of secure IoT clouds is an understanding of the cloud, what it means, and what it's made up of.

Let us start with the set of cloud definitions set forth by the US National Institute for Science and Technology (NIST). NIST defines cloud computing in this way:

> *Cloud computing is a model for enabling ubiquitous, convenient, on-demand network access to a shared pool of configurable computing resources (e.g., networks, servers, storage, applications, and services) that can be rapidly provisioned and released with minimal management effort or service provider interaction.*[3]

As the referenced NIST publication puts it so effectively, cloud computing or any cloud deployment has a few essential characteristics. Some of these are:

- **On-demand self-service:** A consumer can provision computing resources—such as processors, disk storage, and memory—as needed, without requiring human interaction with the service provider.
- **Broad network access:** The resources provided by the cloud must be available over the network and accessible via standard mechanisms.
- **Rapid elasticity:** Resources used by a consumer can be provisioned and released, in some cases automatically, such that it scales rapidly with the demands of the consumer. For instance, if a consumer were to run a series of computations, it should be possible for the actions of that consumer to cause an increase in the number of processors used at points of high-intensity computation, and later the cloud automatically releases extra processors when computation needs are reduced.
- **Resource pooling:** The providers computing resources should be pooled to serve multiple consumers, with different physical and virtual resources dynamically assigned and reassigned according to consumer demand.
- **Measured service:** Resource usage can be monitored, controlled, and reported, providing transparency for both the provider and consumer of the utilized service.

The cloud computing definitions in Table 6.1 further augment our understanding of the cloud, as they describe the various ways that cloud computing resources can be provided to consumers. This table describes how different kinds of clouds can be deployed, depending on who should have access to the computing resources made available via the cloud. The definitions in Table 6.1 are also drawn from NIST's SP-800-145 cloud computing standard.

Table 6.1 Cloud Computing Definitions

Service Models: The manner by which centralized computing resources are made available to a consumer.	
Infrastructure as a Service (IaaS)	The service or capability provided to the consumer is the ability to manage the provisioning and deployment of processing units, storage, networks, and other fundamental computing resources for the purposes of running arbitrary software, which can include operating systems and applications.
Platform as a Service (PaaS)	Software development building blocks are provided to allow consumers to build software applications that run on infrastructure managed by the cloud vendor. The consumer does not manage or control the underlying cloud infrastructure, including network, servers, operating systems, or storage, but has control over the deployed applications and possibly configuration settings for the application-hosting environment.
Software as a Service (SaaS)	The customer is provided with the ability to use a fully developed and centrally managed software application. The software building blocks and underlying infrastructure are completely managed by the cloud provider.
Deployment Models: This defines who has access to the cloud services—IaaS, PaaS, or SaaS.	
Private cloud	The services provided by the cloud are not accessible to the public.
Public cloud	Accessible to the public, in that different customers can host their data and apps on the cloud platform. It is the cloud service provider's responsibility to ensure that the apps and data of each customer are kept isolated from all others.
Hybrid cloud	This includes two or more distinct deployments of cloud services (private or public) that remain unique entities but are bound together by standardized or proprietary technology that enables data and application portability.

Source: Data derived from https://nvlpubs.nist.gov/nistpubs/legacy/sp/nistspecialpublication800-145.pdf

6.3 Cloud Architecture Overview

As a dear friend once put it, "Architecture is the means by which we understand complexity."[*] Architecture provides us with a playground through which we can experiment, analyze, understand, design, and thus, build systems. Through an architectural lens, we can identify required system components and discover or create interactions between components. In this chapter, we will look at the cloud as it pertains to IoT using our architectural lens. To achieve that objective, we will start by exploring the basic component architecture of a generic cloud system.

In Chapter 1, we introduced the IoT architecture and described it as the MGC architecture, with the cloud as one of the three primary components. Then in Chapter 4, we performed a deeper analysis of IoT architecture and introduced a generalized cloud infrastructure architecture based on OpenStack.[†]

As shown in Figure 4.10, any cloud computing infrastructure must contain six foundational components: object storage service, image service, compute service, block storage service, networking service, and identity service. These concepts are applicable to any cloud, public or private. While these components are completely abstracted away from a SaaS or PaaS cloud consumer, they are crucial for anyone operating at the IaaS level. Now, let us explore those foundational cloud components and learn why they are so important.

---

[*] Brook Schoenfield, Director of Advisory Services at IOActive™ Inc.
[†] Retrieved from OpenStack: https://www.openstack.org/software/

6.3.1 Object Storage Service

Object storage is a computer storage system in which data is managed as groups of objects that are unstructured, such that the objects have no precise relationship to one another.[4] Each object typically includes the data itself, some metadata, and a globally unique identifier in the storage system. Since an object is referred to and accessed with their unique identifier, it does not matter which machine the object is located on. All object storage machines are networked together within the cloud infrastructure to constitute the object storage system. Hence, objects can easily be distributed across various storage hardware in the data center. This is a primary advantage (and reason) for the use of object storage in cloud computing. Object storage is suited to data that is read very often and written infrequently—that is, it is suited to data retrieval rather than data modification.

Some examples of IoT use cases for object storage include the persistence of software images for IoT nodes, rulesets for running machine learning systems, and storage of device configuration files and policies.

Cloud infrastructure that supports IoT must provide a means by which data can be stored and accessed as objects. Some examples of cloud object storage systems are OpenStack Swift™, AWS S3™, Google Cloud Storage™, and Azure Blob Storage™.

Cloud object storage services must provide a means by which user access to data objects can be authenticated and user authorization can be verified. This is often done via the identity service, which we shall cover soon. Please refer to our practical security architecture of an IoT cloud in Section 6.6.2—for example, security architecture decisions for cloud storage.

6.3.2 Block Storage Service

Block-level storage in cloud computing involves the emulation of traditional block devices, such as physical hard drives. As such, block storage services provide a traditional block storage device—such as a hard drive—over the network.

Why is this important? With block storage, data is stored in volumes and blocks where files are split into evenly sized blocks. Each block has its own address, but unlike objects, they do not have metadata. Unlike object storage, it is possible to incrementally edit part of a file, which provides performance wins for data access.

Another advantage that block storage provides is read/write speed. Although access to an object in the cloud generally relies on HTTP protocol, block storage systems are mounted as a storage device on the server with underlying file system protocol (such as NFS, CIFS, ext3/ext4, and others) designed specifically for file operations. Object-based mechanisms should not be used for high-activity IO operations such as caching, database operations, log files, etc. Block storage mechanisms are better suited for such activities.[5]

A block storage device can serve as the primary boot device for a server or virtual machine* that stores the operating system, along with necessary system files. It can be used to host databases, support random read/write operations for database files, and keep system configuration files that vary based on the specific running virtual machine.

---

* Retrieved from https://techterms.com/definition/virtual_machine

Some examples of cloud block storage systems are OpenStack Cinder™, Google Persistent Disk™, Azure Virtual Disks™, and AWS EBS™.

6.3.3 Compute Service

Cloud compute services are the means by which cloud service providers enable consumers to create and manage virtual server instances running an OS or virtual machine. The cloud service provider is responsible for the provisioning of the necessary plumbing using hardware virtualization and networked resource pools, making it possible for consumers to set up virtual servers and clusters of networked virtual servers with desired processors, non-volatile storage, memory, and network access restrictions.

Unsurprisingly, compute services are central to cloud computing. They execute the VM images and operating systems from the image service and run the cloud applications which interact with other cloud services to present a complete cloud application. Some of these cloud services are, for instance, identity services used for authentication and authorization, storage services that provide access to data, and networking services to allow consumers to design virtual data centers.

The cloud computing consumer is responsible for architecting their applications and cloud instances to ensure that the IoT cloud application and the cloud environment are secure. For instance:

- Virtual servers (and their hosted applications) must utilize only secure and updated software, free of vulnerabilities; the IoT architects and designers must ensure the software images they are using have been updated, since the image service itself does not normally update the images stored there.
- Every network port must be closed on a server, except for those that are mandatory for the effective operation of the cloud-hosted solution; refer to Section 7.3 for a complete list of network security recommendations.
- Ingress and egress traffic to virtual servers must be restricted via defined network rules.
- User access to server instances must be restricted via authentication.
- Secrets must not be stored on servers, except when encrypted block storage is in use. Other options for storing secrets are tightly restricted object storage groups or cloud-based Hardware Security Modules* (HSMs). A hardware security module (HSM) is a physical device that is specifically designed for the secure storage, usage, and management cryptographic keys. They usually come in the form of a plug-in card for a system that requires it or an external device that is connected to a system or network server.

Some examples of cloud compute services are Google Compute Engine™, OpenStack Nova™, Azure Compute™, and AWS EC2™.

Please refer to Sections 6.6.2 and 6.6.4, as we architect our example IoT cloud to ensure that our virtual servers are configured securely.

---

* Retrieved from https://www.sans.org/reading-room/whitepapers/vpns/paper/757

6.3.4 Image Service

At its core, an image service is a repository of virtual machine (VM) images that can be used to create a new VM. What's a VM image you might ask? It's simply a read-only file containing a preconfigured operating system environment (along with any necessary applications) that can be used to start up a new virtual server. For instance, image services will often have images corresponding to different versions of Windows® or Linux®. Images can also include machine learning software or web development software, depending on the purpose of the image. Please note that these are just examples, and VM images can contain any operating system or software.

Cloud service providers usually host an official VM image service or marketplace. The images available via the marketplace are verified to comply with the cloud service provider's security and quality requirements. However, we must add that some cloud service providers support community-driven image repositories that are not verified by the cloud service provider.

Our point? The responsibility lies with the cloud computing consumer to ensure that the VM images used to deploy their server instances are provided by a trusted source.

6.3.5 Networking Service

Without networking, cloud computing ceases to exist. Imagine a data center and all the networking capabilities that are required to keep it running—routers, firewalls, bastion host subnets, access control lists (ACLs), etc. In cloud computing, all those capabilities must be available in a virtualized form, such that cloud computing consumers can define their virtual data centers to house their virtual servers and cloud applications. The cloud computing consumer is responsible for configuring their network to ensure the security of the virtual data center and all that it contains.

Some examples of cloud networking services are Amazon VPC™, Google VPC™, OpenStack Neutron™, and Azure VNet™.

Please refer to Section 6.6.4 for more details on how we securely architected our virtual data center to support our practical IoT cloud example.

6.3.6 Identity Service

In any cloud computing platform, the identity service is a central pillar that holds up the security of the entire platform. The identity service allows us to manage user access to services and resources hosted in the cloud. We can also use the identity service to authenticate and authorize one service or resource to another service or resource. For instance, ensuring that an application is only able to read from, but not write to, a database. If the identity service is not well designed, it will be hacker Christmas all day every day for such a platform and the applications hosted on it.

The core concepts in identity services are:

- **Users:** Individual user accounts can be created with security credentials such as passwords, access keys, and multi-factor authentication. Permissions can also be granted that define what operations users can perform. Other cloud services are then configured to use the identity service to authenticate and verify authorization for operations or access initiated via user accounts. Users can be administrators who manage the cloud solution and infrastructure, end users who need permissions to access cloud-hosted content, or systems that require automated access to cloud-hosted content.

- **Groups:** It is tedious to manage permissions for every user across a large system. Groups are used to reduce the overhead and maintenance of permissions. First, specific access permissions are defined for cloud-hosted resources. Next, different user groups are created based on the actions or permissions that are required for the users in the group with regard to cloud-hosted resources. Hence, management of user permissions is simplified, since an administrator can manage permissions for a group and thus manage permissions for all associated users simultaneously.

- **Roles:** Roles are similar to user accounts but instead of identifying a particular user, they define a particular function a user would fill, such as a *backup administrator*, or a *database superuser*. Permissions are assigned to roles to control the operations that can be performed by any entity operating in that role. Roles are really different than groups. A user can be assigned to multiple groups at the same time, and any group that includes a particular permission can allow a user to perform actions allowed by that group. This means users can perform actions according to the union of all the groups to which they belong. On the other hand, a user can only have one role at any point in time. Roles operate as the intersection between a user's permissions and the role permissions, restricting the actions of a user to the role they have assumed at the moment. Roles are used in system administration to prevent a user performing a certain role—such as database querying tasks—from accidentally doing something that exceeds the permissions of the role.

- **Permissions:** This is simply a definition of an action that can be performed on a resource. Permissions can be assigned to users, groups or roles. Different cloud services will require different permissions for different actions. For instance, a networking service will have permissions that allow the creation or modification of virtual data centers, subnets, or network-filtering rules.

The identity service is responsible for pulling together users and groups, roles, permissions, and cloud services/entities to ensure that proper usage of the identity service significantly reduces the security risk for cloud-hosted systems and applications. In Sections 6.6.2 to 6.6.4 you will see us making several identity, authentication, and authorization decisions to keep our example IoT cloud secure.

6.4 How the Cloud Enables and Scales IoT Security

In this section, we will take a closer look at what the cloud brings to IoT systems. The benefits of the cloud in IoT systems are easily placed in three buckets: centralized data management and analytics, IoT device management, and multi-presence access. Let us explore each in turn.

6.4.1 Secure Centralization of Data Management and Analytics

An IoT deployment can comprise thousands to millions and even billions of devices collecting data. The cloud provides a necessary centralized means for us to store, secure, aggregate, and analyze that data. With the cloud, we can view IoT data streams as a whole, pick apart and compare individual streams to each other, and produce actionable intelligence to run out systems better and more cost effectively.

Cloud-based object storage services efficiently store and offer up rulesets, programs and AI algorithms for advanced cloud analytics, while block storage volumes are used to store the data that analytics applications crunch on to identify meaningful patterns and trends.

The usage of the IoT cloud for storing and processing data creates three security problems that are rather unique to IoT:

- How can we provide authentication to all devices of an IoT system at scale?
- How can we ingest data from all these devices without the incoming data looking like a DDoS attack?
- How can we provide real end-to-end integrity for data with real data provenance that can be used for data analytics?

Figure 6.1 depicts these cloud-specific issues. Let us consider each, in turn.

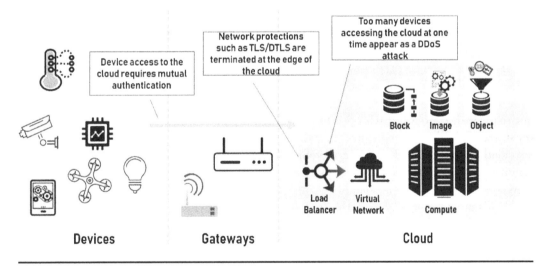

Figure 6.1   IoT Security Issues Created by Using the Cloud for Data Collection and Analysis at Scale

Authentication at Scale

In computer science, a popular maxim is "garbage in, garbage out." This means that flawed input data will cause even the best designed computer programs to produce flawed results. This is true for IoT systems. The information or knowledge produced by an IoT system is highly dependent on the data received by the IoT system. Hence, the system must be designed to ensure that the data processed by the system is received from trusted data producers that are properly authenticated. Additionally, the scale of IoT systems means that any viable authentication mechanism must scale accordingly.

Cloud Authenticates Connected Devices

The primary means for authenticating data producers in an IoT system is the authentication of edge IoT devices themselves, which are the data producers. The cloud must be able to ascertain that it can trust those devices. How? The principles for architecting IoT device trust are described in Chapter 5, Section 5.5.3. It covers how unique identifiers are created and securely provisioned to IoT devices, and how those identifiers are used to authenticate devices to cloud platforms.

As mentioned in Section 5.5.3, the scale of IoT systems means that automation is the only viable means for provisioning authentication identifiers to IoT devices. In an attempt to address the scale problem, some might consider using the same identifier to authenticate multiple devices, but that is WRONG! If the same identifier were used to authenticate multiple devices, compromise of the identifier on a single device compromises the data from all other devices that share that identifier. Authentication identifiers must be unique to each data producer.

An authentication identifier that is often used is an RSA key pair that includes a device certificate containing the RSA public key for that device. To ensure that the cloud is able to validate the device certificate, the approach we recommend is the use of a certificate authority (CA) to sign the device certificate. The CA-signed device certificate and the device private key are pre-provisioned onto secure storage on the IoT device. In this approach, device authentication involves the device performing a registration step by sending its certificate to the cloud inside a signed registration request. The cloud must verify that the certificate is valid (signed by a trusted CA) and has not been revoked by the CA. The registration is validated by its signature made with the device's RSA private key, which the cloud is able to verify using the public key stored in the device certificate. Since the certificate is signed by a CA that the cloud trusts, the cloud trusts the certificate and therefore the device. This solution requires an expensive manufacturing step to provision the device secrets. Other solutions that provide zero-touch provisioning are possible and should be considered for IoT systems.

At this juncture, it is noteworthy that automated re-provisioning of device credentials to an already-deployed system is another hard problem. A solution that tackles both zero-touch provisioning and re-provisioning problem is Intel's Secure Device Onboarding (SDO), which is also described in Chapter 5, Section 5.5.3.

Devices Authenticate Cloud Instances

It is just as important for devices to be able to authenticate the cloud. This ensures that legitimate data is not sent to a rogue cloud. Cloud authentication is usually achieved by securely provisioning devices with a digital server certificate containing the public key associated with the cloud back-end application. The server certificate is used during the authentication step of a TLS or DTLS session establishment, which allows the IoT device to verify the identity of the server or cloud to which it is connecting.

Data Ingestion at Scale

In Chapter 2, Section 2.1, we described how compromised IoT devices were used to mount a successful distributed denial-of-service (DDoS) attack on a cloud service provider, DYN, which caused many major web services and websites to be unavailable for many hours.

Naturally, this begs the question, How can you ensure that your IoT devices or devices belonging to others (since you have secured yours) cannot be used to cripple your IoT cloud? The responsibility for managing data ingestion at scale is shared by the architects of cloud services such as those built into IoT clouds—and the cloud service provider. For IoT architects, managing the flow of incoming data using gateways, as discussed in Chapter 4, and using special protocol and formats as we will discuss in Chapter 7, is vital to preventing data ingestion overload in the cloud. For cloud service providers, managing regional bandwidth requirements and mitigating DDoS attacks are critical. Since the attack on DYN, the use of DDoS attack mitigation has become critical for cloud service providers.

All major cloud service providers, such as Google Cloud Platform, AWS, and Microsoft Azure, publish detailed guidelines for architecting cloud services on their platform to ensure the detection and mitigation of DDoS attacks. As of the time of this writing, some of their primary publications are *Azure DDoS Protection: Best Practices and Reference Architectures,*[*] *Denial of Service Attack Mitigation on AWS,*[†] and *Best Practices for DDoS Protection and Mitigation on Google Cloud Platform.*[‡]

You will notice that the solutions recommended by the different cloud solutions providers listed above are similar. Highlights include:

- Architect your services and servers to intelligently scale out based on demand, using limits to control scaling, and using gateways to collect and upload input from multiple devices to the cloud at approved pre-scheduled times in order to limit device-to-cloud connect times.
- Identify the volume of traffic that is normal for your system. Use this information to configure alerts to system administrators, scaling limits, and automated mitigation options.
  - Some cloud service providers use machine learning to automatically learn the normal volume of traffic for each hosted application.
- Architect for resilience by using specialized firewalls to automatically filter out malformed traffic that is used for some DDoS attacks.
- Minimize your attack surface by limiting the systems and services that are accessible over the internet.

The reader is invited to explore these resources since the same principles are used to provide and protect data ingestion at scale for the IoT cloud. If you use a different cloud provider, they should provide similar guidelines. If you are the cloud service provider—that is, you are building the cloud hosting platform that runs your IoT cloud—we still encourage you to review the resources mentioned previously since they describe DDoS protection approaches that you will need to build into your cloud hosting platform.

---

[*] Retrieved from https://docs.microsoft.com/en-us/azure/security/azure-ddos-best-practices
[†] Retrieved from https://aws.amazon.com/answers/networking/aws-ddos-attack-mitigation/
[‡] Retrieved from https://cloud.google.com/files/GCPDDoSprotection-04122016.pdf

Data Integrity Protection

For the IoT cloud, the primary data integrity concern is that data supplied by edge IoT devices could be tampered with by malicious actors. To prevent this, we shall consider data integrity protection from four major angles.

Network Protection

During transmission of data from edge devices to the cloud, data integrity is usually protected via encrypted communication channels. Two primary ways of achieving this are via the Transport Layer Security (TLS) and Datagram Transport Layer Security (DTLS) protocols. Those mechanisms are described in the prior Data Authentication at Scale section and Chapter 5, Section 5.5.4. In the latter section, we described how to create a secure, mutually authenticated TLS or DTLS channel that involves the cloud platform authenticating IoT devices and IoT devices authenticating the cloud platform.

Data Integrity Tags

Although TLS and DTLS are used to transmit data securely, most cloud platforms terminate those protocols at the edge of the cloud. This means that as the data from the IoT devices enter the cloud, the encryption that protects the information in transit is removed, leaving just the data from the edge device. However, at this point, based on the assumption that prior data protections were applied to the data while both at-rest and in-motion, the cloud platform usually just trusts that the data it is about to process has not been tampered with since it comes from a device that it trusts and the secure tunnel was terminated within the cloud provider's data center. In very sensitive operations, it may make sense to add additional integrity protection over the data prior to sending this data over a TLS or DTLS session, which allows the back-end cloud instance to confirm the data's integrity truly from end-to-end. For instance, an HMAC-SHA-256 integrity tag, or the use of Poly1305 or GHASH to compute an integrity tag over the data, would provide this protection. The tag is created using a secret key and the data, to produce a tag, is usually 128 bits in length. The tag can be truncated to reduce the overhead of integrity, but this should be done with caution—the smaller the tag, the easier it is for an adversary to forge the data or the tag. Normally, tags less than 64 bits are not recommended. It should also be noted that both the device and cloud instance would have to share the integrity key. Adding this tag to the data guarantees data provenance and ensures no corruption—accidental or malicious—can modify the data in transit through the cloud system.

Boot Time Software Identity Verification and Cloud-Based Attestation

Data integrity also depends on the device itself and the software the device is executing. Completing this integrity protection angle involves verifying the source of software components used to start up the device and proving that this software hasn't been modified or corrupted. Normally, this is accomplished with the aid of a hardware-based Root of Trust (RoT) and a cloud-based device attestation service as part of reporting the device's secure boot or measured boot status. RoTs are described in Chapter 5, Section 5.5.3, and management agents that can report on boot status and device integrity are discussed in Section 5.5.2. The way the cloud verifies and uses device status information, such as secure boot information, is covered in the next subsection on device health monitoring via a device management agent running on the operating system. However, if the software that loads the operating system or the operating system itself is compromised, we can assume that the device management agent is also compromised. Hence, the need for the platform RoT to report the secure boot status in an attestation.

There are two primary approaches that can be used to achieve verification of software identity during device boot. They are referred to as secure boot and measured boot.

Verified boot is usually kicked off by running start-up code from a non-writable (and thus protected) component in the processor's memory map—for instance, a ROM or locked flash device.[6] The start-up code starts what is referred to as a Chain of Trust for Signature Verification. This means that in the process of booting up the system, each entity that is started is responsible for verifying the signature of the next component to be started and on and on. For instance, the start-up code verifies the signature of the rest of the firmware and the firmware verifies the signature of the OS kernel before booting up the OS. In closed systems, verification of the chain of trust can be continued beyond the OS to individual applications, and as far as is desired by the system's architects. The security policy set by the system owners dictates what takes place in the event of a failed signature verification during boot. A common and simple policy is merely to halt the boot. For IoT systems, this means that a truck has to roll out and fix the device—an expensive operation. A different policy could be to roll back to a previous version of the operating system stored in Flash, or to boot from a secure network store. An advanced solution is to execute a special recovery operating system in flash and have that operating system communicate with the cloud to repair the system software on the device. Verified boot solutions are normally created to perform some action if the signature verification fails; they never just continue to boot the system.

Measured boot starts the same way as verified boot, but it operates differently. Measured boot involves each component measuring and storing the hashes of the next software component that is to be loaded in the boot process. The hash measurements are usually stored securely via a hardware chip known as a Trusted Platform Module (TPM). The TPM's Platform Configuration Registers (PCRs) are used for storage. An important feature of PCRs is that they cannot be overwritten, only extended. Hence, as the platform boots, each measurement can be accumulated in the PCRs in a way that unambiguously demonstrates which modules were loaded—assuming you know which modules were supposed to be loaded. Once the PCRs have been collected, the next step is to report the values, signed by a key that only the TPM can access. The resulting data structure, called a Quote or Attestation, gives the PCR values and a signature, allowing the Quote to be sent to a cloud-hosted attestation server. The server can examine the PCRs and associated logs to determine if the device is running an acceptable image.[7] For this to work, the cloud must maintain a database of measurements or hashes that correspond to an acceptable system start-up state.

Measured boot (without verified boot) does not involve the verification of signatures of each boot component that is to be executed. As such, measured boot will not prevent booting with a compromised software component, although the questionable system should be discovered afterwards via attestation. Measured boot can be combined with secure boot to provide a more holistic solution.

Runtime Anomaly Detection and Device Health Checks

As described in Chapter 5, Section 5.5.2, device management is usually achieved via the aid of a cloud-hosted device management service and a device management agent running on the device. The device management agent is responsible for processing device updates sent by the cloud, monitoring device health, and transmitting device health data to the cloud.

As an added layer of defense-in-depth, cloud-based services should be used to perform anomaly detection on device health data. For instance, the cloud should be used to define and store expected thresholds such as CPU utilization and network traffic volume, in relation to the routine usage or functions of connected IoT devices or device types. Afterwards, periodic device health analysis via simple rule-based systems or more complex machine learning systems can be used to detect unexpected behavior in IoT devices. As an example, an IoT device that has been measured and tested to consume no more than 50% CPU utilization should trigger an alarm and extra investigation if it's noticed to run at 90% CPU utilization for extended periods. Similarly, if the cloud expects to receive no more than a few megabytes an hour from an IoT device, device health checks that show greater traffic volumes being processed by the network card should cause a system event to be triggered in the device health management system in the cloud, and investigation or remediation of the device should be undertaken.

6.4.2 Secure IoT Device Management

IoT device management is the process of provisioning, authenticating, monitoring, and maintaining IoT devices. Device maintenance usually refers to the process of updating the firmware and software required to keep a device operational and compliant with functional, quality, and security requirements.

The central position of the cloud in IoT systems—the cloud's necessary robustness and the large amounts of compute power available—makes the cloud the ideal control center for IoT device management.

Device Provisioning, Registration, and Re-provisioning

Imagine that you are the CEO of a fridge manufacturing company. Your company has been exploring IoT and has just produced a fridge that is able to sense when food is going bad and notify the owners. Due to the business acumen and customer-first approach of your employees, your product is in demand around the world. You have five million preorders. All your manufacturing plants are in overdrive as your company races to make your release date.

This is your first real IoT rodeo, so it can be expected that there will be a few bumps in the road. In previous pilots with select partners and customers, field engineers have been dispatched to configure the fridges for registration with your cloud back end. However, it is evident that such an approach cannot scale to millions of customers around the world. In a recent meeting with major partners, it has also become clear that your customers expect the fridge to operate securely out of the box, with minimal configuration required.

That is the essence of IoT device provisioning. It involves two major steps:

1. The establishment of the initial connection between the IoT device and the cloud-based IoT solution, and the registration of that device.
2. The application of the appropriate configuration to the device based on the requirements of the cloud solution it was registered to.

Ideally, both steps should occur automatically, once the device is powered on. Provisioning should also occur securely.

To prepare an IoT device for provisioning, there are usually two distinct steps:

1. The manufacturing step, in which the device manufacturer is responsible for preparing the IoT device for provisioning. This usually involves building in the address of the device provisioning service, the private key, and the device certificate that will be used to authenticate the device to the device provisioning service. The private key and device certificate should be securely stored via a Trusted Platform Module (TPM), trusted execution environment (TEE), or other hardware-assisted security component.
2. The cloud setup, in which the IoT system owner/operator configures a centralized device provisioning service that is able to register, activate, and configure IoT devices as they come on. Most cloud computing providers with IoT offerings provide device provisioning services that can be configured by the IoT system owner/operator.

In our smart fridge example, the device manufacturer and the IoT system owner/operator are one and the same. In some cases, they are not. For instance, a smart fridge company may contract out the actual fridge manufacturing to a third party.

Next, we will explore some of the approaches that can be used to provision IOT devices. The approaches described here use RSA key pairs and X.509 certificates for identity and authentication.

1. **Manufacturing-based zero-touch provisioning** (the required steps are depicted in Figure 6.2): The simplest type of provisioning is to set the configuration of the device at the time the device is manufactured. This provides a type of zero-touch provisioning from the perspective of the customer or end-user, since they never have to do anything with the configuration. Of course, this creates devices that only work with a particular cloud or device management system, locking the device to that particular system. As seen in the diagram, all the necessary settings are built into the device, and correspond to a profile or configuration created for the device in the device provisioning service (DPS).

   Most major cloud platforms such as Microsoft® Azure, Google Cloud Platform™, and AWS® support some flavor of zero-touch provisioning based on setting the configuration at manufacturing time.
2. **Bulk provisioning:** At its core, bulk provisioning involves the creation of device configuration data by the customer, and then uploading this data into the device provisioning service for use during manufacturing. In some cases, the generation of unique private keys can be done on the device during manufacturing and the device certificates for

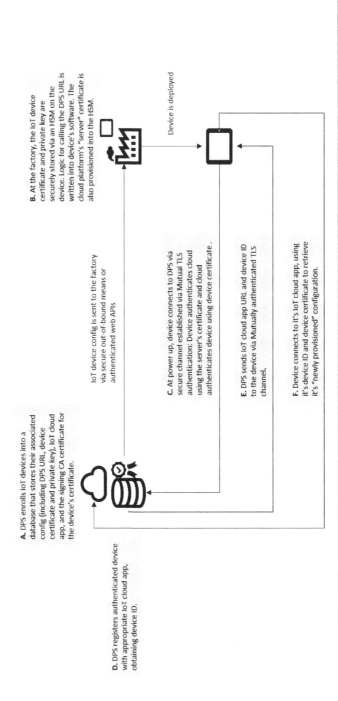

A. DPS enrolls IoT devices into a database that stores their associated config (including DPS URL, device certificate and private key), IoT cloud app, and the signing CA certificate for the device's certificate.

B. At the factory, the IoT device certificate and private key are securely stored via an HSM on the device. Logic for calling the DPS URL is written into device's software. The cloud platform's "server" certificate is also provisioned into the HSM.

IoT device config is sent to the factory via secure out-of-bound means or authenticated web APIs

Device is deployed

C. At power up, device connects to DPS via secure channel established via Mutual TLS authentication: Device authenticates cloud using the server's certificate and cloud authenticates device using device certificate .

D. DPS registers authenticated device with appropriate IoT cloud app, obtaining device ID.

E. DPS sends IoT cloud app URL and device ID to the device via Mutually authenticated TLS channel.

F. Device connects to it's IoT cloud app, using it's device ID and device certificate to retrieve it's "newly provisioned" configuration.

Figure 6.2 Just-In-Time or Zero-Touch Provisioning (also Relevant to Bulk Provisioning)

each device are collected by the manufacturer and sent back to the DPS. The benefit of this approach is that the customer gets the specific configuration of the devices that they want, however the cost per device is usually higher.

3. **The device certificate vending machine (CVM):** This approach is less common and is applicable for IoT system owners who do not have certificates preinstalled on devices and cannot provision device certificates during the manufacturing process. The CVM design pattern allows the IoT device to apply for its own certificate.[8] We assume that the IoT device supports TLS, and to prevent a man-in-the-middle attack when devices and the CVM are communicating, the CVM's TLS certificate should be validated by the device to ascertain the authenticity of the endpoint. Additionally, an IoT device that requests a certificate should have a unique identifier, such as a serial number, client ID, password, one-time-authentication token, or product ID that can be used to verify that the device is valid.

The following is an example of a CVM workflow:

a. The IoT device makes a request to access the IoT cloud application. The cloud verifies the device's ID before sending a command to the IoT device, instructing it to initiate a certificate application to the CVM.
b. The IoT device sends a certificate application to the CVM.
c. Upon receipt of the certificate application, the CVM initiates a CA certificate request to the cloud-hosted Certificate Manager. The Certificate Manager will be a service that allows the automated creation of private keys and associated TLS certificates.
d. The CVM also enrolls the device into a device registration database and associates the device with the CA certificate obtained in (c).
e. The CVM generates a device certificate and key pairs, based on the CA certificate associated with the enrolled device in (d).
f. The CVM returns the device certificate and key pairs to the device.

It is important to note that a potential problem with this approach is device masquerading. The only way the cloud is able to verify a new device is via the device ID, which can be lifted and used by an adversary with access to a legitimate device. As such, it becomes necessary to ensure that device IDs can only be used once for obtaining device certificates. It is also necessary to build solutions that detect and respond to any attempts to reuse a device ID for extra device certificate requests.

For more information on automatic device provisioning, you can check the offerings provided by Amazon* and Microsoft.†

Re-provisioning Devices

Our previous point leads to a necessary mention of the hard problem of re-provisioning devices that have already been provisioned—for instance, if the device was returned by the owner and

---

* Retrieved from https://docs.aws.amazon.com/iot/latest/developerguide/iot-provision.html
† Retrieved from https://docs.microsoft.com/en-us/azure/iot-dps/about-iot-dps

is to be resold by the manufacturer. Pre-provisioned devices are hard to change. Although this is good for security, it drives up the total cost of ownership. Once a device is locked to a cloud management system, how is this linkage reset if the device is resold, or returned?

The IoT device manufacturer is responsible for implementing automated systems for resetting a device such that the steps outlined in Figure 6.2 have to be done all over again before the device can be resold or reused.

The cloud provider should also expose a system and API through which device manufacturers can notify the DPS and the IoT cloud app that a particular device (with a particular ID and device certificate) has been reset. This should automatically trigger a reset in the IoT cloud so that data items such as the device ID and device certificate are invalidated and purged, and any associated user or customer information is also removed.

Device Maintenance, Updates, and Monitoring

IoT devices have a life expectancy of 10 to 15 years. In addition, IoT devices are often deployed in places that are expensive for human technicians to reach - underground, at high elevation, in dangerous spots such as nuclear power plants. Even for IoT devices that are easy to reach, there are often too many devices for manual maintenance to be a viable option.

Hence, it is crucial that IoT devices are designed to support remote maintenance. Once again, the cloud saves the day by providing the capabilities required for designing and running remote maintenance of IoT devices.

As we mentioned earlier, device maintenance is usually achieved via the use of a device management agent. Such an agent implements the logic for receiving commands from the cloud-hosted device management service, as well as the capabilities required to perform the required commands on the IoT device. It is crucial that all information exchange between the device management service and the device management agent occurs over encrypted channels.

The primary device maintenance tasks include:

1. **Device updates (operating system, applications, and firmware):** At the core, device updates involve a few primary things that must be done right:
    - A secure channel of communication between IoT device management service and IoT devices.
    - A secure repository where updates are stored. It should be possible to tag updates with the IoT devices that should get them.
    - The capability for the device management service to map available updates to the appropriate IoT devices.
    - The ability for the device management service to appropriately send the address, Uniform Resource Identifier (URI), of the update package to the device that requires that update.
    - A secure channel of communication between the update repository and all relevant IoT devices.
    - A secure means by which IoT devices can verify the authenticity of downloaded update packages. For instance, digital signature verification.
    - The capability for the IoT device management service to roll back IoT device updates.

In our practical IoT example in Section 6.6.3, we use application containers and container registries to provide over-the-air application updates to our IoT devices.

2. **Device monitoring:** This involves monitoring overall device health, data collection services, the status of ongoing operations (such as sensor operations, software workloads, network health and availability, and the need for software updates), and alerts sent to operators concerning issues that might require their attention.

   It is noteworthy that device monitoring places the cloud in a unique position to detect malicious activity, malware, or otherwise anomalous behavior on IoT devices. To achieve this, device monitoring must be combined with intelligent data analytics in the cloud.

3. **Reboot:** Sometimes the easiest thing to do when a device is malfunctioning is to just restart the device. It removes all the software cobwebs, so to speak, and allows the device to start from a freshly loaded (hopefully) and uncorrupted operating system and set of applications. If the root of the problem is some software glitch, this usually works. If there is a hardware problem or corrupted software, other remediation actions are required.

4. **Factory reset:** Sometimes removing all the updates and returning to the original software load for a device is necessary. This occurs when a software or operating system update goes wrong, or there is a corruption in the flash memory or filesystem of the device. Most devices keep a backup of the original device load, and using the factory reset returns the device this original state. From this point after the factory reset, updates can be provided one by one to bring the device back to a secure and fully updated state.

The Hard Problem of Remote Device Maintenance for Resource-Constrained Devices

Recently, there was an incident where a set of IoT devices had a trust anchor compromise.

What's a trust anchor you might ask? A trust anchor is a cryptographic key that is inherently trusted by the system. Typically, trust anchors are provisioned into the device in a way that prevents or resists attacks that can change them—for instance, being part of a ROM or firmware image. For many systems, the trust anchors are usually root certificates or CA certificates, which are used to sign other certificates. As such, the trust anchors are also used to verify the trust of other certificates and their associated digital signatures.[9] Most operating systems ship with a set of root certificates that can be used to verify the digital signatures for applications on that system. System builders, vendors, or users can also add root certificates to the set of trust anchors in an operating system.

Back to our story: The IoT solution operator, who shall remain nameless, had a trust anchor compromise in the Linux® operating system distro that ran on edge IoT devices. To resolve the problem, a new Linux image had to be pushed from the cloud to all the devices. This was no big deal for the cloud, even though the image was 4 gigabytes. The devices on the other hand, used 3G connections that were prone to significant network errors, and the network operator charged for every byte of data transferred. During regular operation, when the devices only had to upload or download a few hundred bytes of data, the network constraints were not a problem. For the mandatory update of the entire operating system, however, the company's analysis showed that it would cost several million dollars in network costs, plus over a year of downloads, to get all the devices updated.

The Solutions?

The major callout is that solution developers, vendors, and operators must plan ahead. As the saying goes, failing to plan is planning to fail. Here are some solutions for the problem we just described:

1. Since most IoT devices such as sensors and actuators are resource constrained, deploy gateways to the edge, closer to your constrained devices. As described in Chapter 4, Sections 4.2.1 to 4.2.3, gateways reduce the costs in long-haul communication and energy consumption while enabling solution owners to support a high volume of IoT devices.
2. Judiciously select the operating system distros that you base your IoT solution upon. The choice of an operating system means that you are implicitly trusting the correct development, operation, and maintenance of that operating system. You should look into the development and maintenance practices, as well the history of an operating system, before building your IoT solution upon it.
3. Consider the use of containerized distros, which allow you to update sections of the operating system without having to update the whole thing. There are Linux distros that significantly reduce the overhead of updates by using a binary diff file that contains only the differences between images to perform the update. At the time of writing, examples of such operating systems include Intel® ClearLinux, Rancher®OS, and CoreOS®.

6.4.3 Secure Multi-Presence Access to IoT Devices

To illustrate this point, let us briefly consider an IoT solution that is deployed in the offices of LYN, a non-profit charity organization. It allows LYN's employees to adjust the temperature in

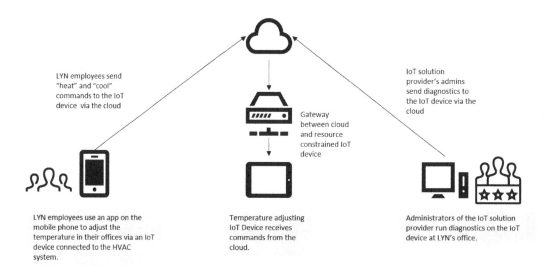

Figure 6.3   The Cloud Provides Multi-Presence Access to Iot Devices

their meeting rooms by using an app on their smartphones. In tandem, administrators of the IoT solution provider are able to run diagnostics using their laptops, at their own office. How does that work? Yes, you're right, it's the cloud. The cloud also ensures that multi-presence access is secured via encrypted communication channels, authentication and authorization.

Figure 6.3 depicts how the cloud provides the central means by which both entities, users and system administrators can access the IoT devices that controlling the temperature in LYN's offices. Thus, the cloud provides, multi-presence access to IoT devices.

6.5 A Summary of Security Considerations for IoT Cloud Back Ends

An attacker who is able to compromise the cloud back end of an IoT solution, has—as security professionals say—the keys to the kingdom. Why? Because of all we have covered in this chapter thus far, which demonstrate the criticality of the cloud to an IoT system.

Before venturing into our practical IoT security architecture, let us summarize the major security concerns for IoT clouds. For each category of security concerns listed below, we reference the attacks and mitigations in Table 6.4 in Section 6.6.6, where we enumerate possible attacks to our practical IoT cloud example and propose mitigations to those attacks.

1. **Secure data collection at scale, data sanitization, and data storage:** It is necessary to ensure that data obtained by the edge devices are transmitted securely to the cloud and stored securely in the cloud. Also, the means by which data is consumed by the cloud must be built to scale securely. For practical examples, please check attack numbers A2, A4, A5, A6, A7, A9, A12, A13, and A14 in Table 6.4 in Section 6.6.7.

2. **Device provisioning, registration, authentication, and authorization:** Edge devices are often vulnerable to physical, local, or replacement attacks. This means that an IoT System must be designed to ensure that all cloud services only communicate with legitimate devices. In addition, the back-end system must be able to prevent data inputs by rogue devices. This is achieved by ensuring that all devices are provisioned with secure identifiers, and that the back-end system verifies the registration, authentication, and authorization of any edge device before processing the data it send. For practical examples, please check attack numbers A3, A19, and A20 in Table 6.4 in Section 6.6.7.

3. **Device management and updates:** Most IoT devices are expected to be operational for at least 10 years. As such, it's important that software (and, in some cases, firmware) on provisioned devices can be updated after deployment. Otherwise, vulnerabilities that are discovered in software components on the device can be exploited to compromise the system. The cloud back-end systems are the primary means through which patches and updates can be seamlessly delivered to IoT devices. Patches and updates must be deployed via a secure channel. Devices must also be able to validate patches and updates before they are applied. The common model is to deliver digitally signed updates via an encrypted channel. The receiving device is then able to validate the signature of the update. For practical examples, please check attack numbers A3, A19, and A20 in Table 6.4 in Section 6.6.7.

4. **Encrypted communication:** Data from IoT devices are usually sent to the cloud via communications channels that are not controlled by the entity that owns or operates the

IoT System. As such, it becomes imperative that any data or information of value that's transmitted over that untrusted channel must be secured via encryption. Encrypted communication is particularly interesting for edge devices as it requires the consumption of extra power and compute resources, which often put constraints on edge devices. For this reason, lightweight encryption protocols are actively being developed and improved to support the provision of encryption capabilities on resource-constrained devices. As part of the security analysis of an IoT system, the security architect and product development team will need to make judgment calls regarding performance and security trade-offs.

Every communication channel—from sensors or actuators to the gateway and on to the cloud—has to be considered for encryption. Such considerations should be based on the sensitivity of the data being transmitted as well as the potential impact of compromise and any other protections designed into the system. For practical examples, please check attack numbers A2 and A12 in Table 6.4 in Section 6.6.7.

5. **Application Programing Interface (API) security:** Cloud back-end systems always consume data through APIs[*] exposed over a network. For systems deployed on public clouds, such APIs could be accessible over the Internet. In addition to the security measures for the concerns we've already covered, all cloud APIs or logical interfaces that accept data or commands must be robustly tested and secured. This involves testing using negative/malformed/unexpected inputs and stress-testing techniques. APIs must also verify and authenticate every access. For practical examples, please check attack numbers A4, A5, A6, A11, A12, A13, and A14 in Table 6.4 in Section 6.6.7.

6.6 Practical IoT Cloud Security Architecture: The "Dalit" Smart City Use Case

The city is Dalit, a fictional city in central Africa. But let us forget that it is fictional so we can enjoy this story, shall we? The Governor of Dalit and the city's Planning and Development Commission are well aware of that all-too-often-mentioned steady pace of technological innovation and the benefits that they can provide. As a result, the city officials wish to build Dalit 2.0, by developing Dalit into a Smart City. Their vision is a city equipped with technologies that improve the lives of city dwellers. They aim to achieve these by deploying systems that help to reduce traffic congestion, traffic accidents, and crime.

As a first step, Dalit's officials decided to build a pilot system that will paint a picture of the activity level at major city intersections. The system being designed presents the city official with real-time data and trends concerning the number of cars, bikes, and pedestrians that are at different city intersections, at different dates and times.

Dalit's smart city pilot makes use of technologies such as cameras for capturing videos and images, edge gateways that process video streams from the cameras in order to produce metadata counts of cars, bikes, and pedestrians, as well as back-end infrastructure and applications.

The back-end systems collect data from multiple edge gateways, store them in a full-text searchable search engine. The data and search engine are made accessible to city officials via a

---

[*] Retrieved from https://www.infoworld.com/article/3269878/what-is-an-api-application-programming-interfaces-explained.html

web application. Configuration and management of the edge gateways is also handled by the back-end systems.

For the purposes of our analysis in this chapter, we shall focus on the security of the cloud-hosted back-end systems.

6.6.1 Introducing ATASM as a Threat Modeling Tool

Chapter 5 provided an in-depth analysis of the art and practice of "threat modeling." Given the importance of threat modeling to the design of secure systems, we will use our Dalit example to explore a different technique of threat modeling. We hope that this will help the reader see different threat modeling approaches, which should improve the reader's understanding of this domain.

At this point, you will likely agree that threat modeling is an art that involves the review of a system's architecture in order to deduce any potential security weaknesses present in the system, determine which of those weakness should be mitigated, and produce a specification of the required mitigations. Threat modeling requires an in-depth understanding of the system, its components and their interactions, the data processed or stored by the system, and the run-time or deployment environment in which in the system runs. Once the system architecture is sufficiently understood, the threat modeler is able to tackle the other aspects of threat modeling, which can be summarized as follows:

1. Identifying the attackers who will be interested in compromising the system
2. Identifying the information in the system that will be of interest to attackers
3. Deducing the attack patterns that are likely to be used by the attackers
4. Identifying the security controls that should be implemented in order to thwart the attackers

In this chapter, we will use an approach called ATASM to threat model our IoT cloud architecture. A brainchild of Brook Schoenfield, ATASM is designed to be a clear and easy way to teach the art of threat modeling. ATASM stands for Architecture >> Threat Agents >> Attack Surfaces >> Mitigations. Here is a brief explanation of each of those four steps:

1. **Architecture:** Digest the architecture. This is almost always done through architectural diagrams.
2. **Threat Agents:** Understand the (human) attackers who will be interested in your system, their capabilities, and their motivations.
3. **Attack Surfaces:** Identify the juicy nuggets in your system, which will be the targets of attackers. Also, identify the attack methods of those annoying attackers and their high-level or system-level objectives for attacking the elements of your system
4. **Mitigations:** Identify the security controls required
   o These become the formal list of security requirements that are the output of threat modeling.

Figure 6.4 illustrates the iterative nature of ATASM.

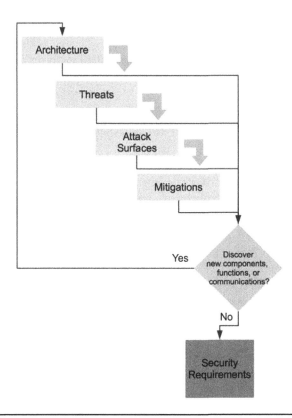

Figure 6.4   ATASM—A Recursive Threat Modeling Process (Adapted from Schoenfield, B. [2015]. *Securing Systems: Applied Security Architecture and Threat Models.* CRC Press/Taylor & Francis Group.[10])

For a deeper dive into ATASM, check Brook Schoenfield's work in *Securing Systems: Applied Security Architecture and Threat Models.*[10]

At this juncture, it is worthwhile to note that there are many ways to threat model. ATASM is only one of them. ATASM's approach, which is an architectural-driven threat modeling methodology, provides clear and concise steps to deduce relevant attack surfaces and their mitigations.

6.6.2 Dalit Cloud Architecture Overview

In this section, we introduce you to Dalit's cloud-hosting environment and the choices the city made that influenced their cloud architecture. In subsequent sections we use different architectural views to present the security architecture of the Dalit cloud.

Choosing the Cloud Provider, Service, and Deployment Model

Dalit had to decide how to host the back-end systems for their smart city pilot. Lacking funds to maintain a highly available data center for a private cloud, Dalit decided to proceed with a public cloud deployment model with one of the well-known and established cloud service providers. This option was additionally attractive since the first deployment was a pilot, and there was not an immediate need to build a data center for the city, even if the funds could be found. But which cloud service provider could they use?

Currently, and perhaps increasingly, there are a number of competent cloud service providers. According to the 2018 Gartner Magic Quadrant for cloud service providers, the top three leaders are Amazon's Web Services (AWS), Microsoft® Azure®, and Google Cloud Platform™ (GCP), in that order.[11]

Dalit decided to choose AWS—although we, the authors, must add with wry smiles that the choice is primarily due to our expertise, and is not an endorsement of AWS over other cloud service providers. Still, this means that our architecture and security analysis will reflect some AWS proprietary technologies or solutions. Even so, most of the security requirements that we will deduce in our analysis will apply to cloud deployments irrespective of the cloud service platform. For instance, user authentication to cloud services can be achieved with different existing technologies. Examples of such identify services proprietary to cloud service providers are AWS Cognito™ and GCP Firebase™, while Auth0® is a third-party solution that can be used with both AWS and GCP. Whenever we address AWS-specific technologies, we will try to mention comparative technologies available with other top cloud service platforms.

Finally, considering the flexibility we want in architecting, designing, and managing our cloud infrastructure and applications, as well as the requirement to configure and manage cloud edge gateways, our chosen service model is an Infrastructure as a Service (IaaS). This also gives us the opportunity to have more fun with security architecture and design! Table 6.2 summarizes the cloud-hosting choices chosen by the City of Dalit.

Table 6.2 Planning Our IoT Cloud: Cloud Service Provider, Service Model, and Deployment Mode

Dalit's Cloud Hosting Choices	
Deployment Model	Public Cloud
Cloud Service Provider	Amazon (AWS)
Service Model	Infrastructure as a Service

Component Architecture

Usually, one architectural view is not sufficient to describe the different elements and components of a system. A complete description requires different views of the system. Multiple views, and in some cases, different levels of abstraction, become increasingly necessary the larger the system is. To illustrate, the different angles from which one gazes upon an elephant provide a piece of the insight into the features of the large mammal. To know the animal even more, one must understand its behavior, its instincts, its physiology, etc.

Examples of architectural views for systems architecture are data flows, component interaction, networking, management, monitoring, and even the business process/interaction. Some of these views can be combined into a single diagrammatic representation.

For the Dalit cloud, we have developed four architectural views. We do not mean to say that those are the only views necessary, but rather, they are sufficient for us to understand the system enough to practice threat analysis of an IoT cloud. Our architectural views can be named and described as follows:

1. **Data Ingestion and Processing View:** This view describes the major cloud components and data flows involved in the ingestion and processing of the data retrieved from edge devices.
2. **Device Firmware and Software Updates Flow View:** Here we depict the cloud components used to update the software on the edge device. The specifics of the mechanism for processing and verifying the update on the edge device is left for the reader to explore, using principles and practices shared in Chapter 5.
3. **Cloud Infrastructure and Networks View:** This view illustrates the infrastructure of the cloud. It depicts the networks, firewalls, and security groups that connect and, in other cases, isolate services running in the virtual private cloud (VPC). A VPC is a virtual data center.
4. **Monitoring and Auditing View:** This describes how Dalit can use AWS components to monitor their usage of cloud computing resources and audit the administrative accounts used to manage the Dalit cloud.

DISCLAIMER: In this section, some of our descriptions and depictions of cloud components and services (such as the Device Shadow Service) are based on components in offerings from AWS and other cloud computing providers. But we do not refer to them as such in order to give us the ability to depict and analyze cloud architectural details that are sometimes hidden by the "user or developer-friendly" web interfaces of cloud computing providers such as AWS, GCP, and Azure.

6.6.3 Data Ingestion and Processing View

In this view, which is depicted in Figure 6.5, there are two numbered flows. One flow depicts data flowing from the edge to the cloud, while the other flow shows how city officials access that data. A good starting point to begin digesting the architecture is the camera at the edge. Cameras are used to process video feeds at city intersections. Video streams from the cameras are transmitted to an edge compute gateway at the intersection. The specifics of the camera and its connection to the edge gateway are outside of the scope of this chapter since it is focused on cloud architecture.

The edge gateway processes the video feeds and uses machine learning to identify cars, bikes, and pedestrians. To avoid storing or protecting highly sensitive data about city dwellers, the edge gateway processes the video feed, counting and storing numeric data, before discarding the video stream. The numeric traffic metadata is sent to the cloud.

At the cloud perimeter, another gateway is deployed: We call it the Cloud Gateway. The Cloud Gateway is a specialized compute service that is responsible for managing all incoming connections. Like any compute service, it leverages block storage for storing and running its operating system and software applications. It is designed to autoscale with traffic.

The Cloud Gateway exposes a Message Queueing Telemetry Transport (MQTT) interface that allows edge-to-cloud data transmission via the MQTT protocol. MQTT is an extremely lightweight Publish–Subscribe messaging transport. It is useful for connections with remote locations, where a small code footprint is required and/or network bandwidth is at a

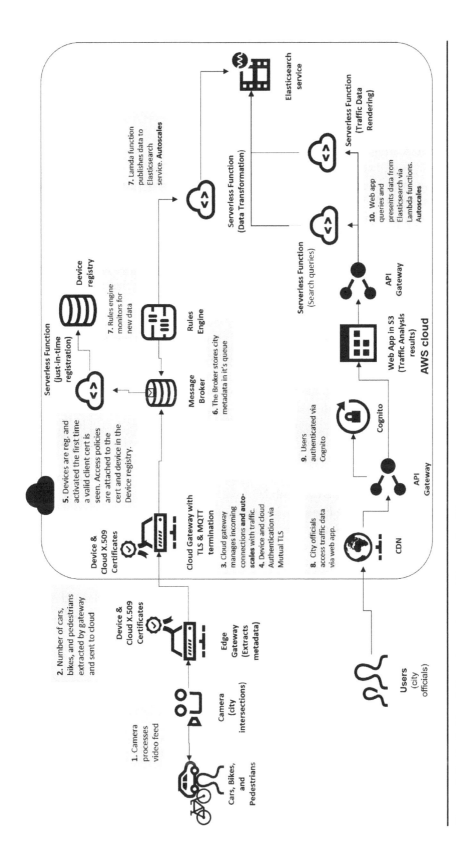

Figure 6.5 Dalit Cloud: Data Ingestion and Processing View

premium—that is, IoT sensors communicating to a cloud broker via satellite link over occasional dial-up connections with healthcare providers, and in a range of home automation and small device scenarios. MQTT is common in IoT cloud architectures to allow the cloud to easily ingest data from many devices. Please check Chapter 7 for more details on the workings of MQTT.

> **What is autoscaling?** This is a capability of a cloud computing provider to automatically increase provisioned computing resources based on the load on the service or application that is using those resources. Autoscaling is normally done with virtual machines (VMs) or with containers. When more traffic enters the cloud than the current set of servers can manage, the cloud management system automatically spins up another server with the VM or container software that current servers are running, effectively adding another server to handle the traffic. Likewise, when the traffic burst subsides, the cloud will spin down these extra servers and maintain a minimal set (possibly only one) to handle the current traffic load. Please check Chapter 11, Section 11.5.5, for more information on autoscaling.

As described in Section 6.4.1's Authentication at Scale, and depicted in Figure 6.5, the edge and cloud gateway devices authenticate each other via mutual Transport Layer Security (TLS). This means that each device holds a X.509 certificate that can be successfully verified by the other device. The TLS protocol allows client/server applications to communicate in a way that is designed to prevent eavesdropping, tampering, or message forgery.[12]

To enable mutual TLS, the edge gateway is configured with an X.509 device certificate that was signed with a CA certificate used in enrolling the device into the cloud. The cloud platform "server" certificate is also provisioned into the device. The device certificate, device private key and server certificate are stored securely on the device via a Hardware Security Module (HSM). The device certificate and its signing CA certificate are also stored in the cloud via HSMs, and are thus accessible to the designated Lambda™ function, just-in-time registration. This flow is described in Section 6.4.2 and Figure 6.2. The cloud-based HSM exposes APIs that the just-in-time registration function is able to use verify the signatures or certificate hashes of device certificates that are sent to the cloud. Our architecture does not go into the details of cloud HSM access or usage. There are multiple resources available on the subject.

> **What is Lambda?** It is the AWS implementation of a newer approach to cloud computing that is referred to as Serverless Architecture or Functions as a Service (FaaS). FaaS is described in more detail in Chapter 11, Section 11.5.2. FaaS allows software developers to write and deploy solutions through event-driven code functions, while the cloud computing provider takes care of the management—setup, scaling, and tear down—of the infrastructure required to run the code. Other examples of FaaS implementations are Azure Functions™ and Google Cloud Functions™.

Using just-in-time registration, the edge gateways are registered and activated with the cloud back end, once they send their first message to the cloud. The cloud gateway contains logic that is used to detect the first connection of a device certificate which causes it to add that certificate info to a specified MQTT registration topic. The just-in-time registration function is subscribed to that topic. The function is responsible for picking up a new client certificate, verifying that it is authentic, before activating it in the device database and attaching authorization policies. Only registered and activated devices certificates can be used to publish city data to the message broker. Device requests to publish or read data cause the cloud gateway to call an authentication lambda function that verifies that the request is by a registered and activated device, with permissions to publish to or read from a particular topic. The response from the lambda function causes the gateway to allow or deny a device's data publish or read request.

The message broker is a compute service that functions as a data bus that houses data that clients can access. It stores data received from edge devices and serves as a conduit pipe through which data can be passed from one component to the other. This is achieved via a Publish–Subscribe mechanism, such that registered components can either publish data to a topic or subscribe to consume data that has been published to a topic. A topic is simply a tag assigned to data. The rules engine is designed to monitor the message broker queue for different message types that would require actions to be taken by the cloud application. In our Dalit pilot, the rules engine subscribes to the message broker topic "city metadata," which contains periodic information concerning the number of pedestrians, cars, and bikes at a particular city intersection. It receives a notification once new city metadata is available and sends the data to a Lambda function, called Data Transformation.

In Dalit's IoT cloud architecture, the data transformation function does some final processing, parsing the data received from edge devices into JSON documents that are constructed to efficiently support future search queries. The JSON documents are persisted in a designated cluster of servers referred to as an Elasticsearch® service. The Elasticsearch service is an auto-scaled compute service that is based on the Elasticsearch software engine built by a company known as Elasticsearch B.V. It provides a distributed full-text search engine that will provide the means for us to let city officials seamlessly search, view, and visualize data streams from city intersections. At this point, the data from city intersections is stored and is searchable via Elasticsearch.

To provide the city officials of Dalit with access to data about pedestrians, cars, and bikes in the city, a static website* is hosted via AWS's object storage service, AWS S3™ (Simple Storage Service). This means that the code files are stored in S3. AWS's identity service is used to grant a select number of cloud administrators the roles and permissions required to update the website's code. All administrator operations are audited. This is done via the AWS CloudTrail™ compute service which logs all administrator activity performed on the cloud infrastructure. AWS CloudTrail supports configurable alerts and log searches to ensure that designated administrators can monitor the actions taken on or via the cloud. It is worthy of note that such a feature is not unique to AWS and is provided by other major cloud providers. Examples are Azure Search™.

Regular website users simply see the website as displayed via their browser. To achieve this, website users have access to read the code files, but they cannot update or otherwise modify them. A Content Distribution Network† (CDN) is used to serve the web application in order to cache content in a cloud server that is geographically closer to the user, thus improving access speed for fetching content.

The web application exposes a user interface through which city officials can query the AWS Elasticsearch service. This is achieved via the usage of two Lambda functions named Search Queries and Traffic Data Rendering. The web application is designed to execute either

---

* Retrieved from https://techterms.com/definition/staticwebsite
† Retrieved from https://www.webopedia.com/TERM/C/CDN.html

of those Lambda functions, depending on the page loaded or the button pressed by the user. In turn, each of those functions search or retrieve data from the Elasticsearch Service. Access to the functionality provided by the web app is restricted via AWS Cognito identity service, which provides user authentication and authorization. AWS Cognito is a specialized service for web-user authentication and authorization that is a component of AWS's identity service. It allows users to access cloud-hosted resources such as the website code files stored via AWS S3.

6.6.4 Device Software (and Firmware) Updates View

As we discussed in Section 6.3, the ability to update software (and firmware) on IoT edge devices is crucial to the success of any IoT solution. Firmware is simply a special class of software designed to control the hardware of a computing system.

Considering an IoT device with multiple software components, we can be certain that, sooner or later, some security researcher or hacker will discover security weaknesses in one of those components. Without the ability to update the software, the device becomes a ticking time bomb.

Dalit's software update mechanism makes use of software containers and the AWS Elastic Container registry compute service.

> **What is a Container?** A computer program running on an ordinary operating system can see all resources (connected devices, files and folders, network shares, CPU power, quantifiable hardware capabilities) of that computer. However, programs running inside a container can only see the container's contents and devices assigned to that container. In addition to isolation mechanisms, the kernel often provides resource-management features to limit the impact of one container's activities on other containers. Containers are usually run via container engines, which use OS-level virtualization to create user space instances that can contain a software application's code, configuration, and all the dependencies that it requires to run. As such, the usage of containers allows software developers to create and deploy their applications without having to worry about supporting different operating systems or environments. This is a big win for portability—especially across an operating system family such as Linux® or Windows®. Please check Chapter 11, Section 11.5.3, for more information on software containers.

In the Dalit Device Software Updates view, the system administrator publishes signed containers to the Container Registry. The container registry is simply a repository of update files stored as containers. Afterwards, the system administrator uses a web application (also hosted via AWS S3), with the aid of a Lambda function, in order to publish a new desired state to the Device Shadow compute service. The Lambda function is accessible to the web app via an API (or URL) it exposes through the Amazon API Gateway™ compute service. The API gateway is simply a service that allows for easy creation and manipulation of REST APIs. The Amazon API Gateway can be used to create an API that is simultaneously connected to disparate cloud-hosted data sources, presenting a simplified interface for a client accessing that API. In our current example case, the REST API is connected to the functions exposed by the "set desired device state" serverless function.

Take a moment to consider what was described in the last paragraph and compare it with the approaches described in "Device Maintenance, Updates, and Monitoring" in Section 6.4.2.

Might there be a smarter way to architect this update system for Dalit? Take a moment to analyze. Yes, you've probably found it. It is possible and perhaps better to architect a system in which device updates are tagged with a device ID or device group ID, such that the system can automatically associate an update with the device or devices. This will eliminate our example step in which an administrator manually updates the device shadow service with a configuration that directs a device to pull down an update from the container registry. Our challenge to you: Go architect it!

Let us not forget that the system administrator's access to the device shadow service and the container registry must be authenticated and authorized via the identity service. Figure 6.6 depicts the admin's authentication to the device shadow service via Cognito. But although Figure 6.6 does not include the admin's authentication to the container registry, that is just as important. How would you accomplish such authentication? There are a few options—utilizing a built-in authentication service such as AWS Cognito, or allowing Secure Shell (SSH)–based direct access to the network and machine(s) that host the container registry. Both approaches make use of the identity service. Also, in either case, the principle of least privilege must be enforced to ensure that only the administrators who are designated to update containers in the container registry have those access permissions.

The Device Shadow service is popularized by AWS but can be implemented by any cloud provider or in a private cloud. It holds records of the last known state and desired future state of edge devices, even when the device is offline. The Device Shadow service publishes new desired states to a corresponding topic on the message broker, where they are be picked up by edge devices that are subscribed to the same topic. To update the edge devices, the future state contains information about the URL of the elastic container registry where the updates reside. Once the device receives this message, it accesses and downloads the update before verifying its signature and installing it. All communication into the cloud must occur over encrypted channels. For instance, system administrators connecting to the admin app used to update the device state, system administrators uploading new containers to the container registry, the edge gateway to the cloud gateway, and the edge gateway to the elastic container registry.

Figure 6.6 depicts the flow for updating software components on the IoT edge device.

6.6.5 Networking View

Figure 6.7 presents the cloud networking view of the Dalit Cloud. Dalit makes use of Amazon's networking service, Amazon Virtual Private Cloud™ to create an isolated network that comprises subnetworks (subnet) and security groups. It is composed of three subnets. Two of the subnets are public networks that can be reached via the Internet. The private subnet can only be accessed via the public networks. Each subnet has a security group, which acts as a virtual firewall to control inbound and outbound traffic.

The bastion host is housed in a public subnet, with a security group that restricts ingress traffic to only those originating from the IT office of Dalit. This network is the administrative subnet that enables Dalit's cloud administrators to manage the various computing systems and services deployed to AWS. Hence the public admin subnet has access to all private subnets. Administrators access the network via the bastion host, which is a server that is hardened to withstand attacks. Access to the bastion host requires SSH key-based authentication. This means that an RSA public, private key pair is generated, with the public key placed on the bastion host. Only clients with the private key can access the bastion host via SSH.

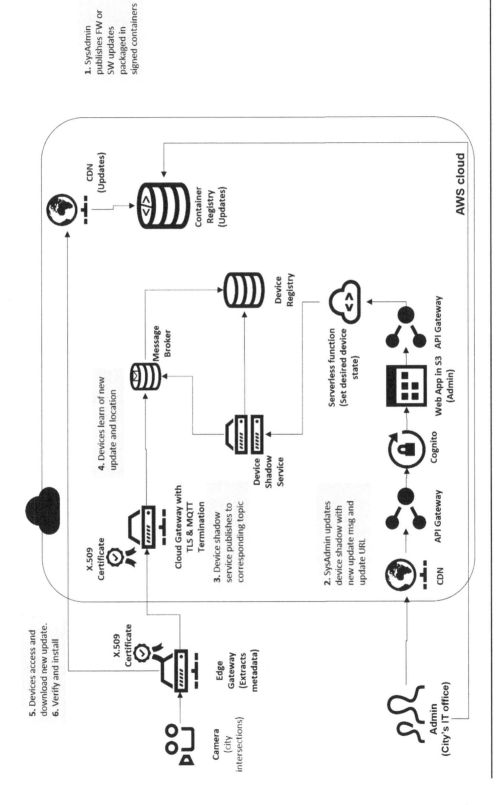

Figure 6.6 Dalit Cloud: Device Software Updates View

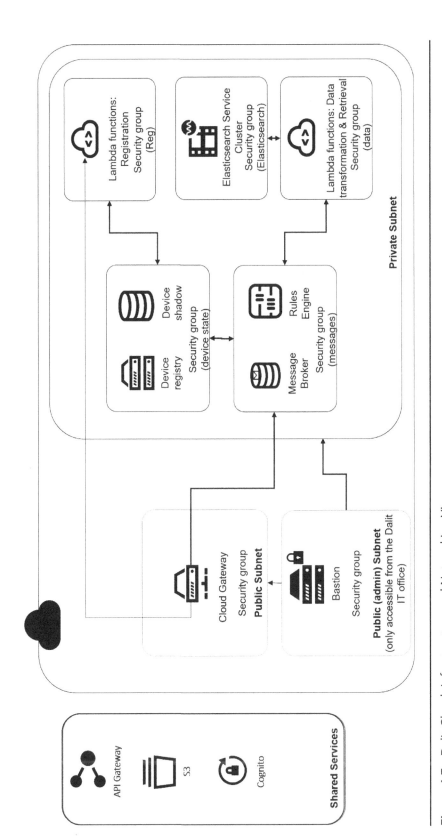

Figure 6.7 Dalit Cloud: Infrastructure and Networking View

All administration of the cloud platform by Dalit makes use of the bastion host. Hence, the admin subnet is able to access the other two subnets in the private cloud. The non-admin public subnetwork houses the device (cloud) gateway and serves as an entry point for all the data coming in from edge devices.

Within the private subnet, there are five security groups. The Registration security group restricts access to the Just-In registration Lambda function. It only allows ingress access from the Cloud Gateway compute service, while two-way traffic is allowed with the device state security group.

The messages security group houses the message broker and the rules engine. This group only allows ingress traffic from the device gateway, the admin subnetwork, the device state security group, and the data security group.

The reader is invited to study Figure 6.7 to deduce the implications of the other connections and security groups that are illustrated.

Finally, there are shared and global AWS services that are used by the Dalit cloud, but they are not located inside Dalit's VPC. They are the API Gateway™, S3, and Amazon Cognito™.

6.6.6 Cloud Resource Monitoring and Auditing View

To keep any system running properly, it is necessary to have oversight and insight into the usage of the system. The monitoring and auditing view in Figure 6.8 depicts how Dalit's IT office uses AWS to monitor the usage of cloud computing resources and administrative accounts.

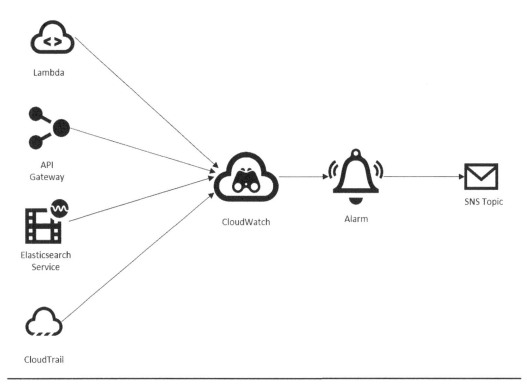

Figure 6.8   Dalit Cloud: Monitoring and Auditing View

Support for the AWS CloudWatch™ compute service is built into every AWS resource or service. It enables administrators to monitor their usage of AWS while using AWS Cloudwatch alarms to trigger notifications (to designated recipients) when preconfigured usage thresholds are exceeded.

The AWS CloudTrail™ compute service makes it possible to monitor and review all actions carried out via an AWS cloud user account. Every action carried out on the AWS platform, such as creating the cloud architecture depicted via all the Dalit cloud diagrams, requires access to AWS via an AWS account. Usually, an organization or entity such as Dalit will create an AWS account and then use that account to create other accounts with varying levels of administrative privileges. All those accounts will have some level of access to the Dalit cloud. AWS Cloudtrail can be used to monitor all of Dalit's AWS accounts.

As depicted in the Monitoring and Auditing view (see Figure 6.8), AWS Cloudwatch receives usage data from Lambda, the API gateway, and the Elasticsearch Service. It is also configured to monitor CloudTrail logs. Thresholds are configured with AWS Cloudwatch Alarms, along with event handlers that send notifications to an Amazon Simple Notification Service™ (SNS) Topic. Amazon SNS is infrastructure created for the mass delivery of messages. SNS topics utilize the Publish–Subscribe method described for the message broker in Section 6.4.2. Subscribers can either subscribe to a topic or be subscribed to a topic owner. Once a publisher sends a message to a topic, it triggers Amazon SNS, which delivers the message to subscribers via the means previously configured into the SNS topic, such as HTTP/HTTPS, email, SMS, etc.

The monitoring and auditing system just described can be used to notify designated people about unexpected usage of AWS accounts and resources—for instance, an aggrieved employee who attempts to delete vital systems or data, or a hacker who attempts to spin up thousands of virtual machine to mine bitcoins.

6.6.7 Threat Analysis

Threat Landscape

In Chapter 2, Section 2.3.1, the current threat landscape was analyzed in great detail. We sought to understand who might want to attack our IoT systems and their motives for attack.

Now, that knowledge is to be applied to the Dalit cloud. Table 6.3 analyses the threat agents who might be interested in attacking the Dalit cloud, outlining their objectives and capabilities.

Table 6.3 Threat Landscape for the Dalit Cloud

Threat Agent	Goals	Risk Tolerance	Work Factor	Methods
Cybercriminal	Financial	Low	Low to medium	Known proven
Script kiddies	Media attention	High	Very low	Known proven
Security researchers	Media attention, financial	Very high	High	Sophisticated
Nation state	Disruption, espionage	Very low	Very high	Highly sophisticated

Attack Surfaces and Mitigations

Cloud architecture veterans will notice that some of the attack surfaces and mitigations described in Table 6.4 do not apply solely to IoT cloud solutions but are also relevant to more generic cloud and web applications. That is to be expected. Certain IoT-specific security considerations come into the mix for an IoT cloud, but the vast majority of "generic" cloud security concerns have to be addressed as well. As mentioned earlier in this chapter, we do not focus too heavily on generic cloud considerations, given the plethora of existing resources that address those topics. But it is still well worth it that we consider and list some of the major cloud security concerns, since you will have to address the cloud as a whole, not just the IoT pieces, in any actual deployment.

Please note that Table 6.4 is not intended to be an exhaustive list of all possible attacks of the Dalit cloud architecture. Instead, it offers an example of various types of IoT cloud security concerns, as well as how to deduce those threats from an IoT architecture. The reader is invited to extend our table of attack surfaces and mitigations.

6.7 What We Learned

As we have seen, the cloud portion of an IoT solution is an attractive, high-value asset to attackers, since it plays such a central role in the life of an IoT solution and the value the solution provides to its users.

In this chapter, we explored the history of the cloud and explained its core concepts. Afterwards, we unpacked how the cloud enables and scales IoT security by providing secure centralization of data management and analytics, secure IoT device management, and secure multi-presence access to IoT devices.

To bring the concepts we covered to life, we capped off the fun by architecting the IoT cloud for a fictional city called Dalit, using state of the art cloud computing approaches and concepts. Dalit is piloting a smart city project that involves the collection of data on the number of pedestrians, bikes, and cars at city intersections. The data collected at the intersections is to be sent up to the cloud for processing and presentation to city officials.

Finally, armed with our knowledge of the Dalit IoT cloud, we created a threat model to identify security concerns and their mitigations. Some of the attack possibilities that we identified apply to any cloud computing solution, whereas some can be mapped to any of the following major IoT cloud security concerns:

- Secure data collection, sanitization, and storage
- Device provisioning, registration, authentication, and authorization
- Device management and updates
- Encrypted communication
- API security

Table 6.4 Attack Surfaces and Mitigations

Dalit Cloud: Attack Surfaces and Mitigations				
#	Specific Attack	Attack Surface	System Objective	Mitigation
A1	Denial of service/ distributed denial of service	• Cloud gateway • Administrative bastion host • AWS API Gateway™	• System disruption	• Configure autoscaling for compute services, including the cloud gateway. • Only the cloud gateway and admin bastion host should be Internet accessible. • The bastion host should only be accessible from the solution owner's network. • Use firewalls to filter out malformed traffic that are used for DDoS. • Throttle requests allowed through the cloud gateway. • Restrict open ports to MQTT interface. Only cloud gateway. • All software and firmware on edge devices must be kept updated and must be free of known security vulnerabilities.
A2	Man-in-the-middle network snooping and network/message data tampering	• Cloud gateway	• Steal confidential data • Corrupt data stored in the cloud	• Mutual TLS authentication: deploy unique digital certificates to each edge gateway and do the same for the cloud gateway. • Restrict open ports on the cloud gateway to the MQTT interface used for communication with the edge device.
A3	Rogue edge device	• Cloud gateway	• Steal confidential data • Corrupt data stored in cloud	• Mutual TLS authentication. • Restrict open ports to MQTT interface. • Sign device certificates with a CA certificate that is registered with the Dalit cloud back end. This enables registration, and activation. • All software and firmware on edge devices must be kept updated and must be free of known security vulnerabilities. • *Implement device attestation.*
A4	Unauthorized access to message queue	• Message broker	• Steal confidential data • Supply invalid and potentially dangerous data to the message queue	• Restrict access to message queue > only the cloud gateway and Device Shadow can publish data. • Cloud gateway authenticates all external access via mutual TLS. • Message broker defines and enforces policies that control connection, subscription, submission, and retrieval of information from the message queue. • Message broker access policies should be associated with registered and activated device certificates, in the cloud.

(Continued on following page)

Table 6.4 Attack Surfaces and Mitigations (*Continued*)

Dalit Cloud: Attack Surfaces and Mitigations				
#	Specific Attack	Attack Surface	System Objective	Mitigation
A5	Unauthorized, malicious message added to message queue	• Message broker	• Cause unintended system behavior • Steal confidential data	• Message broker defines and validates approved data formats or structures. • Message broker subscribers should validate the data retrieved from the queue.
A6	Arbitrary code execution via malformed messages	• Message broker • Cloud gateway	• Control system resources	• Message broker defines and validates approved data formats or structures. • Fuzz testing of APIs exposed via cloud gateway should be used to validate effectiveness of input validation.
A7	Denial of service via flooding of authenticated messages to the message queue	• Message broker • Cloud gateway	• Make system functionality unavailable	• Apply mitigations in A1.
A8	Data theft via rules engine	• Rules engine	• Divert data to unintended destinations	• Only system administrators should be able to configure rules. • All administrative access must pass through a bastion host that restricts access via SSH keys and restricted ingress IP address ranges. • All modifications to the cloud environment must be logged in a manner that's easy to search or audit, via services such as AWS Cloud Trail™.
A9	Code injection and privilege escalation	AWS Lambda™ functions: • Data retrieval • Just-in-time registration • Set device state	• Steal confidential data • Use cloud account owner's compute resources, i.e., for bitcoin mining	• Write secure code and validate with secure code reviews and static analysis. • Robust input validation. • Enforce least privilege: functions only granted access to cloud resources, on a need-to-know basis.
A10	Device info theft Device info corruption	• Device registry	• Compromise database	• Restrict access to the device registry to the message broker's security groups. • Use cloud policies (AWS) to grant necessary read/write permissions to message broker and Device Shadow service.
A11	Privilege escalation	• AWS API Gateway	• Steal confidential data • Compromise database	• The API gateway must authenticate every request to the APIs it exposes. • The API gateway must verify the authorization of every authenticated request before allowing it. • Amazon provides AWS Cognito™ for this purpose.

(Continued on following page)

Table 6.4 Attack Surfaces and Mitigations (*Continued*)

Dalit Cloud: Attack Surfaces and Mitigations			
#	Specific Attack	Attack Surface	System Objective	Mitigation
A12	Man-in-the-middle network snooping and network/message data tampering	• AWS API Gateway • Cloud gateway	• Steal confidential data • Corrupt data stored in cloud	• Mutual TLS authentication between edge devices and cloud gateway. • Server-side TLS for users accessing the API through a browser i.e. access to the Traffic Analysis web application.
A13	Username/password brute forcing (Traffic Analysis web application)	• AWS Cognito • AWS API Gateway	• View confidential city data • Impersonate city officials	• Enable multi-factor authentication. • Define and enforce strong password policies. • Denial-of-service mitigations, mentioned earlier. • Enable CAPTCHAs.
A14	Bad/rogue authentication tokens (Traffic Analysis web application)	AWS API Gateway	• View confidential city data • Impersonate city officials	• API gateway must validate all received authentication tokens before allowing corresponding action.
A15	Cross-site scripting (Traffic Analysis web application)	Web applications via HTTP	• Execute scripts in the victim's browser • Hijack user sessions • Deface web sites • Redirect the user to malicious sites	• Perform input validation on all user-supplied input, to ensure it only contains expected characters. • User supplied input that's to be displayed on the web page must be cleared of all scripting tags.
A16	Cross-site request forgery (Traffic Analysis web analysis)	Web applications and server, in this case, AWS S3™	Leverage a logged-on user's session in order to cause them to send unintended yet "technically" legitimate requests to the web service they are logged on to, i.e., a user logged into a banking app, who is tricked into making an unintended bank transfer via a link or button on a malicious website	• Include a non-predictable nonce in the response to a successful user authentication. The nonce must be present in every request made by an authenticated user. The server must validate the nonce, before allowing the authenticated request.
A17	Rogue cloud administrator (Traffic Analysis web application)	• Administrative account credentials • Bastion host	• Steal or compromise system data • Deploy computing resources for illegal activities, i.e., bitcoin mining	• Cloud admin access must be granted based on a need-to-know basis. • Varying levels of admin access must be granted depending on scope required. • Admin credentials must use SSH keys. • Administrative access must only occur through the bastion. • The bastion must restrict ingress requests to a specific IP address range of the administrative entity. • All cloud administration activities must be logged and monitored.

(Continued on following page)

Table 6.4 Attack Surfaces and Mitigations (*Continued*)

Dalit Cloud: Attack Surfaces and Mitigations				
#	Specific Attack	Attack Surface	System Objective	Mitigation
A18	Network pivoting	Network	Leverage any network/server compromise of (usually in public-facing network subnets) to access data in restricted networks	• Within the virtual network, create separate public subnets for servers that accept external inputs. • All servers must only be administratively accessible via SSH keys. SSH access must only be possible through the bastion. Access to the bastion must also be accessible via SSH keys, and external access must be restricted to the IP addresses of the administrative entity, i.e., the company deploying and maintaining the app. • Within any subnet (public or private), and across public and private subnets, use security groups to restrict which machines/servers are allowed to communicate with each other.
A19	• Malicious updates lodged in container registry used for device updates • Malicious updates processed by edge devices	Container registry	• Compromise edge devices or steal data contained on them • Use edge compute for alternate purposes, i.e., bitcoin mining	• Only administrators should be able to place container images in the registry. • All container images must be signed by the system owner/maintainer. All edge devices must verify the signatures of container images before using them. • All connections (downloads) from edge devices to the container registry must be over the TLS. • All admin access to the container registry must be via SSH keys, restricted IP addresses, and the bastion host.
A20	Compromise device configuration and integrity	Device Shadow service	• Manipulate device configuration • Compromise device integrity	• Restrict access to the Device Shadow service via the bastion host and it's SSH keys, restricted address, and security group. • The Device Shadow service must maintain a unique shadow for each device that's to be remotely updated. • The Device Shadow service must store history-of-device states and support rollbacks. • The Device Shadow service should only receive inputs from the Lambda™ function used to set its state.
	...	...	...	...
	...	...	...	...

References

1. "Internet History of 1970s." Retrieved from https://www.computerhistory.org/internethistory/1970s
2. Amazon Press. (2006, August 24). "Announcing Amazon Elastic Compute Cloud (Amazon EC2)—beta." Retrieved from https://aws.amazon.com/about-aws/whats-new/2006/08/24/announcing-amazon-elastic-compute-cloud-amazon-ec2---beta/
3. Mell, P. and Grance T. (2011, September). "The NIST Definition of Cloud Computing." Retrieved from https://nvlpubs.nist.gov/nistpubs/legacy/sp/nistspecialpublication800-145.pdf
4. Mesnier, M., Ganger, G., and Riedel, E. (2003, August 18). "Object-Based Storage" (PDF). *IEEE Communications Magazine,* vol. 41, issue 8.
5. Kovacs, G. (2017, February 22). "Block Storage Vs. Object Storage in the AWS Cloud." Retrieved from https://cloud.netapp.com/blog/block-storage-vs-object-storage-cloud
6. Zimmer, V. and Krau, M. (2016, August). "Establishing The Root of Trust." Retrieved from: https://uefi.org/sites/default/files/resources/UEFI%20RoT%20white%20paper_Final%208%208%2016%20(003).pdf
7. Fedorkow, G. (2015, September 11). Retrieved from https://forums.juniper.net/t5/Security/What-s-the-Difference-between-Secure-Boot-and-Measured-Boot/ba-p/281251
8. Liu, G. (2018, July 11). "Ensure Secure Communication with AWS IoT Core Using the Certificate Vending Machine Reference Application." Retrieved from https://aws.amazon.com/blogs/iot/ensure-secure-communication-with-aws-iot-core-using-the-certificate-vending-machine-reference-application/
9. Ashmore, S. and Wallace, C. (2010, June). "Trust Anchor Format." Retrieved from https://tools.ietf.org/html/rfc5914.
10. Schoenfield, B. (2015, May 20). *Securing Systems: Applied Security Architecture and Threat Models.* Boca Raton (FL): CRC Press/Taylor & Francis Group.
11. Dignan, L. (2018, May 23). "Google Cloud Platform Breaks into Leader Category in Gartner's Magic Quadrant." Retrieved from https://www.zdnet.com/article/google-cloud-platform-breaks-into-leader-category-in-gartners-magic-quadrant
12. Dierks, T. (2008, August). "The Transport Layer Security (TLS) Protocol Version 1.2." Retrieved from https://tools.ietf.org/html/rfc5246#section-7.4.6

Chapter 7

Securely Connecting the Unconnected

So the whole war is because we can't talk to each other.

— Orson Scott Card, *Ender's Game**

7.1 What Connectivity Means to IoT

Many problems in this world have resulted from a lack of communication. The famous book made into a movie, *Ender's Game* tells the story of a war between humans and an alien race that ensued primarily because the two species could not communicate. Although an interstellar war is unlikely to break out due to communication problems between Internet of Things (IoT) devices, in previous chapters we have described some significant events that have occurred when communications protocols go awry, including the Morris Internet Worm and the Mirai Botnet Attack. In this chapter, we look more deeply at communications in IoT devices and then explore how to analyze communications protocols for security problems, which should equip you to uncover such issues before they take your systems down.

Just as communication is the essence of human existence—the ability to share intentions, exchange information about capabilities, coordinate in the sharing of tasks, warn of dangers, and even receive help when needed—communication between devices accomplishes similar functions. It may be obvious, but without the ability to connect, a device cannot be an Internet-of-Things device.

Communication is so essential to IoT because beyond mere data exchange, the goal of IoT is to achieve real Machine-to-Machine (M2M) communication. This goal is to create the ability within a device to exchange information with another device without any human

---

* Card, O. (1985). *Enders Game.* New York (NY): Tor Books.

intervention, and to accomplish tasks, report conditions and health status, adapt to the changing physical environment around them, and enable the device to survive in order to accomplish its intended mission. In the field of IoT, we often use the acronym M2M to convey this rich idea. And, of course, this requires much more than merely communication. Processing capability, Intelligence (with a capital *I*), and adaption are all required. But communication is a vital first element of that equation. And more than just communication, but a rich language over a communication channel that enables devices to hold a meaningful conversation. The foundation of that rich language is the communication channel created by network protocols. And IoT has many different protocols, which we explore in this chapter.

As we discuss communications, we want to be crystal clear about the capabilities, advantages, and disadvantages with respect to the different communications technologies. Within many IoT articles and marketing materials, it is common to mix different parts of the communications stack. This may result from a lack of understanding of communications protocols, or it may be carelessness, or it may even be purposeful to advocate for a particular proprietary solution. This confusion obscures the boundaries between physical access technologies, transport protocols, and application and presentation layer data-formatting protocols. This goes beyond simplification and creates confusion regarding how to actually build a complete IoT communications stack that effectively meets IoT requirements. Some materials even repackage existing protocols in some *special* way and call it a new protocol. In our opinion, this is more marketing than a technology enhancement. Some examples of this include QUIC, which is really just several user datagram protocol (UDP) ports multiplexed together, and MQTT-SN, which is just a paired-down version of MQTT. In this chapter, we strive to avoid such confusion, and although we cannot cover every protocol or access technology, we provide a reasonably complete coverage of IoT communications, where we separate the technologies and protocols into their different respective layers according to their utility in formal communications model terminology.

As we break down the different wired and wireless communications stacks, it is important to be precise in the language we use to describe the elements of a communications stack; therefore, we prefer to utilize the Open Systems Interconnection (OSI) Reference Model (see Figure 4.1) to organize these elements. The bottom of the stack includes the physical and data link layers; these comprise the *physical access technologies* and determine things such as communication range, power requirements, bandwidth capabilities, number of communicating nodes, and organization of communicating elements. Matching the physical access technology to the needs of a particular device in an IoT system is important. Some physical access technologies are more appropriate for different parts of an IoT architecture, as we will discuss later.

Above the physical access technologies are the network, transport, and session protocols. Although there are choices at these layers, it is nearly universal to use the Internet Engineering Task Force (IETF) Internet Protocol (IP), Transmission Control Protocol (TCP), and UDP. We do not provider further coverage of these protocols in order to reserve our discussion for more IoT-specific topics that, in our opinion, are less well understood and deserving of our (and your) attention.

Above the network, transport, and session layers is the presentation and application layer. There are several very unique and interesting IoT-specific protocols at this layer that anyone designing and developing IoT systems should be knowledgeable about. We provide some brief overviews of these protocols to make you aware of the latest protocols at the IoT application layer.

7.2 Classifying IoT Communication Protocols

As we think about communications protocols, it is helpful to partition protocols into different classifications. Our partitioning of communication technologies is graphically shown in Figure 7.1 and is separated into long-haul communications, wired communications, and wireless communications. Long-haul communications are the protocols used to connect to the cloud. Protocols used to connect with devices are typically shorter range and are split between wired and wireless protocols. These different classifications enable us to better understand the best uses of these protocols within our IoT systems. It may also be helpful to peek ahead at Figure 7.4, which graphically portrays the protocols that will be covered in this chapter.

We are not going to delve into the complete construction of network stacks and the OSI model. There are great books already on this topic, which you will find in our references. And although network stack construction is important to those building communications stacks, for our coverage of communication technologies we only need to differentiate between three different parts of the communications stack: the physical connections, the network and session layers, and the application protocols. Let us review why these three distinctions are important to us.

The physical connections are vitally important to IoT systems because these connections determine the maximum distance IoT stations can be from the other parts of the system, the maximum speed of data exchanges between these stations, and the power expenditure required to execute the communications protocols. This information is necessary to understand so that the best choice of physical communications technology may be made for a particular part of an IoT system.

The network and session layers determine the way devices are found and addressed on the communication media (wireless or wired). With some technologies, the number of devices connected to the system is limited by the underlying communications technology. MODBUS® limits the number of connected devices to only 247; IPv4 (Internet Protocol version 4) is

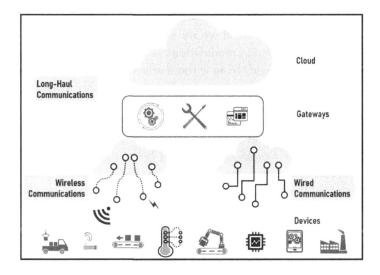

Figure 7.1 Communications Layers of IoT Systems

limited to 2^{32} different stations. There is always some practical limit. Understanding what limitations there are for different technologies and the ways to overcome or deal with those limits is important.

The application protocols determine how different stations communicate. The format, frequency, data size, reliability, among other parameters, are determined by the protocols running above the physical connection. Sometimes these parameters are determined by the session layer, other times by other upper-layer protocols.

We look at each of these three areas in turn, discussing the characteristics of data rate, network size, and maximum transmission distance.

7.2.1 Bandwidth, Bits, Codes, and Hertz

Before we delve into the details of IoT communications protocols, we want to share an important perspective that was shared with Dave years ago by a good friend of his and communications expert, Roger Freeman. When Dave first met Roger, Roger was already an old coot, and also well experienced in communications. As a young man, Roger was a radio operator on merchant marine ships, and he would joke that he was on a ship that heard the Titanic distress call. The Titanic disaster was a bit over a decade before he was born, but to us youngsters it brought home the point that he was older than dirt and had valuable experience we still lacked!

Roger pulled Dave aside one day and said, "Dave, don't ever confuse bits with bandwidth." Then he proceeded to give Dave a lecture on what that means. Dave will admit that at the time he did not fully appreciate Roger's wisdom, but over the years what Roger shared has become clearer and more treasured by Dave. We want to pass this wisdom on to you.

Bandwidth is the width of an electromagnetic wave in a particular frequency band, which we measure in Hertz. That band of frequency is capable of carrying information encoded on the electromagnetic wave. However, we do not measure this bandwidth in bits. Why? Because we can *code* information on that wave using symbols that may represent more than a single bit of information. One way of coding is to apply modulation to a signal—amplitude modulation, frequency modulation, or phase modulation—which increases the number of bits that can be carried on waves in the frequency band we are using. Use of different forms of modulation combined with codes for the different modulations increases the effective data rate of information carried in a particular frequency band without affecting the actual width of the frequency band (the Hertz). Other things can reduce the effective data rate, including error-correcting codes, framing bits, and protocol headers—we refer to these things as overhead. Roger explains this in greater detail in his book.[1]

It is important to bring this up here because the effective data rate one may observe in practice is different from the *bandwidth* advertised by many protocols. For example, in the 802.11 series of protocols, many variants are advertised to have a very large bandwidth (802.11a is reported as using 54 Mbps). This calculation is based on the frequency band, modulation, and coding of that specification without any disclosure of the overhead. This type of inaccurate disclosure of bandwidth has been reported as *marketecuture*.[2] In the following sections we do not attempt to correct the reported bandwidth claims, but we warn the reader here to understand there is a gap between a protocol's reported bandwidth and the actual data rate a particular protocol stack can achieve in an actual deployment.

7.2.2 Physical Layer Communications—Wired and Wireless

In the next two subsections we discuss the physical layer communications protocols commonly used in IoT. We treat the physical and data link protocols together as a single protocol layer, which we refer to here as a Phy—a basic physical and logical connection between nodes created by the combination of the physical and data link layers. Do not confuse this with a PHY, which is an abbreviation for just the physical layer. In this section we want to discuss the bottom two layers of the stack combined. This method of treatment is not uncommon in some communication circles, though many purists may cringe. For our purposes, it is more instructive to treat these two layers together because, as with numerous IoT suppliers, a specific combination of physical and data link protocols is often used together by a supplier's devices. One cannot just combine any physical layer with any data link layer willy-nilly. Those readers desiring deeper details on pure communications theory and protocol stack development are referred to the many good books in our references that are devoted to that topic. So, for our purposes, Phys provide basic connectivity between network stations (devices), media access, framing, and link control. The protocol layers above the Phy allow the network station (a device) to use that connectivity to accomplish different tasks.

The Phys determine how devices are physically connected together—point to point or shared broadcast media. Protocols above the Phy determine the organization of stations and routing of messages—peer-to-peer, star, ring, and mesh.[3,4] Application protocols expand the types of services that can be provided over a connection in a transparent way—for instance, MQTT allows devices to communicate in a store-and-forward manner using a star topology between publishers and subscribers, regardless of how the underlying Phy and transport protocols are constructed.

The next two subsections outline wired and wireless Phys. Wireless Phys are more common in new IoT systems and are, in our opinion, more important to IoT because they enable easier and less costly connectivity between devices. However, wired Phys are still used because they provide robust communications in the harshest environments, the infrastructure to support wired Phys may already be in existence for a particular deployment, or a particular installation may encounter unacceptable interference using wireless communications. Oftentimes, wireless Phys are found in *greenfield* installations (completely new IoT systems), and wired Phys are found in existing systems expanded with IoT capabilities, referred to as *brownfield* installations.

7.2.3 Wired Phys

There are many wired physical protocols in use in IoT systems, but several of them are used more commonly than others. The following descriptions provide a basic overview of the different common wired Phys in use.

Ethernet and **twisted pair** are a type of cabling commonly used in computer networks today. If you plug your laptop in at work or at a hotel, you are most likely using Ethernet. Twisted pair refers to a pair of wires twisted together that provide communications. There is both unshielded twisted pair (UTP) and shielded twisted pair (STP), which is wrapped in a foil or other conductive material to shield the wires from outside interference. UTP is most common in computer networks. A single twisted pair can be used for telephone communications.

Ethernet requires two twisted pairs, or four wires. It is common for sets of twisted pairs to be bundled together and encased in a plastic sheath with plastic connectors at the ends. Ethernet cables are referred to as Cat5, Cat5e, or Cat6 cables, referring to how fast data can be passed through the cables. Telephone cables look similar but actually are not required to be twisted together, though many are. Ethernet cables typically have a plastic end called an RJ-45 connector, whereas telephone cables use a smaller RJ-11.[5]

Ethernet is the term colloquially used for an IEEE 802.3 cabling with 802.3 MAC framing over the wire.[4] The 802.3 specification covers multiple different physical connections, from fiber optic, to twisted pair, to coax cable. But even when restricting our usage to an 802.3 twisted pair configured in a star-type network configuration with a hub or switch, there are multiple versions of the standard that include different speeds. Table 7.1 shows several different common 802.3 twisted pair versions. Although the maximum distance and maximum number of stations allowed by 802.3 twisted pair seem limiting, they are easily overcome using multiport bridges and repeaters.

Table 7.1 IEEE 802.3 Twisted Pair Characteristics[3]

Name	Version	Maximum Distance	Maximum Stations	Transmission Speed
10Base-T	802.3i (1990)	100 meters	1024	10 Mbps
100Base-TX	802.3u (1995)	100 meters	1024	100 Mbps
1000Base-T	802.3bp (2016)	100 meters	1024	1 Gbps

4-20mA loop is commonly called a *twenty-milliamp loop* and is probably the most common connection in industrial IoT and process control environments. This communication is created by a pair of wires between a transmitter and a receiver. The current on the loop of wires is the signal level being communicated. A 0% signal is communicated as 4 milliamps, and 100% as 20 milliamps. Conversion of this signal to some units, such as the liquid level of a storage tank represented as a percentage of the tank's total capacity, or the revolutions per minute of a motor, are done by the receiver. A fault on the loop is easy to detect because the current drops to 0 milliamps, which is not a legal value on the line since 0% would be 4 milliamps. 4-20 mA loops are very reliable, but they can be expensive to set up because each different signal needs a separate pair of wires, and when there are hundreds of loops, the number of wires becomes burdensome, especially to find and fix a fault, because you have to find the one short in hundreds of wires.[6] The maximum length of 4-20mA loops depends on the voltage of the system (which can vary) and the impedance of the connected devices and resistance of the wires. Ohms law must be used to calculate the maximum distance based on the system characteristics, but several kilometers distance is not unachievable.

HART® (Highway Addressable Remote Transducer) is a special physical layer protocol that runs over 4-20mA loop connections to add digital data over the top of the normal information provided on the loop. HART uses a frequency shift keying (FSK) modulation over the wires to encode additional data on the 4-20mA loop so that both digital and analog data can be sent at the same time. This requires a special HART-enabled device at both ends of the wire.[7] HART is part of the IEC61158 Fieldbus specification (please see below for further details).

FOUNDATION™ Fieldbus (FF) has two variants, FF-H1 and FF-HSE. **FF-H1** uses two-wire STP to provide full-duplex digital communications to devices connected via a main bus or daisy chained together. The bus itself must be powered, and the current on the wires is modulated, similar to HART technology, in order to send digital signals to devices; the bus can also be used to power the end devices attached to the network. FF-H1 is an improvement over HART, as it provides a bus organization allowing connections to multiple devices. However, the cabling is more complex and requires junction boxes at join points, but offers savings in the total amount of wiring as well as operational costs. A single bus can reach over a mile with up to 32 devices, but adding repeaters allows this to be increased to almost 6 miles. FF-H1 allows for intrinsically safe (IS) and non-incendiary operations, making it ideal for certain types of manufacturing environments. **FF-HSE** (high-speed Ethernet) is a variant using native Ethernet wiring and is normally used to connect the control network to FF-H1 segments containing the actual devices. The FF-HSE segment operates the standard 10/100 Mbps, allowing it to easily control the lower speed 31.25 Kbps FF-H1 segments.[8,9] FOUNDATION Fieldbus is part of the IEC61158 Fieldbus specification.

RS-232 and **RS-485**, or Recommended Standard 232 and 485, are the basic serial protocols that were very common in computers in the 1990s and were used to connect modems and printers. These serial protocols were adopted under the EIA/ANSI specifications and included both electrical and mechanical specifications. Developed in 1960, RS-232 allows the communication of digital signals, using zeros (+5 volt) and ones (–5 volts), giving more flexibility than a 20 mA loop, as it can communicate a variety of signal data, not just a single percentage or reading value. Basic RS-232 connectors have 9 connections—8 signal pins plus a ground—and include signals for receive data (RD) and send data (SD), which are cross-connected between the master and the slave. Other signals include data carrier detect (DCD) to indicate a good connection to the master; data terminal ready (DTR), indicating that the slave is online; request to send (RTS) is a signal from the slave to the master to indicate that the slave needs to send to the master; and clear to send (CTS) is a signal from the master that the slave can now communicate. The pin ready to receive (RTR) is used in some systems to signal from the slave to the master that the slave is ready to receive traffic. The last signal is the ring indicator (RI), which was used by modems to indicate the presence of an incoming connection. Although nine wires are specified, many RS-232 connections use fewer and loop-back some of the signaling lines in the connector itself to *trick* the station that it received a DCD or DTR signal. There is also a 25-pin connector that includes pins for two independent connections over the same cable and connector.

RS-485 was introduced in 1998. RS-485 is an improvement over RS-232 as it provides increased speed and reliability, and it allows multiple devices to be connected together. RS-485 is still used in some IoT networks because of this ability to connect a series of devices together and provide communication from a master to many devices. RS-485 includes a clock signal, whereas RS-232 requires the receiving station to be manually configured to set the clock speed to match the master. A misconfiguration in RS-232 means that data will be received corrupted, and the device cannot talk to the master. RS-485 provides automatic configuration of devices to the master's settings.[10] RS-232 transmits ASCII text in application-specific (custom) messages, whereas RS-485 typically uses another protocol called MODBUS.[11] MODBUS is covered in the section on upper-layer protocols. Table 7.2 details the characteristics of the RS-232 and RS-485 protocols.

Table 7.2 RS232 and RS485 Characteristics[10]

Name	Maximum Distance	Maximum Stations	Transmission Speed
RS-232	50 feet	1	Up to 1 Mbps
RS-485	4000 feet	32	10 Mbps

I2C (Inter-Integrated Circuit), pronounced *eye-squared-see*, is a serial protocol commonly used in sensors in the Maker Movement, as well as home automation, security systems, and computer peripheral devices. This protocol can also be referred to as two-wire interface (TWI) to avoid trademark issues, though some TWI implementations add proprietary additions to the protocol. Because I2C uses only two wires, instead of a full serial line with eight wires, it tends to be cheaper than RS-232 or RS-485 but also runs at lower speeds and much shorter distances. I2C supports up to 1023 stations, depending on the supported address length of the master controller—127 stations with 7-bit addressing is common. I2C's two wires are a serial clock (SCL) that tells the slave how fast to read signals on the other serial data (SDA) wire. Messages on the SDA are initiated with a start signal from the master, followed by the device address, and then a device command and address for read or write. A write command is followed by the data the master is writing to the slave; a read command requires the slave to respond with the data requested by the master within a specific period of time. The sequence ends with a stop signal.[12] The maximum length for an I2C bus is a few meters, but such a long length requires the transmission speed to be decreased. An I2C repeater is another method to increase the I2C bus length. The bus speed for I2C ranges from 100 Kbps for standard mode to 3.4 Mbps for high-speed mode.[13]

USB (Universal Serial Bus) is a faster serial communications technology that replaced RS-232 and RS-485 serial connections in computers in the early 2000s. USB provides higher speeds than the Recommended Standard protocols and can connect multiple devices to a master hub using a star topology. USB is different from other serial communications as it uses differential voltages across two wires for communication, enabling a higher switching rate and thus higher transmission speeds. USB was invented to provide hot-pluggable (remove and connect cables while devices are powered on), high-speed serial connections to devices. Up to 127 devices can be connected to a single USB hub, with a maximum cable distance of 5 meters. USB allows hubs to be chained together, overcoming the 127-device limit, which effectively creates a star-of-stars topology. The distance from hub-to-device may vary, based on cable quality, since the specification requires no more than a 26-ns propagation delay across the cable. The USB specification currently has three different versions, which all support different speeds. USB 2.0 supports speeds of 1.5 Mbps, 12 Mbps, and 480 Mbps; USB 3.0 supports speeds up to 5 Gbps. Other advantages of USB include the ability to power devices from the cable itself and integrated error detection and recovery for transmitted data. USB also includes a power savings mode where a device's communication hardware is suspended after 3 milliseconds of no activity.[14]

Fieldbus or **IEC 61158** is actually a set of different technologies that were standardized by the International Electrotechnical Commission in 1999,[15] with the most recent update of the standard in 2014.[16] IEC 61158 is composed of 82 specifications that are organized into 6 parts, where the parts cover different layers of the protocol stack, including the physical layer, data link layer, and the application layer. Many different technologies are included in the specification, which organizes the technologies into communication profile families (CPF). We cover only a portion of the IEC 61158 protocols, including HART over a 4-20mA loop, FOUNDATION Fieldbus, MODBUS, and PROFIBUS DP. Other fieldbus protocols we do not cover include CIP™, PROFINET, P-NET®, WorldFIP®, INTERBUS®, CC-Link, Vnet/IP®, TCnet, EtherCAT, ETHERNET Powerlink, EPA, SERCOS, RAPIEnet, and SaftyNet P™.

CAN bus (Controller Area Network bus) is commonly used in automobiles to connect various Electronic Control Units (ECU) together. An ECU contains a microcontroller and is used in the control systems of automotive vehicles. CAN bus is a serial communications technology that supports real-time communication with multiple bus masters and multiple stations connected together with twisted pair.[17] ECUs transmit messages on the bus, and the bus arbitrates messages based on content type, where the content type also determines the message priority. Higher priority messages are transmitted before lower priority messages. For example, a message to activate the braking system or to slow the engine speed is a higher priority message than a message to increase the cabin air flow. CAN encodes bits as 0 volt or +5 (+3.3) volts, but uses a non-return-to-zero (NRZ) encoding scheme, meaning that every bit does not necessarily include a falling or rising edge of the signal. Messages include error codes to detect corrupted messages, and misbehaving ECUs can be removed from the bus. CAN bus was invented in 1986, with the CAN bus 2.0 update released in 1991. In 1993, these specifications were adopted by the International Organization for Standardization (ISO). Figure 7.2 identifies which ISO standards apply to specific CAN bus protocol layers. In 2012, automotive manufacturers introduced a new high-speed CAN bus, CAN FD (flexible data-rate), which addresses the need for higher-speed transmission to accommodate software updates.[18]

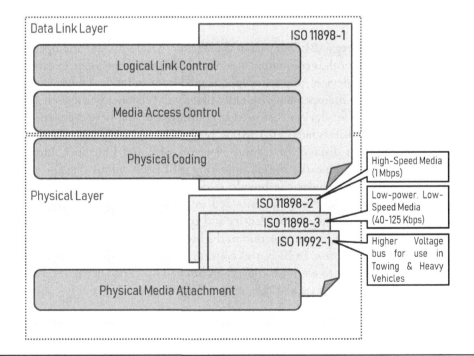

Figure 7.2 CAN bus Physical and Data Link Layer Protocol Diagram

The CAN bus 2.0 specification only defines the data link layer, which controls arbitration, framing of data, error detection, correction, and reporting. The physical layer, which covers how bits are transmitted, was left open to implementers.[19] ISO standardization has partially filled that gap by defining the physical media attachment protocols.

7.2.4 Wireless Phys

There are just as many wireless Phys are there are wired. IoT devices commonly use wireless Phys because they are easier to deploy since no wiring is required, and it also enables the devices to be mobile, which may be required in some installations—for example, maritime shipping, smart logistics, and robotic farming. Of course, from a security perspective, wireless presents several problems because an adversary can eavesdrop on the channel and capture the communications, they can more easily insert data into the communication channel, and they can jam the signal to prevent communications. We discuss more about adversaries and their formal capabilities later in this chapter. The following descriptions cover the most common wireless Phys.

WiFi, also called Wireless LAN or IEEE 802.11, is the radio technology that we use in Starbucks on our laptops and that provides excellent bandwidth and reasonable range. 802.11 has several versions (802.11a, b, g, n, ac).[2] But the 802.11 specifications include other features for WiFi, including quality of service for time-critical networking (802.11e), roaming between access points (802.11f), and self-tuning to select the best channel (802.11h). All these additions are very useful for IoT devices. Security was added in 802.11i, but there have been several breaks and modifications to the security schemes, including the most recent attack of WPA2 in August 2018.[20]

BLUETOOTH® Low Energy (BLE) is part of the Bluetooth 4.0 standard and operates in the 2.4-GHz spectrum.* Bluetooth is commonly used for hands-free phone usage in cars, for wireless keyboards and mouse devices, and for cellphone ear pieces and microphones, but BLE is a different protocol and not interoperable with older Bluetooth versions. The low-energy version of Bluetooth was invented for devices that need to communicate but have even stricter energy profiles than cellphones. BLE is rated for 1 Mbps, but practical throughput is in the 10-Kbps range, with a transmission distance of around 30 feet.[21] BLE and Bluetooth, broadcast out services and capabilities; any nearby station operating a BLE/Bluetooth Phy can receive the broadcast. This broadcast includes the broadcast device's unique 48-bit Bluetooth Device Address (DBA). Because this address is contained in the unsolicited broadcast, any nearby receiving station has the information to identify and connect with that device. The receiving host BLE station can then set up a connection to the broadcasting station. Data is transmitted over this peer-to-peer connection. In BLE, this connection can be set up to be a recurring connection that is automatically re-established every so often. This capability reduces the energy cost of small devices to transmit regular data up to a host.[22]

---

* Both Bluetooth and Bluetooth Low Energy (BLE) operate in the same unlicensed frequency at 2.400 GHz to 2.4835 GHz, but operate on different channels. BLE uses 40 channels each of 2-MHz bandwidth. Bluetooth uses 79 channels each of 1-MHz bandwidth.[20] WiFi also uses the 2.4-GHz band for 802.11b, g, and n, and specifies 14 channels spaced 5 MHz apart, with each channel being 22-MHz wide. This means that many channels overlap.[2]

IEEE 802.15.4™ is often mentioned as an IoT communication solution for sensors and devices, but when it is mentioned it is often difficult to get the details on what that communications technology really comprises. The IEEE 802.15 Wireless Personal Network (WPAN) working group is composed of multiple standards, and this lends to some of this confusion. But even further specifying IEEE 802.15.4 does not define the type of Phy in the device. Working group 4, the low-rate WPAN (LR-WPAN) subgroup of the IEEE 802.15 working group, has published multiple standards under the IEEE 802.15.4 title, and this adds additional confusion. To cut through this confusion, let us take a deeper look at the IEEE 802.15 standards. Table 7.3 provides a listing of the primary 802.15 groups. Although most of these standards are not often mentioned in IoT circles, they still have some relevance to us here because these are all short-distance communications technologies that could be used in IoT. There are multiple versions and revisions of the LR-WPAN specifications, just as there are different versions of the 802.11 standard. The IEEE 802.15.4 standard is composed of 18 different physical layers, along with several different data link protocols that provide different link control schemes and topologies (star, mesh, multiple access). The physical layers cover different radio modulation schemes,

Table 7.3 Listing of the Primary IEEE 802.15 Working Groups and Standards

Working Group	Technology	Published Standards
Task Group 1	Bluetooth v1.1	IEEE 802.15.1-2005 (withdrawn in 2018) https://standards.ieee.org/standard/802_15_1-2005.html
Task Group 2	Coexistence of Bluetooth and WiFi as WPANs	IEEE 802.15.2-2003 http://www.ieee802.org/15/pub/TG2.html
Task Group 3	High-rate (20 Mbs+) WPAN	P802.15.3-2017 https://standards.ieee.org/standard/8802-15-3-2017.html
Task Group 4	Low-rate WPAN with easier security, new frequency allocations, and clarifications	IEEE 802.15.4-2015 https://standards.ieee.org/standard/802_15_4-2015.html
Task Group 5	Mesh topology capability for WPANs	IEEE 802.15.5-2009 https://standards.ieee.org/standard/802_15_5-2009.html
Task Group 6	Body Area Network (BAN) up to 10 Mbps	IEEE 802.15.6-2017 http://www.ieee802.org/15/pub/TG6.html
Task Group 7	Short-range optical wireless communications using visible light	IEEE 802.15.7-2011 https://standards.ieee.org/standard/802_15_7-2011.html
Task Group 8	Peer Aware Communication (PAC) for WPAN	IEEE 802.15.8-2017 https://standards.ieee.org/standard/802_15_8-2017.html
Task Group 9	Transport of Key Management Protocol	IEEE 802.15.9-2016 https://standards.ieee.org/standard/802_15_9-2016.html
Task Group 10	Routing Module Addressing	Draft 802.15.10a-2017 https://standards.ieee.org/project/802_15_10a.html
Task Group 12	Consolidated Link Layer Control	P802.15.12 https://standards.ieee.org/project/802_15_12.html
Task Group 13	Multi Gbps Optical Wireless networks	P802.15.13 https://standards.ieee.org/project/802_15_13.html

Note: No Task Group 11 (TG11) was ever formed, presumably to avoid confusion with the 802.11 group, but no definitive statement was in the IEEE working group notes.

from binary phase shift keying (BPSK) and offset quadrature phase shift keying (O-QPSK) to direct sequence spread spectrum (DSSS) and chirp spread spectrum (CSS), as well as other more unique modulations such as TV white-space narrowband orthogonal frequency division multiplexing (TVWS-NB-OFDM). These get complex pretty fast. Additionally, depending on the chosen physical layer modulation, different frequency bands are allowed, from as low as 169 MHz, various 800- and 900-MHz bands (varying by region), the common 2.4-GHz band (like WiFi), and ultra-wideband in 3.2- and 6-GHz ranges. With all these different options, constructing an interoperable stack is more than just claiming IEEE 802.15.4 support. The IEEE 802.15.4 specifications should be referenced for the specific details.[23]

It is easy to see that with all the options for different radio frequencies, different physical coding schemes, and different link controls, IEEE 802.15.4 is something extremely useful to IoT systems and can be customized to create something useful for different vertical markets. But the primary benefit of the IEEE 802.15.4 specification is the standardization of the MAC and link layer access interfaces, making it easy to create upper-protocol layers that interface with any manufacturer's IEEE 802.15.4 compliant implementation. This also explains why products do not specify anything further than IEEE 802.15.4, because any compliant lower-layer Phy could work with their product. This also then lends to some marketecture, by specifying a unique name for a "protocol" on top of a specific type of IEEE 802.15.4 Phy. Several other IoT wireless protocols do just that and build on top of this standard, including ZigBee® and 6LoWPAN. For interoperability though, you still have to have matching underlying IEEE 802.15.4 Phys.

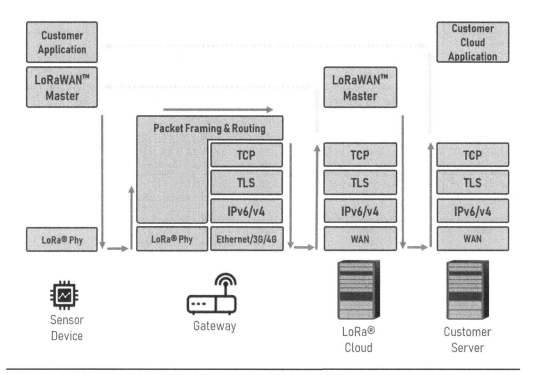

Figure 7.3 LoRaWAN Architecture (LoRa Alliance™)

LoRa® is a long-range proprietary radio modulation technology from the Semtech Corporation that operates in various unlicensed bands below 1 GHz—typically 915 MHz in the United States and 868 MHz in Europe. The physical layer of LoRa uses a modulation that is similar to the CSS technique from the IEEE 802.15.4 specification. LoRa is ideal for long-range radio communications using low power, but it technically only covers the physical layer of communications.[24] The maximum message size is 243 bytes, and transmission speeds top out at 50 Kbps. Longer ranges can be achieved by selecting higher spreading factors in CSS, but this creates a lower transmission speed.[25] LoRaWAN™ is the media access layer protocol managed by the LoRa Alliance™ and is commonly used for LoRa deployments. Figure 7.3 depicts a typical LoRaWAN, where all the communications go back to a central server; this is great for a large public network, say for a smart city, but for multiple coexisting networks, the traffic is intermingled and privacy and security can be a concern. Other MAC layers are possible on top of the LoRa physical layer, with one example being Symphony Link™.[26]

SigFox® is a European company that offers an LPWAN service using proprietary radio technology similar to LoRa. Base stations collect data from IoT nodes and send this data to the SigFox cloud where it is forwarded to subscriber's back-end systems. The radio technology operates as an ultra narrowband in the unlicensed spectrum—868 to 869 MHz and 902 to 928 MHz—depending on the region. The radio modulation uses differential binary phase-shift keying (DBPSK) and the Gaussian frequency shift keying (GFSK). Packets are very small—12 bytes uplink and 8 bytes downlink—but due to the frequency and modulation choices, it can penetrate solid objects providing very good reach. SigFox is working to be available globally and has coverage in Europe, the United States, most of South America, Australia, Japan, and a few other countries in Africa, the Middle East, and the Asia-Pacific region.[27]

Z-Wave® is another proprietary radio technology in the unlicensed 800- to 900-MHz band that is owned by Silicon Laboratories. Technically, Z-Wave uses radio technology from IEEE 802.15.4, but since Z-Wave devices are not interoperable unless they are using the same silicon chips from the sole-source Z-Wave provider, we have chosen to break them out from other 802.15.4 protocol technologies in order to highlight this issue. Z-Wave concentrates on home-automation usage and therefore limits itself to small networks of no more than 232 nodes.[28] Z-Wave includes a full stack up to the application layer and offers developer and system integrator help, including some open-source software for the upper layers. The physical and data link layers are part of licensed silicon.[29]

Weightless-P, **Weightless-W**, and **Weightless-N** are protocols defined by the Weightless™ Special Interest Group (SIG). Weightless-P is an LPWAN technology operating in the sub-GHz frequency bands—depending on the region, one of 138 MHz, 433 MHz, 470 MHz, 780 MHz, 868 MHz, 915 MHz, or 923 MHz. This protocol is designed for small packets less than 48 bytes and a maximum transmission speed of 100 Kbps. Communications from end nodes are sent to nearby base stations that are interconnected through a service operator's base station network.[30] The range is estimated at about 2 kilometers. Weightless-W is a different technology from the Weightless SIG that operates in the TV whitespace spectrum. Weightless-N is a third technology focused on ultra narrowband similar to SigFox. Weightless is only a standards body, so they rely on others to actually build the equipment and set up the service. Weightless technologies do not have the same market penetration as compared with LoRa and SigFox because both of these solutions provide a complete end-customer service.[31]

2G GPRS cellular was the first viable, broadly deployed standard from 3GPP providing data over cellular networks using general packet radio service (GPRS). GPRS provides data speeds from 56 Kbps up to 114 Kbps. Sometimes, 2G GPRS is now referred to as 2G M2M (machine-to-machine) and is still used in some IoT deployments, but there is a strong preference for the newer cellular technologies due to the cost of the cellular devices and battery-life constraints that are better met with the newer technologies.

3G stands for the third-generation mobile radio technology used in cellular phones. Data transmission over 3G can use many different underlying transmission technologies with speeds varying from a few hundred kilobits to a few megabits. Common technologies lumped into 3G include Universal Mobile Telecommunications Service (UMTS) and CDMA2000 EVDO. High Speed Packet Access (HSPA), which builds on the UMTS technology, is a further enhancement for GSM networks in 3GPP Releases 6 and 7. 3G services are more frequently being used for IoT, as the infrastructure is already in place, and most cellular subscribers have moved to 4G services, although 3G is still used as a backup capability for mobile subscribers.

4G/4G LTE is the fourth-generation cellular technology, along with 4G Long Term Evolution (4G LTE), which all refer to the technologies replacing 3G cellular communications. 4G and 4G LTE provide increased reliability and up to 10 times the transfer speeds as compared with 3G. 4G LTE brought web browsing and Internet applications to mobile devices at speeds that are acceptable to most users. 4G LTE is actually a roadmap of capabilities, some of which include the following[32]:

- New evolved packet system (EPS), which is a new flat network running only IPv4 and IPv6, with interfaces allowing easy integration of GPRS and WiFi networks
- New physical layer technologies, including Orthogonal Frequency Division Multiplexing (OFDM) and Synchronization Channel—Frequency Division Multiple Access (SC-FDMA)
- Multiple Input Multiple Output (MIMO) antenna technology, including spatial multiplexing (SM), beam forming (BF), and multi-user-MIMO (MU-MIMO)
- Active antenna systems (AAS) that use multiple transceivers to adjust the radiation pattern (using BF) but do so dynamically, allowing vertical and horizontal splitting of radio beams, cell shaping, and cell splitting, all of which allow for better management of bandwidth demands and current load of subscribers
- Inter-cell interference coordination (ICIC) to coordinate the transmission and reception of the new narrowband techniques in OFDM and SC-FDMA, avoiding interference problems

All of these enhancements, and many more we do not have room to mention, provide higher transmission speeds, more reliability, and better management of the bandwidth. Each 3GPP release adds more capabilities and allows for experimentation with new technologies, with each step moving closer to realization of the full 5G network.

LTE-M (long-term evolution for machines) is the 4G solution for IoT. LTE-M is also known as LTE Cat-M1, or enhanced Machine Type Communications (eMTC). It was part of the 3GPP Release 13 in 2016. LTE-M is a low-power wide-area network (LPWAN) solution that provides lower bandwidth than a regular 4G LTE connection and, therefore, is available for lower cost. LTE-M limits the connection speed to around 300 Kbps with very low latency, in the milliseconds, which is more than adequate for most IoT needs. LTE-M also provides voice communications, which is useful for some IoT installations that might need human communications as a backup—for example, in a smart city–enabled parking garage or kiosk. Battery life for devices utilizing LTE-M is targeted at ten years.[33]

NB-IoT (Narrow Band IoT) is the other 4G LPWAN solution for IoT that is specified by 3GPP. NB-IoT has three modes in which it can operate. The primary mode allows it to reallocate some GSM carrier bands to carry NB-IoT data. The two other modes allow it to use resource blocks in the LTE carrier or guard bands, but then the bandwidth available is limited. Due to the primary usage of GSM infrastructure and the limited availability of GSM in the United States, NB-IoT is rolling out first in Europe. NB-IoT is specially designed for low data-rate only communications at less than 100 Kbps, with a latency up to 10 seconds. NB-IoT is also only half-duplex, meaning that only one side of the communications link can talk at a time, but this is a reasonable limitation for most IoT installations. One of the simplifications of NB-IoT over 4G communications is the removal of station hand-offs. Hand-offs are the protocol support allowing a connection to be transferred from one cellular tower to another, which is required to support mobile (moving) stations. This simplification means that NB-IoT only supports fixed stations.[34]

5G is the next evolution in cellular radio communication, targeted for availability in 2020. 5G provides higher bandwidth using a series of technologies, including virtualized software-defined networking (SDN), network slicing, millimeter wave lengths (30–300 GHz), small cell networks, beam forming, massive MIMO, and full duplex.

7.2.5 Comparison of Different Phys

Table 7.4 brings all the different Phys together, organized roughly by maximum transmission distance.[35–38]

7.2.6 Upper-Layer Protocols

After the detailed explanation of the different IoT Phys in the previous section, we want to complete our review of IoT protocols by looking at the upper layers of the stack. This includes the network and transport layer protocols that determine how a network is organized and how messages are transported from one station to another. We will then review some of the primary application layer protocols. Figure 7.4 provides an organized diagram placing the Phys along with the upper-layer protocols into an organized stack diagram. This diagram will help us to relate the upper-layer protocols to the Phys we have previously covered.

Network and Session Layer Protocols

The most prominent network layer protocol is the Internet Protocol (IP) version 4 (v4) and version 6 (v6). Although IPv6 is now commonly available, IPv4 is still the most-used network

Table 7.4 Comparison of Different IoT Phys[35-38]

Phy Technology	Maximum Distance	Maximum Stations	Transmission Speed	Power Usage
I2C	<10 feet	127	100 Kbps–3 Mbps	Low
USB	<15 feet	127	1.5 Mbps–5 Gbps	Low
RS-232	50 feet	1	Up to 1 Mbps	Medium
CAN Bus	<130 feet	30	Up to 1 Mbps	Low
Bluetooth® Low Energy (BLE)	200 feet	1	10 Kbps	Medium
Bluetooth® 4.0	300 feet	1	25 Mbps (BT)	Low
Ethernet/Twisted Pair	300 feet	1024	Up to 1 Gbps	Medium
Z-Wave®	<100 feet	232	40 Kbps	Low
ZigBee®	<300 feet	65535	250 Kbps	Low
WiFi	<300 feet	10 (b), 50(g/n)	Up to 54 Mbps	Medium
IEEE 802.15.4®	100s of feet[a]	Many[b]	40–250 Kbps	Low
RS-485	4000 feet	32	10 Mbps	Medium
FOUNDATION™ Fieldbus FF-H1	Up to 6000 feet	32	31.25 Kbps	High
2G GPRS	Miles	Massive	10 Mbps	High
3G	Miles	Massive	<2 Mbps	High
4G/4G LTE	Miles	Massive	1–10 Mbps	High
5G	Miles[c]	Massive	>1 Gbps	High
LTE-M	<7 miles	Massive (4G)	300 Kbps	Medium
NB-IoT	<10 miles	Massive (GSM)	<1 Mbps	Medium
LoRa®	<10 miles	Many[d]	<50 Kbps	Low
SigFox®	~30 miles	Massive	<1 Kbps	Low
4-20 mAmp Loop	5+ miles	1	Few bits	Low

[a] IEEE 802.15.4 has varying frequencies, modulations, and coding schemes; although dB limits are specified, exact ranges vary by device and manufacturer and by the configuration of equipment for power usage.
[b] IEEE 802.5.4 can support peer-to-peer and star clusters, and describes network topologies of tree-clusters that can support many nodes. The upper layers define how the network is formed and communicates.
[c] Although 5G connectivity will support transport for miles, the millimeter wave radio technology is short range—a few hundred meters—but the larger network provides long-range communications.
[d] LoRaWAN supports a star-of-stars topology, and LoRaWAN gateways can control the frequency of end-station communications, resulting in a gateway supporting many stations.

protocol on the Internet today. Above IP, the TCP for connection-oriented sessions, and UDP for connectionless sessions, are most common. There are plenty of good books on these protocols,[41-43] so we will not cover them in any detail here. Instead, this section focuses on transport and session layer protocols that are less common and used more exclusively in the IoT space.

Z-Wave® was one of the Phys covered in the previous section, but Z-Wave includes a full protocol stack, up to and including the application layer. This is due to the fact that Z-Wave is a full solution for home automation and includes the applications to run certain types of home services, including applications that open and close door locks, control lighting, access the home alarm system, and monitor and control the thermostat. The network and session layers

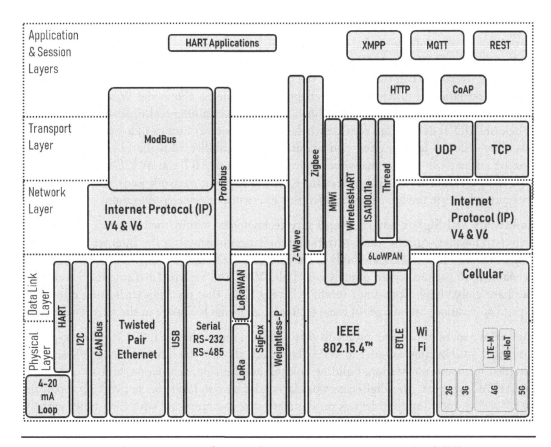

Figure 7.4 Logical Arrangement of Protocols in an IoT Communications Stack.[39,40]

of Z-Wave create a mesh network between devices.[44] This means that any device can talk with any other device, although there is a single primary controller device that normally has access to the Internet and through which all commands flow. The primary controller receives commands, for example, from your smartphone, and delivers those commands to the devices in the local Z-Wave network. The primary controller handles registration of new devices onto the network as well as creating and managing a routing table on how the mesh network is constructed. Multiple secondary controller devices are also allowed. Z-Wave includes some security, although the open-source releases of the specifications and source code (called open wave) do not include the security layer. There has been some analysis of the Z-Wave security, covering source-origin-authentication and confidentiality using AES encryption, and a weakness was found in the implementation of door locks, which was subsequently fixed.[45] Although Z-Wave does not use IP as a network layer, it does provide a specification, Z-Wave over IP, that allows a Z-Wave network to receive commands from a gateway that were received over an IP network.[46]

WirelessHART® is a specialized protocol for fieldbus deployments in the industrial IoT (IIoT) that is standardized in IEC62591-1. It is based on IEEE 802.15.4 LPWAN technology but provides a unique MAC and network layer on top of the standard radio physical layer. It uses time-division multiple access (TDMA) and channel hopping to provide a robust and real-time communications channel. The network and transport layers provide security, routing with redundant paths, over a self-healing mesh network, with the ability to send large messages segmented across multiple packets. A network manager maintains the network, plans routes to devices, and distributes information on time slots to schedule connected devices so that they know when they can transmit over the network. Adding additional access points into the network, operating at different non-overlapping frequencies, allows the WirelessHART network to be expanded and support more devices while maintaining real-time communications. WirelessHART is designed to provide wireless connectivity to various industrial field devices, including 4-20mA loop devices, and maintains compatibility with the HART applications that are currently in use for the wired devices.[47] WirelessHART claims 128-bit AES security with integrity and device authentication. Rotating keys for network joins and per-message encryption keys are used to limit the effects of a network or device penetration.

LoRaWAN and **SigFox** rely on proprietary technologies within their network for message delivery. However, current standardization efforts are underway in the Internet Engineering Task Force (IETF), the standard body of the Internet, in the *IPv6 over LPWAN* working group to enable IPv6 packets to run directly over LoRaWAN and SigFox.[48] Because these technologies have global (and expanding) reach, it is very likely that this standardization effort will expand applications and usages of these technologies in the IoT space in the next few years.

ZigBee® is a set of upper-layer protocols that builds on top of the IEEE 802.15.4 Phy, typically uses the 2.4-GHz range (similar to WiFi), but it can also use the sub-GHz range Phys. It was designed for home automation, building control, and heating, ventilation, and air conditioning (HVAC) systems. The ZigBee network layer adds a mesh local area network on top of the 802.15.4 Phy, meaning that devices on the mesh can communicate with each other without having an actual radio link to the other device by hopping from one device to another.[49] Figure 7.5 shows an example of a ZigBee mesh network. Although ZigBee is a mesh, it uses a controller node to set up the network. This controller establishes what frequency will be used and sets up a beacon message allowing other nodes to join the network. Router nodes are

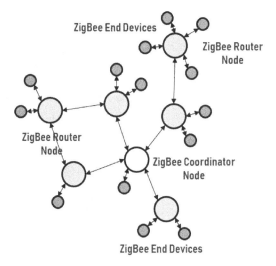

Figure 7.5   ZigBee Mesh Network

nodes on the ZigBee network that can route messages to devices on the mesh network; device nodes are on the mesh network but may power down when they do not need to communicate, so they are not part of the mesh routing. ZigBee uses a type of ad hoc routing (RFC 3561 Ad Hoc On-Demand Routing) that can change and update as the network changes, just in case a node moves, loses radio communication with another node, or must power down.[50] There are two complaints with ZigBee. First, ZigBee nodes cannot directly communicate with other IP devices, but need a gateway to translate at the application layer. The second complaint is interoperability; many ZigBee devices from different manufacturers can form into the same network, but their upper-layer protocols and commands are not interoperable, so the devices cannot really work together. The ZigBee Alliance is working to correct this through additional conformance criteria.

6LoWPAN is short for IPv6 over Low-Power Wireless Personal Area Network. It is a light-weight low-power wireless protocol that provides low data rates and is frequently used for home automation and building control systems. 6LoWPAN uses encapsulation and header compression to directly send IPv6 packets over the small data frames defined in IEEE802.15.4 wireless networks. 6LoWPAN is defined in RFC 6282 and defines a mesh network, much like ZigBee, but the primary difference is that a simple edge router device on the 6LoWPAN network directly compresses IPv6 packets and delivers them into the wireless mesh network to the smallest of IoT devices. The 6LoWPAN devices use an autoconfiguration protocol defined in RFC 6282 to self-generate a unique IPv6 address, which can be exposed through the edge router.[51]

Thread® is a mesh networking protocol built over 6LoWPAN that adds commissioning (setting up the network), routing and security. Applications built on Thread use the UDP protocol to send and receive messages.[52] Thread is a standard offered by the Thread Group with backing from some large companies, including Google, Arm, NXP, Schneider Electric, Siemens, Qualcomm, and Silicon Labs. Because of the significant players influencing this standard, it is important to watch.

MiWi™ protocol is a proprietary offering by MicroChip based on IEEE 802.15.4 mesh networks in the sub-GHz and 2.4-GHz bands. MiWi was specifically designed to compete with ZigBee but provides a significantly smaller code size for the network stack, allowing MiWi to run on the smallest microcontroller units (MCUs).[53]

ISA100.11a is a wireless networking standard designed for industrial process automation and published by the International Society of Automation (ISA). ISA100.11a uses an IEEE 802.15.4 Phy—in the 2.4 GHz band using DSSS—with some customizations at the data link layer, and it interfaces with the IP at the network and transport layer using 6LoWPAN. The data link layer modifications provide for clock synchronization, better message delivery, adaptive channel hopping, local link addressing, and message timing and integrity. ISA100.11a is very similar to WirelessHart, but a standardized interface to process control application protocols is missing from WirelessHART, which could create interoperability problems. Additionally, ISA100.11a's use of 6LoWPAN provides better connectivity to standard network infrastructure equipment.[54]

MODBUS® is a common serial communications protocol used for programmable logic controllers (PLCs) in industrial and Supervisory Control And Data Acquisition (SCADA) systems. MODBUS was originally designed for serial communications using RS-232 (in 1979) and later for RS-485, but it supports a variant that runs over TCP/IP, meaning it can use Ethernet or even a wireless Phy. MODBUS serial allows a master to control up to 247 slave devices,[55] whereas the TCP version supports an unlimited number of devices.[56] For MODBUS serial, each device is addressed by the master using the device's unique device ID (a number between 1 and 247, inclusive), allowing a maximum of 247 devices connected to the master; 2-byte addressing is also possible, which would address up to 65535 devices. MODBUS controls devices by reading and writing from a table of registers on the device, which are mapped to device-specific functions; each device has its own application-specific (custom) definition of what a particular register does. MODBUS defines four different register types: bit-addressable read-only *discrete inputs*, bit-addressable read-write *coils*, 16-bit word-addressable read-only *input registers*, and 16-bit word-addressable read-write *holding registers.*

A message on the MODBUS protocol includes several elements, including an identifier for the slave device the message is intended for, a function code to define the operation (read or write from/to a particular table entry), and the address and data of the affected elements. The message also includes a cyclic-redundancy-check (CRC) value to detect corruption of the message.[57]

PROFIBUS DP (Decentralized Peripheral) is a term used to describe a standard set of industrial communications protocols that originally operated over RS-485 as a multi-master serial bus protocol. Recall that we identified this protocol as one of the fieldbus technologies in IEC-61158. It includes versions for process automation (PROFIBUS PA), for safety critical applications (PROFIsafe), and for high-speed communications (PROFIdrive). PROFINET defines these same protocols over industrial Ethernet. PROFIBUS supports multiple masters on the bus using a token; masters can send and receive messages from their slave devices only while they have the token. This is similar to a token ring system, but only the masters have access to the token, which limits the amount of time that any one master might not be able to send and receive traffic. The masters are limited to how much time they can hold the token before they must pass it on. This make PROFIBUS suitable for real-time control loops in industrial systems. PROFIBUS also limits the RS-485 specification to increase the possibility of interoperability between different manufacturers, as well as by defining basic instrumentation inputs and outputs, which makes it easier to integrate different instruments into the network. PROFIsafe limits the amount of power on the bus, making it intrinsically safe and cheaper to install and operate in a hazardous area.[58]

7.2.7 Application Layer Protocols for IoT

The previous sections provide a good overview of the different Phys and network protocols used in IoT systems. Application protocols run on top of these protocol stacks to perform various functions in an IoT system. The number of standard application protocols are large, and the number of proprietary application protocols may be even larger. This section cannot even attempt to do justice to that number of protocols. However, there are several standard protocols that come up over and over again in IoT systems, and these deserve some attention. Anyone developing or analyzing IoT systems should be familiar with this set of application protocols. The following sections provide an introduction to these protocols, with references allowing you to find additional details.

MQTT is the message queue telemetry transport protocol that provides asynchronous delivery of messages in constrained networks. MQTT was designed by IBM and originally standardized by the OASIS group.[59] Later, it was submitted and standardized in ISO as ISO/IEC 20922.[60] Google's Cloud IoT Core™ and AWS IoT™ both use MQTT as a primary protocol for device communication.

MQTT implements the publish–subscribe design pattern, allowing authentic publishers to submit messages to a broker, which then delivers those messages to subscribers as they query the broker. This construct is used quite often in IoT systems because it allows control systems or other devices to queue up messages in the broker for delivery to end devices when it is convenient, or when the end device wakes up and checks in with the broker. This allows the IoT system to conserve energy and scale to a very large number of devices. This conforms very nicely to the MGC architecture, where the gateway or cloud maintains the brokers, and the devices check in with the cloud regularly to receive messages waiting for them. This description sounds a lot like email. But MQTT is very different because the messages are structured in a way to reduce the storage and message-processing requirements of the broker.

An important concept in MQTT is that every message has a title, called a topic. Topics are like filenames or pathnames on a filesystem and are composed of words or groups of words separated by some character—like a slash (see Figure 7.6). There is a special limited version of MQTT designed specifically for sensor networks, called MQTT-SN, with the primary benefit being shorter topic IDs that can be used in place of full topic names.[61]

As shown in Figure 7.6, MQTT topics are arranged hierarchically by the broker. Message subscribers can subscribe to messages by topic, or by some prefix of the topic, and then receive everything under that hierarchical branch. Using this structure, the broker can organize

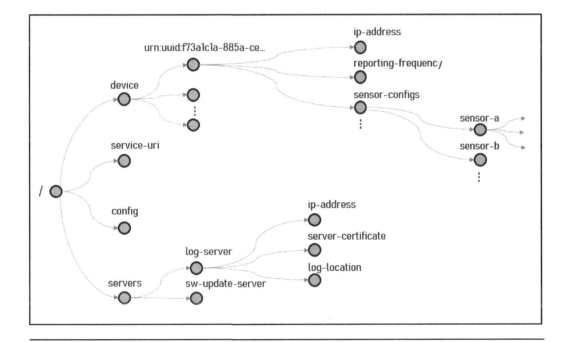

Figure 7.6   MQTT Topic Hierarchy Example

submitted messages into a database and categorize the messages by the parts of the topic. When a device checks in to see what messages it has waiting, the broker can easily use pattern matching from the device's subscription record to find messages in the database that should be delivered to that device. Of course, some method must be employed by the broker to ensure that it only delivers such messages to a device only once. Subscribers request—or subscribe to—messages by the full topic name, or they can use wildcards to replace any part of the topic name with any word, allowing that subscriber to receive any message that is categorized under a specific topic prefix. For example, from Figure 7.6, a device topic that retrieves all the configuration for "Sensor A" would be: `/device/URN:UUID:F73A-1C1A-885A-CE4D/sensor-configs/sensor-a/*`.

MQTT supports basic security features to protect the messages. The broker authenticates publishers and subscribers when they connect to the broker using passwords. Message confidentiality can be provided by using transport layer security (TLS). It is possible to configure mutual authentication on TLS and customize the TLS certificate verification process so that the client certificates are used to authenticate the publishers and subscribers, providing even stronger authentication than just passwords.

MQTT is often used in two places. First, it is used in the gateway layer and cloud as a way to deliver messages to the devices layer at the convenience of the end device—asynchronously. This usage of MQTT is over the network and must include appropriate network security—TLS for confidentiality and integrity, appropriate authentication and authorization controls, and network protections such as firewalls and intrusion-detection systems.

The second way MQTT is used is locally within a gateway as a message-orchestration layer. In this configuration, MQTT operates with local domain sockets on a single machine, not over the network. As shown in Figure 4.9, the orchestration layer allows different components within a gateway to be loosely coupled through the orchestration layer but retain high cohesion through the use of supported topics. This is extremely important in IoT systems design, as it allows different subsystems within the gateway to plugged into the orchestration layer, or removed, without affecting the rest of the design. This improves extensibility and specialization of the gateway without destabilizing the software design or the platform. Message orchestration can be used on small IoT devices as well, although it is most common in gateways or end devices that employ sensor fusion.

The benefit of MQTT is its asynchronous nature and flexible message format. Messages can be posted to the broker in any format that the subscribers are capable of reading, and the publisher can post messages whenever it is convenient for them. The publisher and subscriber never have to communicate, and the publisher can reach thousands of subscribers with a single post because the subscribers and broker do all the hard work. The message topic is used to identify and deliver the message whenever the subscribers happen to check in with the broker. It eliminates a lot of complexity in the code for publishers and subscribers as it removes the requirement that they be available when the other party is also available. Only the broker needs to be always awake and operating. This identifies the weak point of the system as well. The broker becomes the single point of failure in this system—if the broker crashes or is attacked, all the messaging in the system stops. Having redundant brokers is a solution to this problem, but getting all nodes to contact the secondary broker becomes a difficult issue. This setup works really well in a mesh network where the routing protocols can be leveraged to update the location of the broker, regardless of what node that happens to be at the moment. Of course, the other problem with this setup is synchronizing all the message content between the primary broker and the secondary broker nodes so that no messages are lost if the primary broker goes down. None of this is handled by MQTT but can be built on top of the protocol.

AMQP (advanced message queuing protocol) is an application message protocol that supports message queues such as MQTT but is much more generic as well as complex. AMQP was first published as an OASIS specification[62] and then later as an international standard and documented in ISO/IEC 19464.[63] AMQP is similar to MQTT in that it can provide a publish–subscribe type architecture with application-specific messaging. However, AMQP differs from MQTT in message format and message delivery. AMQP has a detailed message format that includes a specific binary encoding of message headers, properties, delivery instructions, and data. There also is no concept of a hierarchical topic structure, though an application could build this on top of an AMQP queue node. AMQP has a complex view of message delivery and a very specific terminology for protocol elements, defining *nodes* as the endpoints of communication (producers, consumers, and queues), and *containers* as the broker or client devices. Messaging between *containers* is performed over an AMQP *Connection*. Each AMQP *Connection* contains multiple *Channels* that are half-duplex; two half-duplex *Channels* between *endpoints* defines an AMQP *Session*. AMQP provides special message properties, like guaranteed-delivery, deliver-only-once, and deliver-to-multiple-parties.

AMQP is a very complex protocol and not used on simple sensors. However, it is used in cloud and gateway communications to build very robust messaging. The drawback to AMQP is its complexity and large code size. Some feel its binary-formatted messages are also a drawback because developers have a hard time reading messages during development and when investigating system failures. But others see the binary message size as reducing message overhead for bandwidth-restricted environments. Binary messages make sense in pure M2M communications when human-readable messages just add redundancy that rarely (if ever) provide any value.

XMPP (extensible messaging and presence protocol) is defined by the IETF in RFC 6120[64] and is used to exchange data formatted in the extensible markup language (XML) between two or more network stations. XMPP was originally designed for instant messaging (IM) but has been found to be very useful in many situations, including IoT. XMPP does not define message formats other than to stipulate that the data exchange is intended for small XML-formatted messages, referred to as XML stanzas. Although XMPP defines peer-to-peer messaging, RFC 6120 points out that the XMPP message flow is typically from a sending client to a server, then to another server, and finally to the receiving client. The servers act as intermediaries for the XMPP messages to allow communication between the clients. This architecture works very well for IoT devices, which communicate to gateways and cloud servers.[64]

The primary benefit of XMPP is its wide usage in the Internet and the proliferation of tools and frameworks that can construct and parse XML messages. The drawbacks to XMPP are the large message size that is typically associated with XML messages and the inherent complexity involved in parsing XML messages, especially on small devices. The complexity commonly leads to security problems when integrated into protocols, and the large size is exacerbated when combined with the requirement for a complete TCP stack.

REST is short for representational state transfer, which is an hypertext transfer protocol (HTTP) architectural style that uses the basic HTTP operations—GET, PUT, POST, and DELETE—to operate on web objects using uniform resource identifiers (URI). A URI can either be the unique name of a resource or object—uniform resource name (URN)—or it can be the unique name and location of a resource or object—uniform resource location (URL). AWS IoT® allows the use of a REST interface for publishing data to IoT systems.

REST is really the way IoT devices and gateways can interact with a logical object model created in the cloud. The objects in the model are URNs, and the IoT devices interact with the objects using HTTP commands. For example, a temperature-sensor device wakes up and performs a GET operation on its configured gateway. The GET operation returns a JSON* object containing a command string to be executed:

```
{
    "command-list" : [
    {
      "cmd"   : POST
      "who"   : Urn-for-self
      "what"  : latest-temp-readings
    },
    {
      "cmd"    : SLEEP
      "who"    : Urn-for-self
      "period" : 8
     }
}
```

The device interprets the JSON object and formulates a new REST query from the first command in the JSON: POST /device/URN:UUID:F33D-6870-A5D5-661A/temp-readings/latest. After completing the POST, the device performs the command to sleep for 8 hours. Part of this example is a REST protocol—the GET and POST are RESTful commands on the gateway. The other elements are implementation-specific commands that are interpreted by the JSON parser and are built into the device software. This is just a simple example to give you an idea of what is possible.

The REST portion allows the device to drive actions without the server—the gateway in this case—having to remember any state or implementing any state machine. Each REST command is self-contained. If the device fails in the middle of the operation, the gateway does not have to clean up any state or sessions variables. It makes programming the server or gateway very easy.

The problem with REST, as can be seen by this example, is that there often is a strong coupling between the client device implementation and the data objects returned. Any upgrade to the server data objects must be correlated to a code change on the clients, so they know what do with the data they parse out of JSON. Any client application that is more than just a trivial example has tight coupling to the server application. Then, when something needs to be upgraded or changed, a whole new set of URNs needs to be created. You cannot just change the existing URNs and JSON files because it is impossible to upgrade every device at the same time. Thus, there will be clients that exist with new code that interprets the URNs one way, while there are also old clients that interpret the URNs a different way. It is impossible to change every device on the IoT system all at once. The end result is that REST applications

---

* JavaScript Object Notation—a popular format for web messages.

that live a long time tend to be cluttered with old code that responds to URNs that are not used anymore. This creates a bit of a maintenance headache and can also be a source of attack by hackers that find a way to use that old code to execute buffer overflows or other remote code-execution exploits. Our recommendation is to update the client more frequently, removing the URN code that is no longer in use.

There are wonderful benefits to REST. One benefit is the ease of adding services and the ability for humans to read the JSON or XML or other messages that are sent between the device and the sever or gateway. Also these protocols are used in many mobile and enterprise web services, so there are plenty of reliable and well-tested open-source software packages around that one can use to build such services.

CoRE (Constrained RESTful Environments) is not a protocol but a working group in the IETF that is defining a set of protocols to enable constrained network nodes to take advantage of REST and to use RESTful protocols to realize machine-to-machine (M2M) communications. REST has become so important on the web and for cloud environments that it is extremely important to be able to utilize REST in IoT systems all the way down to the smallest devices. The CoRE working group has published several RFCs to make this happen and has several more in draft form. In the remainder of this section, we will look at two of the more important approved RFCs from CoRE—CoAP and SenML—and two supporting RFCs—CBOR and COSE.

CoAP (Constrained Application Protocol), defined in IETF RFC 7252,[65] is a web transfer protocol that optimizes HTTP REST interactions specifically for very constrained nodes, such as those running 6LoWPAN, enabling M2M interactions. Beyond just making the REST protocol messages smaller, CoAP adds additional features that are needed by constrained devices, including discovery, multicast, and asynchronous message exchange. Figure 7.7 depicts a CoAP

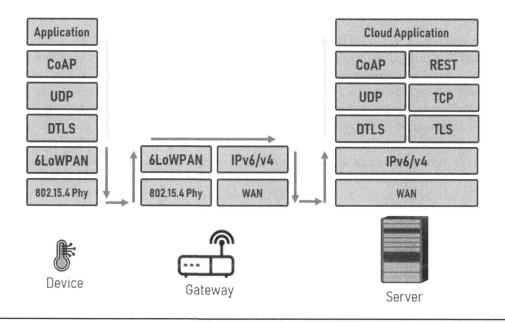

Figure 7.7   CoAP Web Architecture Diagram

stack with security using Datagram Transport Layer Security (DTLS) over a 6LoWPAN 802.15.4 wireless mesh network.

As depicted in Figure 7.7, there are three important protocol transitions that occur within the protocol stacks that should be noted. The first is CoAP and HTTP. The diagram shows CoAP being used on the device client, and both CoAP and REST/HTTP on the server. While it is possible for cloud servers to implement CoAP, it is also possible to create a proxy layer that translates between REST/HTTP and CoAP. The CoAP specification defines a stateless mapping allowing an HTTP client to access resources on a CoAP web server, or for a CoAP client make use of simple REST/HTTP resources through a simple proxy. CoAP does not allow all HTTP interfaces or REST APIs to be reachable through CoAP because HTTP is too complex. However, the CoAP specification defines basic mappings between HTTP and CoAP, and RFC 8075 defines the specific rules to be implemented in a CoAP-HTTP proxy.[66] All of this translation is possible because CoAP has defined its headers and parsing requirements in a way that reduces the workload for constrained devices. This is the first and primary benefit of CoAP.

The second important transition shown in Figure 7.7 occurs in the gateway between 6LoWPAN and IPv6. The gateway performs a network layer bridge function, allowing network connections between stations on the 802.15.4 wireless network so that hosts on the IP network can be reached. RFC 4944 defines the basic format for IPv6 packets over IEEE 802.15.14 wireless networks that allows for this translation, and also includes several other features that enable this protocol translation to work effectively, including specific protocol rules for compression of packet headers (RFC 6282), a paging service used to locate and wake up wireless stations (RFC 8066), and a discovery service enabling IP hosts to find stations on the wireless network (RFC 6775). Although these protocols are more related to 6LoWPAN, they are important services that enable the use of CoAP; otherwise, a CoAP client on an 802.15.4 wireless network would be completely cut off from Internet hosts.

The final transition shown in Figure 7.7 is the security layer. This is actually a replacement of TLS with DTLS, not a translation as the other two transitions were. DTLS allows the use of TLS over the UDP. DTLS is defined in RFC 6347.

CoAP has three other important features that are needed for IoT: asynchronous message exchange, multicast capability, and URI and content-type support.

CoAP is designed to run over UDP, which is a connectionless datagram service. The CoAP protocol allows messages to be sent as either confirmed or unconfirmed. CoAP claims asynchronous message exchange because a CoAP server can delay its response to a CoAP client. This is not the same as the asynchronous message delivery provided by MQTT or AMQP, which allows messages to be queued in a queuing server until the destination wakes up and retrieves its messages. However, this type of asynchronous service allows a very constrained device to take as much time as necessary to construct its reply message.

CoAP supports the use of multicast groups, allowing a CoAP message to be sent to multiple stations. This would be useful when a gateway wants to collect readings from multiple sensors. CoAP supports multicast using specially-designated all-CoAP-node multicast address groups. For IPv4, the assigned address is 224.0.1.187. For IPv6, the address FF0X::FD is assigned, but this is a variable multi-scope address, where X can be assigned any valid scope. Recommended scopes include both the Link-Local (X=2) and Site-Local (X=5), allowing these multicast messages to be limited within a particular network or network segment.[67]

CoAP defines a new URI scheme enabling access to the resources on CoAP servers. Just like the familiar HTTP URIs that are constructed from the protocol—*http* or *https*—and the hostname and optional port address, CoAP uses the same structure but uses *coap*, or *coaps* for secure access using DTLS. The general structure of a CoAP URI[*] is:

```
coap-URI = "coap:" "//" host [ ":" port ] path-abempty [ "?" query ]
```

The CoAP specification is very well written and complete, but there are numerous additional RFCs that add to it and expand or complement other services. CoAP is becoming an essential protocol for IoT and should continue to gain in prominence, especially given the next set of data format standards that can be layered on top of CoAP.

CBOR (Concise Binary Object Representation) is a data format defined in IETF RFC 7049 that is being adopted by IoT frameworks in place of XML and JSON. The problem with XML and JSON is the size of the produced messages and the code libraries necessary to generate messages in those frameworks. CBOR is specifically designed for small message sizes, extremely small code size, and extensibility without the need for explicit version negotiation. CBOR is a binary data encoding similar to ASN.1 (Abstract Syntax Notation, version 1, ISO/IEC 8825-1:2015) but without the accompanying complexity requiring a special data compiler and large code library to generate and parse messages. CBOR was designed to be easily translated into JSON and vice versa.[68] Although CBOR is technically a presentation-layer protocol, not an application protocol, because it can make other protocols more useful for IoT—for example REST/CoAP and MQTT—we felt it was important to include it in our discussion.

COSE (CBOR Object Signing and Encryption) is a companion IETF specification for CBOR, defined in RFC 8152 and describing how to create signed, encrypted, and integrity-protected messages in the CBOR format. This is an important addition to CBOR, as it allows messages—such as in CoAP and MQTT—to be independently protected apart from their transport mechanism and to achieve actual end-to-end protection.[69] COSE includes support for the following algorithms:

- Edwards-curve Digital Signature Algorithms (edDSA), the asymmetric digital signature algorithm using an Edwards elliptic curve
- Hash-Based Message Authentication Codes (HMAC) using the SHA-2 algorithms, SHA-256, SHA-384, and SHA-512
- AES[†] Cipher-Block-Chaining (CBC) Message Authentication Code (AES-CBC-MAC)
- Encryption with AES-GCM (AES in Galois Counter Mode)
- Encryption with ChaCha20 and Poly1305

Although these algorithms seem a bit limited, the benefit of COSE is the definition of how to apply hash, encryption, and public key signing algorithms across CBOR objects. Future extensions or additional standards can add new algorithms while applying the same formatting and ordering rules to the data elements.

---

[*] The elements of the URI are defined in RFC 3986, including "path-abempty."
[†] Advanced Encryption Standard (AES), defined in FIPS-197.

SenML (Sensor Measurement Lists) is an application layer format for carrying arrays of simple sensor data in HTTP or CoAP. Defined in IETF RFC 8428 by the CoRE working group, SenML defines a data model for sensor readings and creates new media types to encode the data model in JSON, CBOR, XML, and Extensible XML Interchange* (EXI). A SenML *Pack* is a single array carrying a set of records wherein each record represents a sensor measurement. A sensor measurement is composed of a set of optional base fields and a set of regular fields. Base fields apply to every record in the array up to the next record that redefines that base field; this allows the array to list the fewest number of fields as possible. An example of a SenML record format is as follows:

```
[
  {"bn":"urn:uuid:f73a1c1a-885a-ce01-080063cc82f110",
      "u":"Cel","t":1.276020076e+09,"v":23.5},
  {"u":"Cel","t":1.276020091e+09,"v":23.6}
]
```

In this example, the base name (bn) is defined as a universal identifier for a sensor and applies to both records in the SenML *Pack*. Both readings (v) are in the units (u) Celsius with the time (t) recorded.

SenML is useful for gateways and servers to collect and exchange sets of values between elements. Although SenML is designed for sensor readings, it can be applied to the exchange of other data sets as well.[70]

7.2.8 Protocols Summary

This section has provided a broad survey of many of the most popular IoT protocols, covering both wired and wireless Phys, all the way up to some of the more common application protocols that are transforming IoT today. Although we were not able to delve deeply into any one protocol, the broad coverage of many protocols should enlighten you about some protocols you may not have been aware of and provide enough information to guide you in choosing the correct protocol for your IoT system. In the remainder of this chapter, we refocus on security, covering the steps you should be taking to protect your IoT network and how to analyze the protocols you use for security flaws.

7.3 Network Security for IoT

The primary adversary for IoT systems is the network adversary. Whether that adversary is a nation state, a group of cybercriminals, or just a curious hacker looking to explore or exploit your system, the network ports and protocols are the most common avenue of attack. And because this is the case, we must offer some advice on how to protect your IoT systems from such network attacks.

---

* EXI is standard developed by W3C defining a condensed binary format for XML encoding (https://www.w3.org/TR/exi/)

7.3.1 Protecting the Little Ones

The small devices in IoT systems are the most difficult to protect. As we discussed in Chapter 4, the small devices do not really have the resources to run much in the way of protection software. This means that the small devices need to be protected by the gateways and other network infrastructure devices. However, the small devices themselves can follow certain design principles that make them harder to attack. The following list provides general guidance on the design and installation of small IoT devices to make them more resilient to network attack:

- **Turn off all unnecessary ports**—As we mentioned in Chapter 2, an open port is a weak-point in the system because it creates an avenue of attack, just like gates in a castle. Open as few network ports as possible. This means services such as simple network management protocol (SNMP) and Telnet should not be used. Well-constructed devices can perform all operations using a single open port for accepting management commands—either HTTP (port 80), HTTPS (port 443), MQTT (port 1883), or MQTT over TLS (8883). Since outgoing communications don't need an open port—they open a port when they start communicating—only the management port needs to be open. Remember that protocols can be configured to run over different ports, and it is not uncommon to run HTTP over some other port, such as 8080; so the port number isn't as important as the number of ports.

- **Never run network code under a privileged user account**—Whatever software is listening on the network port, or sending packets on a network port, should be given the lowest possible user privileges and should never be run as root. Because code that directly owns the network port—whether sending or receiving—is likely the first code to encounter an attack, this code should not be executing as a privileged account on the platform. If an attacker is able to activate a fault in the network code due to malformed data, and leverage that fault to execute arbitrary code, the attacker's code will be executing with the privileges of the code it attacked. You don't want that arbitrary code running as root or admin on the system. Sometimes it is the lack of privileges that helps contain the malware and prevent it from doing damage to the system.

- **Perform fuzz testing on all open ports**—When you do have an open port, run a software fuzz tester—for example, Peach,[*] American Fuzzy Lop (AFL),[†] or OWASP Zed Attack Proxy (ZAP)[‡]—to detect improper handling of malformed packets. Any software crash created by a fuzzing tool needs to be investigated for a potential software defect because crashes indicate that an arbitrary code execution bug is latent in your code.

- **Perform static code analysis (SCA)**—The entire codebase of the device should be analyzed for potential software defects, such as buffer overflows, integer over or under runs, heap errors, format string errors, and other common software defects that can be turned into remote attacks by hackers. Common static code analysis (SCA) tools include Klockwork®, Kiuwan®,[§] and Checkmarx®. Although SCA tools will not catch everything, and some personal analysis that Dave has done shows most tools can only detect simple

---

[*] Retrieved from https://github.com/MozillaSecurity/peach
[†] Retrieved from https://github.com/mirrorer/afl
[‡] Retrieved from https://www.owasp.org/index.php/OWASP_Zed_Attack_Proxy_Project
[§] Retrieved from https://www.kiuwan.com/

buffer overflow attacks, performing SCA on your code is good software hygiene. SCA scans that show problems can indicate a software development team that needs to re-examine their codebase for good coding discipline.

- **Use a secure boot mechanism and reboot often**—During operation, devices should be booted using a secure boot mechanism that utilizes cryptographically sound digital signatures or encryption across the software to ensure that the software has not been changed maliciously or accidentally. By performing a reboot and reverifying the software used to boot the platform, including the operating system, any memory-resident malware that might have leaked in through the network gets purged, and the device starts fresh with none of the bad stuff running. Of course, if a network attacker was able to compromise a system once, they will do so again, so detecting attacks and then updating the device's software to remove vulnerabilities is vital.

- **Provide and utilize a secure update mechanism for all platform software and applications**—All devices, no matter how small, must have a mechanism to securely update all the software and firmware running on that device. In this case, "securely" means that the software is digitally signed to ensure that the software cannot be modified (integrity), and that the signature proves the original source of the software (non-repudiation, proof-of-source). The verification key should be held in the device in a tamper-resistant location so that an attacker cannot spoof valid software by attacking the location in which the trusted verification keys are stored.

- **Authenticate all incoming connections; protect outgoing messages**—In Chapter 4 we discussed the fact that connections with small devices may not be able to be protected with TLS because that protocol is very heavy and interactive. We did not mean that all attempts for secure communications should be abandoned when working with small devices. DTLS can be used with a pre-shared key; this is lightweight enough for many small devices. Alternatively, outgoing messages can be directly protected instead of using a complete security protocol such as DTLS. Such an approach would protect messages with a simple elliptic curve signature using ECDSA,[*] or the message can be encrypted and integrity protected using standard COSE formats.

- **Protect small devices behind a firewall**—Since most small devices cannot perform much of their own network protection, placing these devices behind a firewall is the best option. The firewall in this case would be a gateway providing connectivity back to the cloud instance for the IoT system. Clearly, this works for the wired protocols that we discussed, but what about the wireless? Wireless is a bit harder to protect, but wireless protocols with integrated encryption provide a solution. Devices joined to a gateway organized in a star network (BLE, IEEE 802.15.4, LoRaWAN, LTE-M, NB-IoT) with security during link establishment and at least integrity, if not full packet encryption over the data link frame, provide good separation from network adversaries. An adversary would have to break the encryption or somehow gain admittance into the network through the gateway. In a star formation, the gateway can filter traffic from unauthorized stations and drop packets that fail integrity checks. This doesn't mean that an attacker couldn't flood the radio channel with bogus packets, but for the packets to get through the integrity checks of the data link layer, the adversary would have to break the integrity/encryption scheme, which, if the encryption is done properly, is a pretty difficult task for the attacker.

---

[*] Elliptic Curve Digital Signature Agorithm is specified in FIPS 186-4 (https://csrc.nist.gov/News/2015/FIPS-186-4-RFC-NIST-Recommended-Elliptic-Curves)

7.3.2 Additional Steps by the Bigger Devices—Self-Protection Services

More capable devices and gateways need to run services to protect themselves. These include preventative services that thwart attacks, as well as services that detect and report attacks. Of course, all the advice that small devices should follow also apply to these more capable devices, but larger devices can and should do more. The following list includes the additional baseline services that should be considered for gateways and more capable IoT devices.

- **Run a firewall to protect yourself**—A firewall service to protect the downstream devices is important, but the gateway and larger devices should also run a firewall to protect their upstream communications to the cloud. Why is this important, since these communication channels should be running TLS already? Mostly, this is a defense-in-depth protection to ensure that if anything is misconfigured on the device, the firewall will prevent an unprotected port from being exposed. It only takes a small configuration error to change a web server that is configured with HTTPS on port 443 to also include a second listener on an unprotected port 80, or port 8080. You might be surprised how often an operational mistake like this takes place. A high-priority change or response to some customer-reported problem gets the operations team in fire-fighting mode to resolve the issue. Tests are performed in a preproduction environment, and it is every hand on deck with folks pulling all-nighters to root cause and correct the problem. Great teamwork! Solution found! Push it out to production. But everyone is tired, and what actually gets pushed to production is the debug configuration that includes a second web server connection in the clear, or maybe a special telnet port for engineers to peek into the device and run test scripts on the device. Without a firewall on the actual device in the field serving as an extra level of protection that no unauthorized ports are opened onto the *Big Bad Internet*, that HTTP or Telnet port will be exposed to any network adversary that happens to perform a port scan on the device. But you are convinced that your team doesn't make such rookie mistakes? OK, what about the network adversary that happens to find a zero-day attack on the operating system? They can install any software they want, and that software can open any port they want. Doesn't it make sense to add a firewall to deter such occurrences? A firewall does take some cycles, but it can save you from untold and unexpected damage to your system.
- **Filter incoming traffic to weed out known attacks**—Even with a firewall, devices and networks get penetrated all the time. Overcoming a firewall isn't as easy as we see on some television shows, where the script goes something like this: "Oh, no! The system is protected with a firewall. Wait a second . . . let me just <cascade of furious keyboard clicks> . . . OK, I'm in!" However, firewalls can be bypassed. Applications like snort and suricata[*] can be used to detect and warn of such attacks and respond automatically.
- **Maintain a trusted list of endpoints**—The device should keep a list of trust anchors. A trust anchor is a public key that is used to authenticate communications over TLS or other secure channels. The trust anchor list is the set of known trustworthy stations that a device can communicate with for management, software updates, reporting incidents and problems, and just reporting data. Applications, and especially the device management service, should validate that connections and commands are coming from entities

---

[*] Retrieved from https://suricata-ids.org/

that can demonstrate use of a trusted key in the trust anchor list. Demonstrating this can be performed through TLS or DTLS, or by verifying an ECDSA signature on a received MQTT message.

- **Run host intrusion-detection software**—Gateways must be running software that can check for known bad programs and execute memory scanners to look for compromised applications in RAM.
- **Always use a secure channel**—TLS is your friend. As we will talk about in the next section, many protocols have been found to have latent defects, but running them within a secure channel like TLS prevents those defects from being used by an attacker.
- **Leverage hardware protections on your platform**—Many platforms today include sophisticated hardware protections. Services such as trusted execution environments (TEE), secure boot, secure elements (SE), trusted platform modules (TPM), and control flow protections protect software running on the platform from hackers and exploits. Some of these technologies have reported attacks, but even in light of these attacks, these protections are not without merit. Ensure that you are familiar with the advanced security protections on your platform and use them to the greatest extent possible.

7.3.3 System Protect and Detect Services

Individually protecting each device is important, but there are protections that should span the entire system. Threat detection and monitoring systems are becoming common place in the market today, and many of them are using artificial intelligence (AI) to improve their effectiveness. These systems should be integrated into your IoT system and used to monitor traffic across as much of your network as possible. Logs and activities from all your devices should be fed into an advanced threat-detection system for analysis in order to detect threats before they completely overtake your system.

7.4 Security Analysis for Protocols

In Chapter 5, we described the process of security architecture, including threat modeling, and we graphically depicted that process in Figure 5.1. Performing security analysis on protocols is similar to the process we used for security architecture. However, instead of an architecture picture, an understanding the protocol itself is the first step. The rest flows pretty much the same—except for that really tricky part where we attempt to discover the actual threats on the protocol. That part is actually really hard. There are many papers written about protocols that were thought to be secure, and then flaws were discovered. You are probably aware of many of them. Remember WEP? The Wireless Equivalent Privacy protocol that secured WiFi and was published as part of IEEE802.11i. When WEP was broken, it was replaced by Wireless Protected Access (WPA). But if you have been paying attention, you will recall that most everyone is now using WPA3, meaning that there have been a few new problems found and corrected since WEP. Lest you think this is just a problem with WiFi, take a quick look across the published academic literature and you will find it is rife with papers written about breaking a protocol and presenting a fix, only to have someone else publish the next paper that breaks the "fixed" protocol and presents a new solution. And this cycle seems to continue like a never-ending game of musical chairs in Purgatory. In this section, Dave presents a technique based on formal security models that he has used to help with this protocol threat analysis, and if done carefully, it avoids the back-and-forth build-break-rebuild cycle of protocol design.

Dave has been working in communications and building protocols for many years. He invented a key management protocol called the Security Association Management Protocol (SAMP) in 1992 for the US government and spoke nationally on key management and in standards groups. And Dave participated in some of those build-break-rebuild cycles. These experiences helped him to adopt a loose model of protocol adversary capabilities,* which he used to analyze and develop protocols. But at Intel, Dave had the privilege of being mentored by Jesse Walker. Jesse introduced Dave to the published works in formal protocol analysis before Jesse retired from Intel. This section draws heavily from the material[74–78] that Jesse shared with Dave.

7.4.1 The Preliminaries and Definitions

Before getting into an explanation of the model, it is best to build a vocabulary for us to use when discussing protocols and the elements that comprise them. So, to start out, let's define a few terms.

First, let's define a *protocol*. A protocol is a series of *steps* that two or more parties must follow to accomplish some specific goal or goals. For each *step*, the protocol defines

- The ***sending party***, which is the entity that sends a message in this step
- The ***receiving party*** or parties,† which are the entities that accept messages in this step
- The **message format**, which defines exactly how a specific message is constructed by the sending party
- The ***message processing rules***, which defines what the receiving party must do with the message upon receipt
- The ***shared state***, which is the set of data elements used by the parties in the protocol, some of which are known to each other (e.g., next message ID, shared symmetric keys, public keys, previously exchanged messages, etc.) and some of which are secret to a particular party (e.g., passwords, private keys, etc.)

In each step of the protocol, the sender and receiver may be different. Typically, the sender and receiver swap roles in each step of the protocol, allowing the receiver to reply to a message that was just received from the sender, but this is not required. In some protocols, both parties can be a sender and receiver during the same step, though this is usually more difficult to

---

* This approach was birthed from the Dolev–Yao model,[71] but Dave's understanding and use of this informal model came primarily from two books.[72,73] The Stalling book, especially "Chapter 1.2 Security Attacks," has a very concise treatment of the adversary actions on a protocol.

† From this point forward, we will describe the protocol-analysis process from the perspective of a single receiving station. It should be understood that this process can be easily extended to any number of stations by appropriately extending the definitions and analysis to multiple stations. This statement should not be construed to trivialize multi-party protocol design, as the complexity of multi-station interactions is more complex.

model. The party that begins or initiates a protocol exchange is referred to as the ***Initiator***, and the initial receiver of the protocol messages is called the ***Responder***. It is important to remember that during some steps of the protocol, the Responder may send messages to the Initiator (the swap roles thing). The names Initiator and Responder reflect the roles of the parties at the very beginning of the protocol and do not change; the terms sender and receiver may change with each step of the protocol.

Next, let's define some common protocol goals. These are things that a protocol attempts to accomplish through the operation of the protocol, and usually these goals are tied to some type of cryptographic operation to make them secure. This is not a comprehensive list, but the most common goals include the following:

- Establish a secure channel (***session key establishment***) providing ***confidentiality*** and perhaps ***integrity*** of exchanged messages—the primary goal of the Transport Layer Security (TLS) protocol
- Establish the identity of another party (***entity authentication***)—the goal of several well-known protocols, including Kerberos, Challenge Handshake Authentication Protocol (CHAP), and Entity Authentication Protocol (EAP)
- Establish the origin identity of a data object (***non-repudiation–proof-of-origin***)—this is one of the purposes of Online Certificate Status Protocol (OCSP)
- Establish a secure receipt for an operation or data object (***non-repudiation–proof-of-receipt***)—one of the goals of Secure/Multipurpose Internet Mail Extensions (S/MIME)

Finally, let's define the ***adversary*** and ***adversary actions***. The adversary is the evil guy (or gal!) who wants to break our communications. Their objective is the opposite of our communications goals—confidentiality, integrity, authentication, etc. In our model, the adversary has complete control of the communications channel and can do anything they want to the messages transmitted by others—you can think of the adversary as your Internet service provider (ISP). Your ISP receives all the messages and data packets you transmit and delivers to you all the messages and data packets you receive. You cannot communicate without your ISP. So what if your ISP were evil and trying to mess with your communications? In this case, your ISP—the adversary—could perform many actions, which include the following:

- Message observation or eavesdropping
- Message modification
- Message insertion
- Message deletion
- Message delay
- Message reordering
- Message replay

In our model, the adversary is powerful, but their actions are only on the messages. The adversary doesn't have your cryptographic keys, so if you encrypt a message, then the adversary can observe your message, but cannot read it. The cryptographic key is part of your shared state, and nothing in the adversary's toolbox allows them to look into your shared state—unless there is a defect in your protocol that allows the adversary to guess your cryptographic key or parts of your shared state from messages that you have sent!

There are some models where the adversary not only has control of the network but is given the ability to spin up new parties to engage in multiple protocol sessions at the same time. For these other parties, the adversary is given the ability to attack them and expose their short-term session keys or replace their long-term identity keys at will—these are called *reveal* and *corrupt* attacks, respectively. The adversary isn't allowed to perform these actions on you, the test subject, who we are using to prove the security of the protocol. But the adversary is allowed to do this on anyone else.

This more extreme ability is required to model perfect forward secrecy and special attacks like a BORE attack. A BORE attack—break once, run everywhere—is an attack where an adversary compromises one party and is then able to gain a secret that allows them to compromise everyone. Without modeling multiple parties at the same time, testing for the existence of a BORE attack couldn't be done in our model; the *reveal* operation enables the adversary to attempt a BORE attack. Forward secrecy is the property where a compromise of your long-term secret does not compromise past secure sessions. The *corrupt* operation enables the adversary to attempt to break forward secrecy on your (the test party's) sessions by compromising another party's long-term secret.

In order to resist these attacks, you must apply cryptographic operations to your messages. This book does not cover cryptography, but there are plenty of good, practical books that cover that subject.[79]

7.4.2 An Informal Analysis Model for Protocol Design

Recall that in our discussion of security architecture threat analysis, we used our knowledge of the system and previous attacks that we have observed, or we had previously read about, or attacks we were able to construct ourselves, as a way to assign threats to parts of the system. We could use attack trees as a way to walk through potential attack scenarios, and we discussed security design principles as a way to understand the properties of a system that should not be violated.

In an informal model of protocol design, we use similar guides to inform and mold our protocol, and we use the adversary's actions as way to attack the protocol objectives. In the formal models,[75–78] we just use a more rigorous methodology wherein we reduce the protocol to basic cryptographic operations and then assess the cryptographic strength of protocol operations using complexity theoretic games used to test cryptographic algorithms. We are not going to do that here—we are going to stick to the more informal approach. The informal approach is not as rigorous, but is reasonably effective and is used for many protocols designed today. Mainly, this is because performing the rigorous formal approach requires someone who is intimately familiar with cryptographic complexity theory, and there aren't a ton of those people around. Let's go through an example using this informal approach to make it more clear.

7.4.3 An Informal Analysis of a Digest Authentication Protocol

Let's take a look at an authentication protocol over a REST API using digest authentication.[80] The initiator and responder (client and server) share a state that includes a username and a shared secret—this might be a symmetric key but is usually a password. The protocol exchange

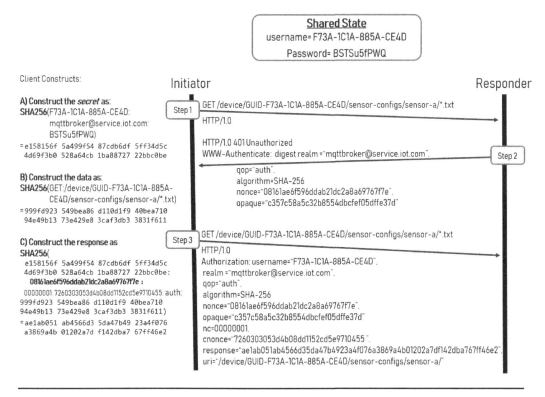

Figure 7.8   Detailed Example of the HTTP Digest Access Authentication Protocol

is shown in Figure 7.8. In the three-step exchange, the initiator first sends a request to the responder for a particular resource. In our example, we show the initiator asking for a URI representing configuration files for a particular sensor. These were published in the server's MQTT broker, but the client is leveraging a REST API instead of an MQTT to access them.

The server requires that the initiator be authenticated before releasing the resources, so in Step 2, it sends back the "Unauthorized" response with some parameters that tell the initiator how to respond with proper authentication. The nonce and opaque parameters are used by the server (responder) to match this message with a future authentication request message that comes from the initiator.

In the final step, the initiator follows some protocol rules to construct a response to the server that proves the initiator knows the secrets in the shared state with the responder. The initiator constructs these values using the hash function SHA256, which the server had requested as the authentication algorithm. Construct "A" uses the secret from the shared state elements—username and password—along with the realm parameter that the server (responder) provided. Then the initiator creates Construct "B," which specifies the HTTP request method and the URI being requested. Constructs "A" and "B" are then mixed with four values to create construct "C": the server nonce, a nonce use count (nc), a client nonce (cnonce), and the quality of protection value ("auth" in our case). Construct "C" is the response value returned to the responder. This full initiator message is shown in Step 3.

Analyzing the Digest Authentication Protocol

So how do we analyze this protocol using an informal methodology? First, we look at the protocol's objective. This is an authentication protocol, so the objective is to perform authentication. That was easy. Sometimes it isn't so easy. Sometimes protocols have multiple objectives. Think about TLS, which includes confidentiality, integrity, and mutual authentication. Things get trickier as the protocol has more goals. For this example, we stick to just authentication.

Now we have to put our thinking caps on. Does the protocol complete the objective? Yes. The responder can take the elements passed to it by the initiator and create Construct "A," "B," and "C" from the elements in the message and the shared state, and then compare the SHA256 that it calculated and ensure that it matches the "response" parameter sent by the initiator. Since SHA256 is a cryptographic hash algorithm, which means it is a one-way function and is pre-image and collision resistant,* no one except someone possessing the shared secrets could have created the "response" parameter! Are we done? Not just yet. We need to make sure the adversary cannot break the protocol. We do this by reviewing the adversary actions.

Clearly, the adversary can perform a denial-of-service (DoS) attack by deleting all messages and never letting the initiator talk with the responder. We ignore this condition for our protocol, mainly because we have given a huge advantage to the adversary in being able to delete all messages, so this DoS attack isn't really fair. But there are other attacks.

In protocol analysis, we attempt to use the adversary's capabilities in different combinations to attempt to reveal the secret shared state or cause one of the protocol's guarantees to be violated. What we basically need to do is to look at each adversary action and determine if the adversary can break the protocol using that action. Looking back at the adversary's list of actions, some actions are trivial to analyze (eavesdropping, deletion, delay, reordering), and they have little effect on this protocol, but some other actions are not so easy to evaluate—especially message modification. In the informal model, we need to look at each message parameter and analyze what modifications could cause a problem. And what about message replay? Can an adversary just replay the message sent by the initiator in order to gain access to private resources?

Let's look at replay attacks for a moment because it is interesting, and it is covered in the RFC. If an adversary captured the message in Step 3 and replayed it to the responder (server), the responder should accept the message. RFC 7616 mentions two protections. The responder-created nonce value is identified by the RFC as protecting the exchange because they can carry things such as the IP address and a timestamp to prevent Message 3 from being replayed by an adversary. The other protection is the recommendation to use HTTPS or TLS.

Clearly, if the exchange between the initiator and responder were encrypted and integrity protected with TLS, then the adversary could not see any of the messages and could not mount any attack. This is the recommendation of the RFC. However, what if someone were not using HTTPS or TLS because, for example, the exchange was occurring within a protected LAN segment, inside an enterprise computing environment, a data center, or on your home network? Let's try out our model and see how this plays out.

Say that the adversary captures a Message 3 from some initiator. The adversary cannot use that message directly because the responder's nonce and opaque value would detect that

---

* Hash functions have three characteristics: they are one-way functions, meaning they are pre-image resistant; they are 2nd pre-image resistant; and they are collision resistant. For more details on what this means, refer to a good cryptography primer.[79]

the adversary is requesting this from the wrong IP address, and perhaps even the embedded timestamp had expired. So instead, the adversary sends a new request to the responder for themselves and uses the same URI as in the original GET request. Explained another way, the adversary sends a Message 1 to the server from themselves, effectively starting a new authentication sequence. The responder follows the protocol perfectly, returns a new Message 2 to the adversary, with directions on how to authenticate, and includes a new server nonce and opaque value—*but this one includes a fresh timestamp and the IP address of the adversary!* The adversary modifies the Message 3 that was captured from the initiator and modifies it by including the server nonce and the opaque value that was received from the server in the adversary's Message 2. Now this modified Message 3 contains the authentication response from the initiator—something the adversary could not generate themselves because they do not know the secret password—and the server nonce and opaque values issued to the adversary. However, when the responder (server) attempts to validate this message, it will use the server nonce value supplied by the adversary in their Message 3, not the nonce value used by the initiator when they calculated the actual *response* parameter. The server's calculated response will be different from the one originally calculated by the initiator because the initiator used the server nonce sent to them, but the adversary sent a different server nonce. Because the server nonce is bound into the response parameter, the adversary cannot forge a message using cut-and-paste between message conversations.

The protocol properly protects against a common type of attack on protocols wherein an adversary mixes messages or parts of messages to create a new conversation in order to trick the responder. It can also work on an initiator, in some cases. Because the nonce value is bound to the authentication token in the response parameter, the attack is thwarted. Pretty cool—even the informal security analysis helps us look deeply at protocols.

Now, assume a team you were working with decided to modify the server implementation, or that there was a bug in the server-side implementation. The RFC states that the server nonce should be constructed with an IP address and timestamp that identifies the client to protect against replay attacks. If this was not done, or there was a bug in this code that made the nonce just a random value not tied to the IP address or timestamp, an adversary could perform the "cut-and-paste" attack we discussed above. This could be problematic. But let's say you perform defense-in-depth and use TLS to protect your exchanges. It would be perfect then, right? Let's talk about a *corrupt* attack. If the device authentication exchanges go through a gateway that logs the exchange, or the server itself logs the exchange, and an adversary is able to *corrupt* the server and steal the logs, then if the nonce didn't include a timestamp and IP address, the protocol could be attacked. You might say, if the adversary is in my server, the game is over, so why care? Logs can be attacked at many points, including backups and log-collection servers. The production web server might be completely secured. The adversary isn't going to play by your rules. The point here is to think like the attacker: Where can my protocol be compromised? And be careful about modifying little things, even the construction of a simple nonce value.

Although we are going to stop here, the analysis isn't done. You would continue to look at all the different parameters and inspect what interactions might occur between them. Using the informal model, you need to try a lot of different things and think craftily about what an adversary might do.

In the end, it is experience with protocols and an understanding of attacks that help us come up with a good list of threats and attacks. This can be a bit ad hoc to engineers new to the process, and there is certainly a bit of art and experience that enters into the process. Experience in any endeavor is always beneficial. But even experienced engineers and protocol designers miss things. Just look back at the history of protocols such as IEEE802.11i for some examples. We are not poking fun here or criticizing at all—protocol design is hard and fraught with dangers.

What Formal Models Are We Talking About?

The first formal model for protocol analysis was the adversarial model created by Dolev and Yao (DY model) in 1983,[71] and much of the informal model we discussed has its roots in this model. But the DY model lacks complete treatment of cryptographic operations, resulting in some cryptographic properties—the protections or lack of protections afforded by a cryptographic operation—not being accounted for properly.

Bellare, Rogaway, Canetti, and Krawczyk wrote several papers constructing different models to prove the security of key exchange protocols.[75-78] These models can be adapted to analyzing other protocols beyond just key exchanges and secure sessions and are frequently used for authentication protocols as well.

The first model that provides a proof of security under the standard cryptographic assumptions is the Bellare–Rogaway Model of 1993 (BR93),[75] which was quickly followed by an updated model for multi-party protocols in 1995 (BR95).[76] After these first two models, there were a number of other papers describing modifications to these models, including a Bellare–Canetti–Krawczyk model in 1998 (BCK98)[77] and a Canetti–Krawczyk model of 2001 (CK01).[78]

The difference between these models centers around the adversary actions and how the security objective of the protocol is proved through the model. More information on the models and a comparison between them is found in Choo's PhD thesis.[74]

7.4.4 The Formal Security Models

Protocol analysis has been studied for a long time—decades in fact. And in part, the informal model we discussed in the last section draws from this research—especially the details on the adversary capabilities. But the difference between the informal approach and the formal approach has to do with creating a formal proof of security based on computational complexity theory and reducing the protocol down to atomic cryptographic actions that fulfill the formal protocol objectives.[73] This is very heavy stuff that goes beyond the scope of this book, but bear with us for just one paragraph for a brief discussion that will help guide your informal analysis.

In the presence of an adversary, the only thing that can be trusted are cryptographic operations that have a foundation in computational complexity theory. Computational complexity is just a fancy term meaning there is a proof showing that the base cryptographic operation is so complex that no adversary can break it in a reasonable amount of time. And our concept of time in computational complexity theory is based on many, many computer resources, which are finitely bounded by some polynomial equation. You can think of the time bound being something hugely unreasonable, as if every atom of the universe were a computer, and if the adversary could control all those computers to work on breaking your cryptographic algorithm, then those computers would need to work on the problem for ten billion years in order to be successful. This is where the concept of 128-bit security comes from. Think about how many computers would be needed to try 2^{128} different combinations. The number is really large, but it is finite. And although it is finite, it is completely unreasonable to be able to mount those resources to break a cryptographic algorithm.

It is unlikely that you will be creating computational complexity proofs for protocols you are using. But, if you use well-known and accepted cryptographic algorithms, such the Advanced Encryption Standard (AES), the Rivest–Shamir–Adleman (RSA) public key algorithm, or the Secure Hash Algorithm version 2 (SHA2), then these algorithms have already been proven to be secure under reasonable computational complexity conditions. You just have to ensure that you are using them properly.

In our informal security model we do just that. We ensure that each operation in our protocol is protected with a cryptographic operation that provides the guarantees our protocol needs—confidentiality, integrity, indistinguishability, non-repudiation, etc. This is difficult, and without experience and proper training, there are many pitfalls that can make you go astray in your logic. The informal security model we discussed is useful to analyze how you are putting together existing protocols and ensure that you are not violating any of the conditions a protocol depends on. It is very useful for systems architecture. We don't expect you to be able to properly analyze brand-new security protocols or create your own security protocols after only reading this section.

So, if this is so hard, why did we spend any time on it at all? It is dreadfully important for you to understand how difficult protocol analysis actually is. Too many teams make up their own protocols, or tweak an existing protocol "just a little" and think they are going to be OK. In IoT teams, some protocols are pared down to make them fit in the small devices, or have features or capabilities removed in order to work well in the IoT realm. This is really dangerous. We discuss security analysis of protocols so that you will understand how complex protocol changes actually are, and if you are using a protocol that you really must depend on (e.g., key exchanges or authentication), you will understand how important it is that the protocol be analyzed in a formal security model. What does all this mean: Don't roll your own security protocols.

7.5 IoT Protocol Conclusions

In this chapter we looked briefly at a lot of different protocols. While we didn't go deeply into the details, we provided a very broad view of the numerous protocols used in IoT systems. You can use this survey to identify the protocols most applicable to your systems, and use the long set of references at the end of this chapter to delve deeper into the details. We also talked about how to analyze protocols for security, and for protocols you really need to be secure, how those protocols should be analyzed in a formal security model. But we introduced you to techniques that allow you to evaluate protocols on your own and understand the reasons behind why some parameters and security protections are included in the respective protocols. You should use the informal security model approach to verify that you are using reviewed protocols correctly, and that your usage doesn't violate the security protocol in some way. There is both an art and a science to secure protocol design, so don't create your own security protocols, but instead use protocols that are well published and standardized.

References

1. Freeman, R. (2001). *Practical Data Communications*, Second Edition. New York (NY): John Wiley & Sons.
2. Earle, A. (2006). *Wireless Security Handbook*. Boca Raton (FL): Auerbach Publications/Taylor & Francis Group.
3. Freeman, R. (2005). *Fundamentals of Telecommunications*, Second Edition. New York (NY): John Wiley & Sons.
4. Tanenbaum, A. (1996). *Computer Networks*. Upper Saddle River (NJ): Prentice Hall.
5. Dummies. (2018). "Network Basics: Twisted-Pair Cable." Retrieved from https://www.dummies.com/programming/networking/network-basics-twisted-pair-cable/
6. Paonessa, S. and McDuffee, B. (2018). "Back to Basics: The Fundamentals of 4-20 mA Current Loops." Retrieved from https://www.predig.com/indicatorpage/back-basics-fundamentals-4-20-ma-current-loops
7. Field Comm Group™. (2018). "HART Explained." Retrieved from https://fieldcommgroup.org/technologies/hart
8. Yokogawa Electric Corporation. (2012, March). "Fieldbus Technical Information." 4th Edition. TI 38K03A01-01E. Retrieved from https://web-material3.yokogawa.com/TI38K03A01-01E.pdf
9. Rockwell Automation. (2011, June). *FOUNDATION Fieldbus Design Considerations*. Reference Manual. Retrieved from https://literature.rockwellautomation.com/idc/groups/literature/documents/rm/rsfbus-rm001_-en-p.pdf
10. National Instruments. (2018, April 17). "RS-232, RS-422, RS-485 Serial Communication General Concepts." Retrieved from http://www.ni.com/white-paper/11390/en/
11. Weis, O. (2018, February 22). "A Quick Tutorial on RS485 and MODBUS." Retrieved from https://www.eltima.com/article/modbus-vs-rs485/
12. I2C Info. (2018). "I2C Bus, Interface and Protocol." Retrieved from http://i2c.info
13. Embedded Systems Academy. (2017). "I2C FAQ." Retrieved from http://www.esacademy.com/en/library/technical-articles-and-documents/miscellaneous/i2c-bus/frequently-asked-questions/i2c-faq.html
14. Anderson, D. (2001). *Universal Serial Bus System Architecture*, Second Edition. Boston (MA): Addison-Wesley.
15. Fieldbus Inc. (2011, October 14). "IEC61158 Technology Comparison: State of the Bus." Retrieved from http://www.fieldbusinc.com/downloads/fieldbus_comparison.pdf
16. IEC. (2018). "IEC 61158-1:2014: Industrial Communication Networks—Fieldbus Specifications—Part 1: Overview and Guidance for the IEC 61158 and IEC 61784 Series." Retrieved from https://webstore.iec.ch
17. CiA. (2018). "CAN Lower- and Higher-Layer Protocols." Retrieved from https://can-cia.org/can-knowledge/
18. CiA. (2018). "History of CAN Technology." Retrieved from https://can-cia.org/can-knowledge/can/can-history/
19. Bosch. (1991, September). "CAN Specification Version 2.0." Retrieved from: http://esd.cs.ucr.edu/webres/can20.pdf
20. Steube, J. (2018, August 5). "New Attack on WPA/WPA2 Using PMKID." Retrieved from https://hashcat.net/forum/thread-7717.html
21. Bluetooth SIG. (2016). "Bluetooth Core Specification." Retrieved from https://www.bluetooth.com/specifications/bluetooth-core-specification
22. MikroElekronika. (2016, March 26). "Bluetooth Low Energy—Part 1: Introduction To BLE." Retrieved from https://www.mikroe.com/blog/bluetooth-low-energy-part-1-introduction-ble
23. IEEE Computer Society. (2015). IEEE Std 802.15.4™.
24. Semtech. (2018). "What Is LoRa®?" Retrieved from https://www.semtech.com/lora/what-is-lora
25. Mekki, K. (2018). "A Comparative Study of LPWAN Technologies for Large-Scale IoT Deployment." ICT Express. Retrieved from https://doi.org/10.1016/j.icte.2017.12.005
26. LinkLabs. (2018, June 26). "What Is LoRa? A Technical Breakdown." Retrieved from https://www.link-labs.com/blog/what-is-lora

27. SigFox. (2018). "SigFox Technology Overview." Retrieved from https://www.sigfox.com/en/sigfox-iot-technology-overview

28. Z-Wave® Alliance. (2018). "About Z-Wave Technology." Retrieved from https://z-wavealliance.org/about_z-wave_technology/

29. RF Wireless World. (2012). z-wave physical layer | zwave PHY layer basics. Retrieved from http://www.rfwireless-world.com/Tutorials/z-wave-physical-layer.html

30. Weightless SIG. (2017, November 7). "Weightless-P System Specification. Version 1.03." Retrieved from http://www.weightless.org/about/weightless-specification

31. LinkLabs. (2015, November 23). "What Is Weightless?" Retrieved from https://www.link-labs.com/blog/what-is-weightless

32. 5G Americas. (2017). "Wireless Technology Evolution Towards 5G: 3GPP Release 13 to Release 15 and Beyond." Retrieved from http://www.5gamericas.org/files/3214/8833/1313/3GPP_Rel_13_15_Final_to_Upload_2.28.17_AB.pdf

33. GSMA. (2018). "Long Term Evolution for Machines: LTE-M." Retrieved from https://www.gsma.com/iot/long-term-evolution-machine-type-communication-lte-mtc-cat-m1/

34. Baby, R. (2018, June 15). "Specifications and Applications of Narrow Band IoT (NB-IoT)." Retrieved from https://www.rfpage.com/specifications-and-applications-of-narrow-band-iot/

35. RS Components. (2015, April 20). "11 Internet of Things (IoT) Protocols You Need to Know About." Retrieved from https://www.rs-online.com/designspark/eleven-internet-of-things-iot-protocols-you-need-to-know-about

36. Glow Labs. (2018). "Table Comparing Wireless Protocols for IoT Devices." Retrieved from http://glowlabs.co/wireless-protocols/

37. Vidales, M. (2017, May 16). "802.15.4 Wireless for Internet of Things Developers." Retrieved from https://blog.helium.com/802-15-4wireless-for-internet-of-things-developers-1948fc313b2e

38. Chua, J. and Yang, D. (2018, October 31). "An Examination of LPWAN Technology in IoT." Retrieved from https://www.electronicdesign.com/industrial-automation/examination-lpwan-technology-iot

39. Al-Fuqaha, A. (2015). "Internet of Things: A Survey on Enabling Technologies, Protocols, and Applications." *IEEE Communication Surveys & Tutorial*, Vol. 17, No. 4, Q4-2015. Retrieved from https://fardapaper.ir/mohavaha/uploads/2018/10/Fardapaper-Internet-of-Things-A-Survey-on-Enabling-Technologies-Protocols-and-Applications.pdf

40. Hardwood, T. (2018, August 20). "IoT Standards and Protocols: An Overview of Protocols Involved in Internet of Things Devices and Applications." Retrieved from https://www.postscapes.com/internet-of-things-protocols/

41. Stevens, W. (1994). *TCP/IP Illustrated, Volume 1: The Protocols.* Reading (MA): Addison-Wesley Publishing.

42. Wright, G. and Stevens, W. (1995). *TCP/IP Illustrated, Volume 2: The Implementation.* Reading (MA): Addison-Wesley Publishing.

43. Miller, M. (1998). *Implementing IPv6: Migrating to the Next Generation Internet Protocol.* New York (NY): M&T Books.

44. RF Wireless World. (2012). "z-wave protocol stack | z-wave protocol layer basics." Retrieved from http://www.rfwireless-world.com/Tutorials/z-wave-protocol-stack.html

45. Fouladi, B. and Ghanoun, S. (2013). "Security Evaluation of the Z-Wave Wireless Protocol." Retrieved from http://www.neominds.org/download/zwave_wp.pdf

46. Silicon Labs. (2018). "FAQ: What Is Z-Wave Over IP." Retrieved from https://www.silabs.com/products/wireless/mesh-networking/z-wave/specification/faq

47. Field Comm Group™. (2018). "HART without the Wires." Retrieved from https://fieldcommgroup.org/technologies/hart

48. Weissberger, A. (2017, October 25). "LoRaWAN and Sigfox Lead LPWANs; Interoperability via Compression." Retrieved from http://techblog.comsoc.org/2017/10/25/lora-wan-and-sigfox-lead-lpwans-interoperability-via-compression/

49. ZigBee Alliance. (2017). "Why ZigBee PRO?" Retrieved from https://www.zigbee.org/zigbee-for-developers/zigbee-pro/

50. RF Wireless World. (2012). "ZigBee Tutorial | ZigBee Protocol, Frame, PHY, MAC." Retrieved from http://www.rfwireless-world.com/Tutorials/Zigbee_tutorial.html

51. Olssen, J. (2014, October). "6LoWPAN Demystified." Texas Instruments. Retrieved from http://www.ti.com/lit/wp/swry013/swry013.pdf

52. Thread Group. (2018). "FAQ: What Aspects of the Wireless Network Does Thread Address?" Retrieved from https://www.threadgroup.org/support#faq

53. Microchip. (2018). "Low-Cost Embedded Wireless Connectivity for Commercial and Smart Home Networks." Retrieved from https://www.microchip.com/design-centers/wireless-connectivity/embedded-wireless/802-15-4/software/miwi-protocol

54. Nixon, M. (2012, September 23). "A Comparison of WirelessHART™ and ISA100.11a." Retrieved from https://www.emerson.com/documents/automation/white-paper-a-comparison-of-wirelesshart-isa100-11a-en-42598.pdf

55. Schneider Electric. (2018). "What Is ModBus and How Does It Work?" Retrieved from https://www.schneider-electric.us/en/faqs/FA168406/

56. MODBUS. (2006, October 24). "MODBUS Messaging on TCP/IP Implementation Guide V1.0b." Retrieved from http://modbus.org/docs/Modbus_Messaging_Implementation_Guide_V1_0b.pdf

57. MODBUS. (2012, April 26). "MODBUS Application Protocol Specification V1.1b3." Retrieved from http://modbus.org/docs/Modbus_Application_Protocol_V1_1b3.pdf

58. Powell, J. (2013, October). "Profibus and Modbus: A Comparison." Retrieved from https://www.automation.com/automation-news/article/profibus-and-modbus-a-comparison

59. MQTT. (2014, November 7). MQTT. Retrieved from http://mqtt.org/

60. ISO. (2016, June). "ISO/IEC 20922:2016 Information Technology—Message Queuing Telemetry Transport (MQTT) v3.1.1." Retrieved from https://www.iso.org/standard/69466.html

61. Stanford-Clark, A. and Truong, H. (2013, November 14). "MQTT For Sensor Networks (MQTT-SN) Protocol Specification Version 1.2." Retrieved from http://mqtt.org/new/wp-content/uploads/2009/06/MQTT-SN_spec_v1.2.pdf

62. AMQP. (2012, October 16). "Advanced Message Queuing Protocol." Retrieved from https://www.amqp.org/resources/specifications

63. ISO. (2014, May). "ISO/IEC 19464:2014 Information Technology—Advanced Message Queuing Protocol (AMQP) v1.0 Specification." Retrieved from https://www.iso.org/standard/64955.html

64. IETF. (2011, March). "Extensible Messaging and Presence Protocol (XMPP): Core." Retrieved from https://tools.ietf.org/html/rfc6120

65. IETF. (2014, June). "The Constrained Application Protocol (CoAP)." Retrieved from https://tools.ietf.org/html/rfc7252

66. IETF. (2017, February). "Guidelines for Mapping Implementations: HTTP to the Constrained Application Protocol (CoAP)." Retrieved from https://tools.ietf.org/html/rfc8075

67. IANA. (2018, November 26). "IPv6 Multicast Address Space Registry." Retrieved from https://www.iana.org/assignments/ipv6-multicast-addresses/ipv6-multicast-addresses.xhtml

68. IETF. (2013, October). "Concise Binary Object Representation (CBOR)." Retrieved from https://tools.ietf.org/html/rfc7049

69. IETF. (2017, July). "CBOR Object Signing and Encryption (COSE)." Retrieved from https://www.rfc-editor.org/rfc/rfc8152.txt

70. IETF. (2018, August). "Sensor Measurement Lists (SenML)." Retrieved from https://www.rfc-editor.org/rfc/rfc8428.txt

71. Dolev, D. and Yao, A. (1983, March). "On the Security of Public Key Protocols." *IEEE Transactions on Information Theory,* Vol. IT-29, pp. 198–208. Retrieved from http://www.cs.huji.ac.il/~dolev/pubs/dolev-yao-ieee-01056650.pdf

72. Russell, D. and Gangemi, G. (1991). *Computer Security Basics.* Sebastopol (CA): O'Reilly & Associates.

73. Stalling, W. (1999). *Cryptography and Network Security.* Upper Saddle River (NJ): Prentice Hall.

74. Choo, K. (2006, May). "Key Establishment: Proofs and Refutations." Thesis. Information Security Institute, Faculty of Information Technology, Queensland University of Technology.

75. Bellare, M. and Rogaway, P. (1993). "Entity Authentication and Key Distribution." Retrieved from http://cseweb.ucsd.edu/~mihir/papers/eakd.pdf

76. Bellare, M. and Rogaway, P. (1995). "Provably Secure Session Key Distribution—The Three Party Case." *Proc. 27th Annual Symposium on the Theory of Computing (STOC 95)*, pp. 57–66, ACM. Retrieved from http://web.cs.ucdavis.edu/~rogaway/papers/3pkd.pdf

77. Bellare, M., Canetti, R., and Krawczyk, H. (1998, January). "A Modular Approach to the Design and Analysis of Authentication and Key-Exchange Protocols." Retrieved from https://eprint.iacr.org/1998/009

78. Canetti, R. and Krawczyk, H. (2001, June). "Analysis of Key-Exchange Protocols and Their Use for Building Secure Channels." Retrieved from https://eprint.iacr.org/2001/040

79. Aumasson, J. (2018). *Serious Cryptography: A Practical Introduction to Modern Encryption.* San Francisco (CA): No Starch Press.

80. IETF. (2015, September). "HTTP Digest Access Authentication." Retrieved from https://www.rfc-editor.org/rfc/rfc7616.txt

Chapter 8

Privacy, Pirates, and the Tale of a Smart City

By JC Wheeler

8.1 Shroud for Dark Deeds or Fortress for the Vulnerable

At the end of the movie "The Circle," Emma Watson's character Mae seems to peacefully accept the surveillance culture evolving around her—resigning herself to the inevitable erosion of privacy in a digitally interconnected world. Perhaps you feel the same, and who can blame you? The conveniences offered by ubiquitous devices, smart homes, and smart cities provide us with innumerable benefits, from tailored strategies to manage our health and well-being to customized home environmental settings. If we have to sacrifice a bit of privacy for the quality of life offered by intelligent Internet of Things (IoT) systems, then so be it. The surveillance side effect isn't so bad for society as a whole anyway. As Mae's boss, a social media mogul, reasoned, people behave better when they know they're being watched. Although his ideology is ironically self-serving, he's not wrong. Those familiar with the Hawthorne effect[*] know this to be true. Why, then, is the preservation of privacy so important? If the right to privacy does little more than enable unethical or illegal behavior to go unpunished, then by all means let us do away with it. However, when we recall historic moments of human compassion and courage, many of them—for example, the Underground Railroad, the French Resistance during WWII, and the hiding of the Jews from the Nazis—were galvanized by whispers in secret. In these cases, privacy meant protection for the vulnerable. So, like many things, privacy is complicated. On the one hand, it shrouds dark illicit deeds, whereas on the other hand, it is a fortress for the vulnerable. Rapid advancements in IoT and artificial intelligence (AI) have robbed us of much-needed time to determine the level of privacy necessary for a healthy society. But, determine it we must—and soon before it's too late. In an interconnected world, run in no small part by algorithms and fed by data collected courtesy of IoT devices, the preservation

---

[*] The Hawthorne effect, according to en.oxforddictionaries.com, is "the alteration of behavior by the subjects of a study due to their awareness of being observed."

of privacy takes on an expanded role. It becomes the key defense, the last fortification against discrimination and denial of service for the most vulnerable of society and beyond.

This significantly elevates the role you, as an architect or planner of smart IoT systems, have to play in preserving life, liberty, and the pursuit of happiness for all citizens. It is you who will determine whether smart cities impair or enrich society. You are that important, and your understanding of digital ownership, your willingness to challenge your assumptions, and your creative mind are all critical as we engineers venture forth as the new change agents of this millennium.

8.2 Chapter Scope

This book divides the topic of privacy into two chapters. This first chapter analyzes two IoT scenarios from a privacy perspective. Both the "Safe Driver's App Meets Smart Fridge" scenario and the "Tale of a Smart City" culminate with examples of discrimination and denial of service—medical and socioeconomic, respectively. The "Tale of a Smart City" centers around an autonomous transportation system and the commerce and smart city management that develop around it. Also included in this chapter is a discussion of deep fakes and their relationship to IoT. The chapter concludes with a Privacy Playbook to assist in the exposure and mitigation of privacy concerns at the product planning and architecture level. The follow-on privacy chapter describes algorithms, architectures, and software techniques for applying privacy mechanisms. In addition, privacy guidelines and regulations pertinent to IoT are discussed.

8.3 AI and IoT Unite—Amplifying the Engineer's Significance in Society

At the end of the 1990s, I was one of hundreds of engineers all scrambling to make their own personal imprint on Motorola's latest brain child—the development of what could now be considered a primitive version of the modern smartphone. The project was called Aspira, and I was working with a team of engineers to dream up services and features that took advantage of the small GPS processor that was to be embedded in the cellphone. Location Services was a popular buzzword in the office. My colleagues and I were excited to be a part of something new to the world of telecommunications technology.

To design something groundbreaking that will make a positive impact on society is what most young engineering students dream of when they study the likes of Hertz, Tesla, and Steve Jobs. The group of engineers working on Aspira were no exception. The atmosphere was filled with energy as engineers brainstormed on the myriad ways Location Services could be used to enrich the cellphone user's experience, assist in emergency situations, and be monetized. Few stopped long enough to contemplate how Location Services could be abused or misused to the detriment of an individual, a society, or, as reported recently, to our nation's soldiers.[1] If we had, we would have put more safeguards in place and thought more about when location data is collected, who has access, and how to provide easily accessible user controls.

The potential for technology innovators to unknowingly and unintentionally cause harm to individuals, economies, companies, and countries is very real in a digitally interconnected, algorithm-driven, world. In the case of privacy, information is power, and it is becoming all too easy to convert discombobulated pieces of data into valuable information that can be misused by special interest groups, governments, and for-profit titans. The small amount of innocuous data collected by IoT devices and applications will and already is being sold or shared with third parties, where it is then combined with other innocuous data, churned on by AI algorithms, and correlated into a profile

> Innocuous data collected by IoT devices . . . will be used to determine employability, insurability, whether you are worth the time of a sales clerk, whether the self-driving vehicle you requested picks you up first . . .

of you, your child, your friends, and your parents. It will, and already is on a small scale, be used to determine employability, insurability, whether you are worth the time of a sales clerk, whether the self-driving vehicle you requested picks you up first, and if you will be one of the premium customers whose car is authorized to take the fastest route to your destination. Socioeconomic status, age, religious affiliation, and health may all be factors in the algorithms' decision processes. And, most of this information will be derived from harmless non-personal data that has nothing to do with any of these factors. *It is a brilliant deduction that rivals anything Sherlock Holmes® ever did, and it's discriminatory in nature.*

Please understand, I'm not proposing that we all quit our day jobs and build a tiny house off the grid—quite the contrary. In the last decade, study after study has demonstrated that the global distribution of cellular technology was the single largest contributor to the reduction of poverty in history.[2,3] Many engineers in IoT today participated in this philanthropic effort of Herculean proportions. This is something we can be immensely proud of, and it demonstrates that technology can be powerful medicine for societal ills. IoT smart systems need not be the exception, provided there are mechanisms in place that strategically address misuse and abuse.

> Technology can be powerful medicine for societal ills. IoT smart systems need not be the exception.

Do you feel the power you have to impact the evolution of our society? "With great power comes great responsibility."[†]

8.4 The Elephant in the Room

The technology and credit card titans pretty much know everything there is to know about us already, don't they? So, isn't the privacy debate already lost?

Granted, search engines, smartphone providers, on-line retailers, social media haunts, credit card companies, and data brokers[4–6] know more about the average American than they would feel comfortable telling a close friend. This is nothing new. For the moment, however, we can still retain relative anonymity if we choose to. We can still pay cash at certain medical establishments or labs to keep medical conditions or diagnoses confidential from employers or insurance companies. And, we are not forced to have a social media account that could be

---

[*] Sherlock Holmes is a fictional character by Sir Arthur Conan Doyle.

[†] The Quote Investigator presents a *marvel*ous synopsis of the possible origins of this quote, citing numerous sources such as the Bible, the French Revolution, Winston Churchill, and Teddy Roosevelt, and, of course, culminating with the one you are perhaps thinking of—Stan Lee. See https://quoteinvestigator.com/2015/07/23/great-power/ and https://www.marvel.com/characters/uncle-ben-ben-parker

used to make an employment determination. We can still pay cash at the store, and we can use search engines that don't track our searches. But, it's getting harder to do so.

The eroding of an individual's ability to retain some semblance of anonymity and a say in the distribution of their personal information is finally getting its day in the halls of Congress and will spark much debate in the next several years. However, the larger concern is the use of personal information in deep learning, to train neural networks to make resource allocation decisions. These decisions can be detrimental to the data donors themselves—which includes you and me. So, in this age of machine learning and deep learning, digital privacy regulations and mechanisms should not be limited to the data realm but should also include learning algorithms. This significantly expands the role of digital privacy governance. The impetus for this expanded role will be demonstrated in the scenarios depicted within this chapter.

8.5 Scenario: Safe Driving App Meets Smart Fridge

Let's switch gears for a few minutes and work through an exercise together:

> **EXERCISE:** *Read the scenario below and then answer the following question. What could a third-party entity deduce about a person by simply having access to data collected from both their safe driver's app and their smart fridge?* Contemplate the public health uses and misuses of this data.*

You may be thinking, "Timeout. How would a third-party entity such as a healthcare insurer or researcher get access to data from these two independent sources?" Great question. Chances are they will pay a data broker[4–6] for the information. Perhaps the data broker is an affiliate of the car insurance company and or the appliance company. If so, the user has likely given them access via their user agreement—somewhere in the fine print, affiliate relationships will be mentioned. Third-party common cloud providers could also be affiliates, along with parent companies and various partners that the insurance or appliance companies have agreed to sell data to.

Scenario

Ravi had a great year. He purchased a smart fridge with all the bells and whistles, to include a control console that enables voice-activated management of his family's smart home, meal planning, energy monitoring, entertainment, and event scheduling. The refrigerator also takes a picture of its contents each time it closes. These pictures are stored in the cloud, and Ravi can access them using an app that he downloads to his smartphone. This allows Ravi to view the contents of his refrigerator anytime and anywhere—very helpful when at the grocery store. In addition, Ravi saved $400 on car insurance by agreeing to use a safe driving monitoring service—requiring the use of a dongle and an app. A dongle is plugged into the ODB II port under his car's dashboard, and a safe driving app is downloaded to his smartphone. In addition to monitoring and providing driver-safety statistics, the service provides Ravi with useful information such

---

* This is a fictitious scenario. Although it is plausible, given current technology, it is not meant to implicate any specific company or individual.

as travel logs, fuel economy, and a car-locating feature. It further alerts Ravi to any diagnostic trouble codes that are derived during periodic diagnostic scans performed by the dongle. The dongle uses 4G LTE to send all collected stats to the cloud and provides a WiFi hotspot feature for the car's occupants. During the app download process, Ravi registers with both the smart fridge manufacturer and the car insurance company—both requiring, among other personal data, his email and smartphone MAC addresses.

The insurance company and appliance maker both use the same cloud provider* and have an agreement to expose some of the user data to that provider in exchange for discounted cloud storage fees. This agreement makes the cloud provider an affiliate of both companies so that the user agreements are not violated. The appliance maker shares images of the refrigerator's contents, its location within an urban zone but not the address, the MAC address of the smart fridge console, the WiFi MAC of all WiFi-enabled devices in the home, and Ravi's email and smartphone MAC addresses. The refrigerator's images include date and time stamps. The car insurance company shares time and location information for frequent stops; car make, year, and model information; WiFi MAC addresses of all WiFi-enabled devices in the vehicle—with date and time; and Ravi's email address and smartphone MAC addresses. This is all summarized in Figure 8.1.

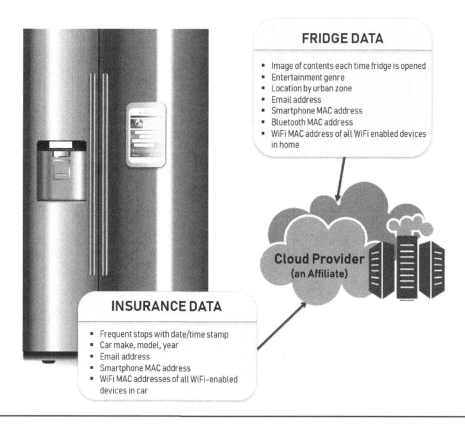

Figure 8.1   Data Shared with Cloud Provider

---

* The data from both parties might just as easily have ended up with a data broker.

The cloud provider has a data center with rack upon rack of state-of-the-art servers and GPUs specifically optimized to execute deep learning algorithms on a colossal number of data sets per hour. The cloud provider thus quickly discovers that Ravi is the common denominator between the fridge and safe-driving data sets, then uses the combined data to derive a profile for Ravi or adds to an already-existing profile. The cloud provider further makes use of all provided WiFi addresses to link Ravi to other devices and individuals, adding to profiles it may have on said individuals.

Take a moment and jot down what you could learn about Ravi from the information provided in Figure 8.1. Make a mental list and we will compare notes shortly.

In this exercise, I am deliberately choosing to ignore the treasure trove of data that the smart refrigerator manufacturer could obtain from voice conversations, the calendar, and other apps on the console. My editor gave me a page limit, but kudos to you for thinking of it. I am also choosing to assume that certain pieces of information such as name, address, phone number, and age are anonymized. This is because I want to demonstrate that Personally Identifiable Information (PII) can be easily derived using non-PII data. The implications of this statement should make us pause for a moment. It

> PII can be easily derived using non-PII data.

means that governments can attempt to regulate PII, but this will, at best, be effective in

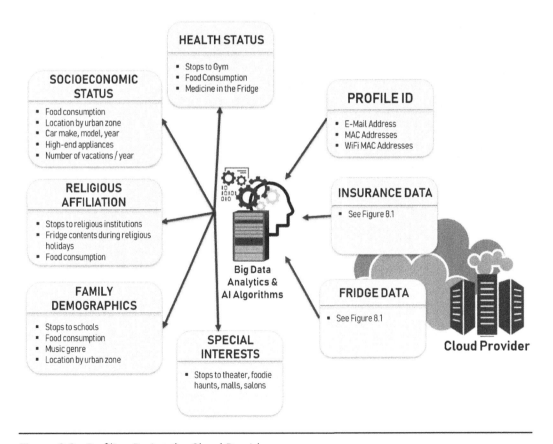

Figure 8.2   Profiling Ravi at the Cloud Provider

limited applications because data analytics algorithms will still be able to derive PII from non-PII data.* So, try as they might, lawmakers will have a difficult time mitigating PII exploitation via regulations alone. Engineers, however, can meet lawmakers halfway by implementing mitigations such as encryption, differential privacy, and other forms of obfuscation and separation that, while imperfect in and of themselves, can complement, enforce, or audit regulatory compliance. More on these mitigation strategies in Chapter 9.

Okay, let's compare lists, mine is depicted in Figure 8.2 and summarized below.

Socioeconomic Status: We can derive socioeconomic status from the value and age of the car, the high-end appliance, and the general location of the home, as well as from the places frequented, such as malls and restaurants. The types of food in the refrigerator can also hint at financial status and discretionary income—for example, organic foods and gourmet brands.

Religious Affiliations: We might be able to guess religious affiliation, if any, as well as dedication to beliefs or lack thereof. For example, if there are frequent stops at a church, a temple, a mosque, a private religious school, or philanthropic organizations run by religious institutions, then these behaviors can be correlated across religious categories to find other common behaviors and predict still others—called micro-targeting.[7] With images of the refrigerator contents, we can also correlate refrigerator use with religious observances. Types of foods might also help here, for example, vegan or kosher.

Family Demographics: The types of foods in the fridge might indicate a family with young consumers, for example, hot dogs, milk, type of yogurt and juice, lots of fruit, and chicken legs versus an empty nest, for example, boutique wines, lots of greens, low-fat dressings, and fish. Recurrent images of fridge contents late at night, say 10-ish, could possibly indicate that a ravenous teenage boy lives in the house or someone is cheating on their diet or perhaps even someone is becoming prediabetic. Frequent destinations involving schools, karate dojo, and dance studios all are indicative of families with kids. Certain urban zones appeal to more mature couples, some to young childless couples, and others to young families. These data points, in and of themselves, don't tell us much, but big data analytics is all about correlating volumes of data sets, over time, to draw conclusions.

> The data points in and of themselves don't tell us much, but big data analytics is all about correlating volumes of data sets, over time, to draw conclusions.

Health Status: If the gym is a frequented destination, and the fridge is stock full of whole foods, health brands, power drinks, maybe organics, and minimally processed foods, then Ravi is likely a health-conscious person. This doesn't mean he is healthy, but it does mean that the probability of good health increases. Conversely, if there are lots of sweets, soft drinks, and processed foods or take out in the fridge, and Ravi does not frequent the gym, then the probability of good health decreases.

---

* Vadhan, S. (2017). "The Complexity of Differential Privacy." In *Tutorials on the Foundations of Cryptography,* edited by Y. Lindell (Chapt. 7, p. 348). Cham (Switzerland): Springer International Publishing.

Special Interests: Over time, Ravi's driving habits will provide some understanding of his interests. How often he drives to arenas, parks, museums, campgrounds, and malls can help determine a level of interest in sports, travel and leisure, art, home and garden, and shopping— all valuable for targeted marketing.

Profile ID: While the data provided to the cloud affiliate does not include Ravi's name, address, or phone number, it does include email and MAC addresses, which are unique identifiers. The cloud provider or a data broker may already have this information from one or more other entities, like those listed in Section 8.4. This allows the combining of data sets associated with Ravi from many different sources. Note that, over time, the cloud provider could deduce who Ravi is even without being provided with email and MAC address information—through the correlation of location information with the other data categories listed in Figure 8.2. It takes more data, but the IoT will provide that.

What we can learn about Ravi from a refrigerator and the digital equivalent of a backseat driver, evokes a feeling not unlike that experienced when we have a psychologist for a friend. It would be a bit uncomfortable if we suspected our friend was psychoanalyzing us. But, most of us would tolerate such discomforts if there were other redeeming factors in the relationship with our friend. There are certainly redeeming factors to both consumers and industry alike in the collection of personal data. The obvious benefits of the present scenario include targeted marketing and safer driving habits. Safe-driving habits conceivably lead to fewer car accidents, reductions in car insurance costs, and reduced fatalities and injuries on the road. I have to admit, that's a win-win for society and a sigh of relief for parents of teenagers everywhere. On the contrary, a primary disadvantage to the collection of personal data specific to this scenario is the potential for the delay of medical services and denial of employment, health insurance, or healthcare altogether.

If the smart fridge app were to provide Ravi with encrypted cloud storage options, and or "opt out" options, then Ravi could determine whether the advantages of data sharing outweigh the potential risks. Shouldn't it be his choice to make?

8.5.1 IoT Saves Our Bacon,* but Tattles if We Eat Cured Fatty Pork

According to the Centers for Medicare & Medicaid Services (CMS.gov), the National Health Expenditures for 2017 totaled $3.5 trillion. This includes federal and state spending, insurance spending, and consumer out-of-pocket spending.[8] That's roughly one-sixth of the value of all goods and services produced in the United States. That's a lot of money going to healthcare. So, the government, insurers, and employers have turned to data analytics and AI to reduce the healthcare price tag. The combining of electronic medical records data with data collected from wearables and other IoT products, such as our smart refrigerator, will provide insurers and researchers with petabytes of data to analyze for this colossal effort. Medical care costs are already being reduced through detecting costly distribution channels,[9] predicting future

---

* For my engineering comrades who may not be familiar with the phrase "saved our bacon," it is slang for "saved us from harm." Bacon is also cured fatty pork. So, the title is a play on words, get it? This brings back fond memories of my perplexed look when the Hindi proverb, "a finger held straight cannot hook ghee" was once explained to me.

health issues,[10] and incentivizing good health habits.[11] Insurers and employers are partnering to incentivize employees to exercise regularly, make good food choices, and participate in a preventative care program. In exchange, employees are promised a reduction in insurance premiums and or deductibles. These incentives require the monitoring of peoples' behavior. Such monitoring is currently in its infancy, but as IoT ventures into health and wellness monitoring, the data will be available to better predict, for example, cancer risk, heart attacks before they happen, and the potential for opioid addiction. There is the possibility, however, especially in these early stages of AI, for the deep learning algorithms to make incorrect predictions that end up penalizing individuals unjustly. And even if the algorithms were spot on, ethical questions loom around, for instance, the issue of genetic predispositions for diseases. How accountable should an individual be for their DNA?

There are many challenges to work through, one being who decides what is and isn't healthy and how much food is too much? What if the decision makers are wrong? They've been wrong before. In my lifetime, eggs have been out, then in, then out, then finally in again. Fat used to be bad and whole grains good, but now, paleo and ketogenic dietary professionals say fat is healthy and grains perhaps are not. Nutrigenomics is a relatively new science. Its supposition is that genetics plays a large role in determining the types of food you should eat and exercise you should engage in. This, of course, requires an analysis of your DNA. If employers or insurers somehow obtain access to this DNA, enabling them to search for markers for diseases such as ALS,* a massive opportunity for discrimination or denial of service results.

8.5.2 Smart Algorithms to the Rescue

What if Congress approves a single-payer healthcare system? A single-payer system, and Medicaid/Medicare for that matter, can't use lower premiums or deductibles to incentivize healthy habits because recipients don't pay into the system. So, what about using prioritization of care as an incentive? What if the average wait time for a biopsy was one month? Is it okay to prioritize the exercising, food-conscious individuals first—giving them a better chance of survival? It seems reasonable to this pragmatic engineer, but then I stop for a moment and think about the sole surviving parent of three who works and puts food on the table and is just too exhausted to exercise. If he were to die, his kids might be relegated to an already overfull foster care system, which itself would cost more than treating the cancer. What about the VP of a non-profit who, instead of devoting time to self-care, would rather expend her energy, time, and resources caring for young teens who are homeless? Will the AI algorithms in our IoT clouds be intelligent enough to incentivize those who enrich society by caring for others more than themselves?

> Will machine learning algorithms in IoT clouds be intelligent enough to incentivize those who enrich society by caring about others more than themselves?

What about those individuals in a higher socioeconomic bracket. They tend to pay more in taxes, and the government needs the revenue right? Should tax revenue have an impact on deciding who is first in queue for that biopsy? Data analytics and machine learning are forcing these difficult questions. We can throw such decisions over the fence to the policy makers,

---

* Amyotrophic Lateral Sclerosis or Lou Gehrig's disease.

or we can perhaps make some of them obsolete by finding work-arounds that meet customer requirements while reducing the risk of discrimination.

Protecting privacy and user choice will go a long way to minimizing such ethical dilemmas in the first place. For example, if the goal of a wearable is to monitor activity—HR, EKG, PH, mineral analysis of sweat (why not?), and sleep patterns—as well as warn the user of possible health problems, I can think of several ways of accomplishing this goal without the threat of insurance companies, cloud affiliates, or employers acquiring this information. If I can think of them, so can you. The Playbook and the next chapter will help.

> It won't be a satisfying win, if we punish the most vulnerable members of our populous to accomplish it.

It's exciting to think that technology advances have the potential to significantly reduce the costs of healthcare and increase longevity and quality of life. Much like the cellphone's impact on poverty, the impact of IoT on the reduction of our federal Medicare/Medicaid budget could be history making. But it won't be a satisfying win, if we punish the most vulnerable members of our populous to accomplish it.

8.6 From Autonomous Vehicles to Smart Cities

Eating pancakes, dreaming of flying cars, and watching "The Jetsons" cartoon were my favorite Saturday morning pastimes as a child. I wondered if I would live long enough to drive a flying car. If the progress made in autonomous driving over the last five years is any indication, then I just might live to see that day.[12] It's hard to fathom, though, a city's transportation grid being dominated, if not exclusively run, by AI. True, we have driverless vehicles in small quantities on the roads today, but there are a number of reasons to be skeptical that they will ever dominate the roadways. Perhaps the most notable is humanity's innate fear of losing control to the AI systems we create. Let's face it, a lot of us think about the movie "The Terminator" when we contemplate AI. The thought of being trapped in a vehicle, driven by Arnold's evil AI doppelganger, scares the wits out of people. How much more so an invisible AI algorithm that we can't at least hit with a baseball bat if it puts our life in jeopardy? We humans like to be in the driver's seat (pun intended).

Like it or not, autonomous transportation networks will become the dominant mode of transportation in the near future. Why is it inevitable despite our fear of AI? It solves some challenging issues faced by cities around the world, it's better for the environment, and it's big money, which will be injected into economies across the globe. Also, it's well underway. Within the next decade, a rich set of autonomous features, if not full autonomy, is on the roadmap of every major car manufacturer in the world, and big tech is in chest deep.[13-15] Further, governments are embracing autonomous driving with open arms. National guidelines, regulations, and investments in Europe, the United States, Japan, and China are all supporting efforts toward autonomous transportation systems.[13] Goldman Sachs has projected growth in the autonomous vehicle industry from $96 billion in 2025 to $290 billion by 2035,[14] and that doesn't include monetary gain from ancillary services such as data brokering for targeted marketing and consumer transactions occurring within the autonomous transportation ecosystem—both of which we discuss in the scenario below.

Not only will vehicles be driving themselves, but they will also receive and process data from other vehicles, traffic systems, advertisers, and service providers—making 5G a hot commodity. Thus, autonomous vehicles will evolve into autonomous interconnected transportation networks that are managed by AI traffic control software in the cloud. This AI traffic control software will not only manage the autonomous vehicle fleet but also entire metropolitan transportation grids, to include traffic lights, subway systems, and commuter trains—making it the linchpin of a smart city. The grid will be continuously optimized to meet the demands of the moment. Sounds futuristic I know, but the wheels* already seem to be set in motion. In January 2018, Ford's CEO, Jim Hackett, announced plans to develop a "Transportation Mobility Cloud" service through its newly acquired subsidiary Autonomic[SM].[†] Autonomic's website provides the following description of their open cloud–based platform:

> The Transportation Mobility Cloud will connect the diverse components of urban mobility systems—connected vehicles, mass transit, pedestrians, city infrastructure and service providers—with the goal of orchestrating a safer, more efficient and sustainable transportation network.
> It is a flexible and secure platform that provides the necessary building blocks for smart mobility applications, such as routing self-driving cars, managing large-scale fleets or helping residents plan transit journeys.

Source: https://autonomic.ai

Imagine what your daily commute would be like if you no longer had to navigate through a swelling river of people—all flooding into the city at once. Instead, a virtual traffic chauffeur, looking nothing like Arnold, provides you and your fellow commuters with your personalized commuter itinerary—guaranteed to get you to your destination in the least amount of time and hassle possible. So, effectively, the river of people is reduced . . . to multiple steady, but manageable, streams of traffic. Calming, isn't it. If we take it a step further, your virtual chauffeur arranges for the delivery of your custom soy coffee mocha with cinnamon sprinkles during a red light enroute—piping hot and paid for through the transportation app. Now that's smart.

There are a number of good reasons to pursue autonomous vehicle systems to be sure—reduced road rage, increased fuel economy, independence for the elderly, and reduced traffic accidents, to name a few. However, emphasis on *however*, friends, there will also be, if we are not careful, some distinct societal consequences impacting personal choice and fair and equitable treatment. Whether the advantages outweigh the disadvantages will depend, not in small amount, on the engineer's ability to preserve aspects of digital privacy.

8.6.1 Scenario: The Tale of a Smart City

Here we step through a scenario involving a large hypothetical US city surreptitiously named Myopia. The scenario starts in 2030 AD and spans a 15-year time period. The scenario is based on existing technology, technology roadmaps, and the challenges faced by US cities around

---

* Just could not resist the pun—it was right there for the taking.
† Please note that *The Tale of a Smart City* is my vision of what an autonomous transportation grid might look like in the near future; it is not based on anything produced or planned by Autonomic[SM] or any other company.

the country today. So, a version of it just might become reality in a city near you in the not-so-distant future. The background below sets the stage. Bear with me, it all ties together in the end.

Introduction to Myopia and Timeline Summary

Myopia, a large sprawling metropolis in the southwestern region of the United States, found itself faced with the same problems mulled over by mayors and city councils all over the country today:

- A densely populated downtown area
- Traffic congestion and above national average commute times into downtown
- Sizable public debt, above USD 10 billion
- Inadequate public transit footprint
- Old and under-maintained public infrastructure
- Poor air quality (above national average for NOAA's Air Quality Index)

  With the need to overhaul its public transportation network, and with limited funds to do so, Myopia turned to public–private investment options. In 2016 (14 years prior), the state legislature approved regulations that made the state a friendly environment for autonomous vehicle testing. Deployments of small autonomous fleets in lower population centers within the state had been wildly successful. This gave Myopia's City Council the political capital to push an initiative to its residents for approval. The initiative awarded a sole contract to a major car manufacturer to pilot a Level 5* Autonomous Vehicle Fleet (AVF) program within a 10-mile radius of the downtown area. The car manufacturer itself was the front facing arm of an autonomous vehicle consortium, predominantly composed of big tech companies, all bringing intellectual property together to provide a fully autonomous fleet of vehicles, along with the cloud infrastructure and service management interfaces needed to maintain the fleet.

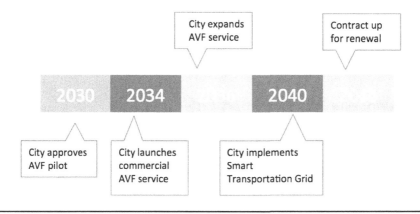

Figure 8.3   Timeline for Autonomous Vehicle Fleet (AVF) Service Evolution

---

* Fully automated vehicle.

The agreement, an incremental approach, started with a small number of prototypes, then expanded over time to provide a fleet of autonomous cars, passenger vans, and buses to augment the existing commuter transit system. A second agreement was made in 2036 to expand the fleet size; then, in 2040, the monitoring and control of the city's diverse public transportation systems was converged into a single smart transportation grid (see the timeline depicted in Figure 8.3).

2030: City Approves AVF Pilot

In 2030, the city engaged in a 15-year exclusive Service Level Agreement (SLA) with the Autonomous Fleet Consortium. The consortium funded the upfront and operating costs of the fleet in exchange for the right to:

- Collect passenger data
- Charge passenger fees based on a pricing schedule set by the city
- Advertise and provide other digital services to passengers for direct and indirect compensation

The consortium agreed to:

- Provide the city with traffic flow data and passenger data collected from user registrations and commute patterns
- Assist the city in implementing a Public Affairs Customer Relationship Management (PA-CRM) service for data analytics on AVF data
- Work with the city to optimize traffic flow through the downtown area
- Add a "VIP" service level that prioritized pick-up times and routes for top-level city officials and those they designate

The traffic-specific data collected by the AVF is depicted in Figure 8.4.

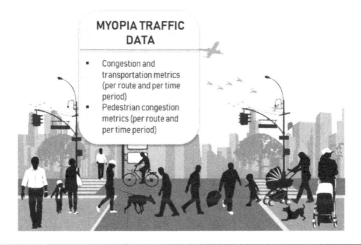

Figure 8.4   Traffic Data Collected by Autonomous Vehicle Fleet (AVF) Service

In addition, the city agreed to:

- Offer tax incentives to downtown residents, small businesses, and corporations that use the AVF
- Reduce available public parking structures by half in the downtown area
- Incrementally eliminate its existing dilapidated bus network, replacing it with AVF passenger vans and buses
- Guarantee tax incentives, at both the state and local levels, to the AVF service provider

All AVF vehicles are equipped with interior and exterior camera systems and voice-activation software, in addition to radar, Lidar systems, WiFi, 5G, and GPS. Each vehicle maintains cloud connectivity as well as connectivity to other fleet vehicles within its vicinity. This provides each vehicle with real-time traffic flow information and enables data to be sent to the cloud for additional data analytics. This data is used to model and anticipate future traffic patterns and will be used to optimize fleet distribution and traffic light sequencing. In addition to the traffic data collected by the AVF, the AVF vehicles and user registration system collect the user data shown in Figure 8.5.*

Figure 8.5   User Data Collected by the Autonomous Vehicle Fleet (AVF) Service

---

* AI algorithms eat through massive datasets faster than teenage boys eat carbs, so each autonomous vehicle collects thousands of data elements, but only the elements that apply to this scenario are depicted here.

Passengers must download and register with the AVF app to use the AVF service. The app's registration process collects personal data and user preferences that assist the AVF service in providing a pleasant user experience. The user is also asked to agree to the terms of use, which includes a privacy agreement that mentions data sharing with affiliates. These affiliates are not specifically named, but they include the municipality of Myopia, all members of the consortium, local downtown businesses, and various special interest groups paying for user-profile data. The user also gives permission to access Location Services in the phone. The app maintains a continuous active log of the user's location, regardless of whether the user is actively seeking transportation. This log is periodically sent to the cloud and is used in refining the user profile and in resource planning. Once the user registration process is completed, the commuter uses the app to make AVF transportation requests.

The interior cameras and a voice-activation system within each vehicle monitor passenger activities and conversations. This helps the AVF service determine how best to interact with passengers for the purpose of drawing their attention to products and services tailored to their lifestyles and interests. The AVF service uses this data to partner with local businesses along users' commute routes, matching businesses with prospective clients.

The AVF service takes the user data it collects from each registered passenger and creates a user profile. Figure 8.6 depicts the user profile. The profile data comes from a combination

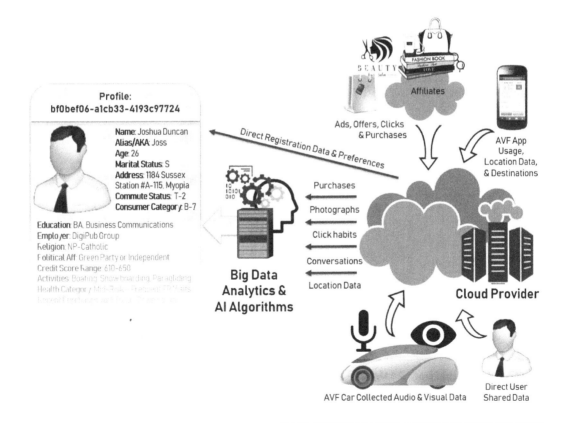

Figure 8.6   AVF User Profiles

of sources to include registration; user preferences; data the app collects from user interactions with ads and travel requests; and data the vehicle fleet collects from interior cameras, exterior cameras, and the voice-activation system. The profile also includes information that is inferred from the raw data. For example, a visit to the cardiac doctor could infer a possible heart condition or at least an interest in heart health, whereas a visit to a pet shelter would imply an interest in animals.[*]

While the AVF pilot fleet is busy collecting the mass volumes of data needed to fuel its AI algorithms, the consortium's data analytics arm has taken on a side project, consulting with Myopia's IT staff to assist in the implementation of a Public Affairs Customer Relationship Management (PA-CRM) system. With the PA-CRM, city officials can easily access and analyze data from a number of disparate municipal databases, to include the traffic and user data provided by the AVF. Dashboards and an easy-to-use custom-reporting interface enable employees, pending access-control privileges, to correlate the data from AVF user profiles with data collected from various city agencies—all with a few mouse clicks. Knowledge is power, and it is now at the fingertips of career politicians (see Figure 8.7).

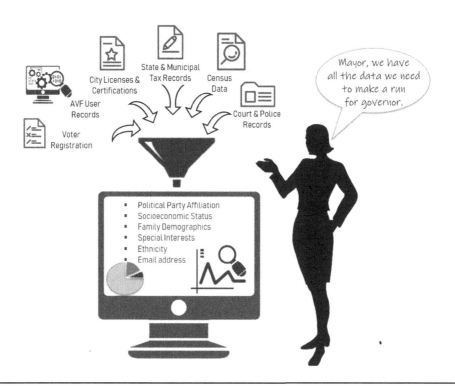

Figure 8.7 Myopia Data Analytics—A Treasure Trove for Politicians

---

[*] Other locations, such as Planned Parenthood Clinics and gun stores, can imply political affiliations and are valuable information to special interest groups—and their opponents.

2034: City Launches Commercial AVF Service

Up to this point, the fleet is small. For the commercial launch, the fleet size expands and the app is updated. Further, the AVF service now boasts a sizable number of local business partnerships. The updated app now features multiple end-to-end commute options, if available, along with estimated arrival times for each. These options included routes that combine the AVF service with other modes of public transportation and pedestrian segments. So, a passenger may elect, for example, to use the AVF for the entire commute or walk to an AVF multi-passenger hop-on/hop-off zone to save money, or perhaps take the commuter train into the city and have an AVF vehicle meet them at the station to take them the last mile to work. Each option is presented along with estimated arrival times and costs.

Another new feature of the app is the addition of a digital wallet and digital currency payment options. The consortium creates a private cyber currency off of a well-known blockchain-based cryptocurrency. They set up their own cryptocurrency exchange as well. This means that they control all mining operations for new coin, which is done to ensure that the value of the currency does not fluctuate.

> All the while the AVF service provider gets commissions on every purchase, from both the businesses and the digital currency transactions, as well as revenue from the sale of user data to third-party affiliates, and, icing on the cake, gets to charge the passenger for transportation. But wait, there's more, the profit may just end up being tax-free.

Soon after the AVF service is in full commercial mode, Myopia's downtown businesses experience an uptick in revenue. In interacting with the ads, deals, and pre-ordering options on the AVF app, commuters are able to order goods from local businesses and have them waiting for them enroute—temperature-perfect foods and beverages, groceries, dry cleaning, and toiletries all packaged just moments before needed, as if the commuter took the time to shop and wait in line themselves—all brokered and purchased through the app.

Use of geofencing triggers real-time offers on services enroute to enable local restaurants, nail salons, massage clinics, and other service providers to draw additional clients and reduce employee idle time. The lift to the local economy represents a big win for Myopia's coffers and its inhabitants.

Myopia's traffic engineers use the traffic-flow data and recommendations from the AVF service provider to modify traffic light configurations, determine the best times for road maintenance, propose plans for street widening/narrowing, new routes, and one-way road configurations, and to augment other public transportation services.

The Myopia City Council and other local agencies use the passenger data provided by the AVF service as input to urban planning initiatives, such as new housing developments, parks, and trendy retail hotspots. They also now have the email addresses of all residents using the service. This provides the city with an inexpensive and convenient way to reach its citizens. In addition, AVF passenger and traffic data is provided to the state for similar activities at the state level, while a subset of traffic data is provided to the EPA and Federal Transportation Agencies.[*]

---

[*] Perhaps a subset of data is sent to an undisclosed government agency, one that enjoyed the shadows prior to Snowden's digital shot heard round the world. I dare not speculate on such things.

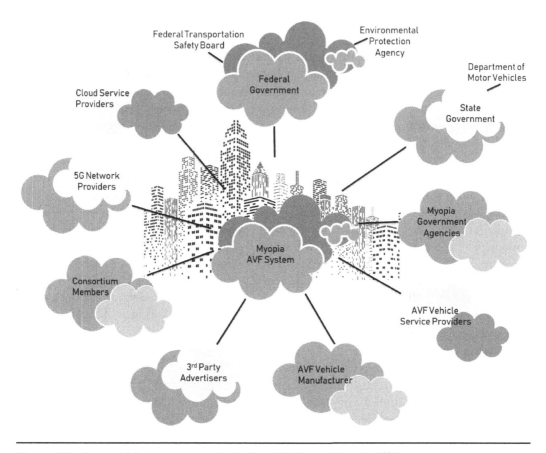

Figure 8.8　Potential Autonomous Vehicle Fleet (AVF) and Myopia Affiliates

Figure 8.8 depicts the various possible partners and agencies that the AVF service provider and Myopia share information with.

2036: City Expands AVF Service

Within a 24-month period after commercial rollout of the AVF, the city is able to rezone several downtown parking garages for much-needed residential and commercial space, reduce traffic flow congestion by 21%, increase air quality by 16%, and reduce traffic-related accidents by 33% in the downtown area.* Downtown residents and businesses view the AVF favorably. The program is so successful that the city council wants to expand the AVF. The consortium agrees, provided the following provisions are added to the SLA:

- Executive- and premium-level transportation options—pricing not regulated by the city; the city's VIP level would still be the highest priority service level.

---

* These are, of course, fictitious metrics. Real metrics will depend on a number of factors, such as fleet capacity, city layout, and population density, but I think we can agree that measurable improvement will occur in these areas.

- Reduction of downtown public parking by another 25% to further increase demand for the AVF.
- Addition of several park-and-ride parking lots 10 miles from the downtown center, to be paid for by the city.
- Control of downtown traffic lights is relinquished to the AVF service provider.
- Taxi services, already limited due to lack of demand, would be eliminated in the 10-mile downtown area.

These new provisions enable the AVF service to better target upper middle class to wealthy individuals for additional revenue. The executive and premium services use upgraded vehicles and boast faster pickup and commute times.

With yet more downtown parking garages being converted into residential spaces, and new park-and-ride lots opening on the outer reaches of the central city corridor, the demand for the AVF increases. The AVF app becomes a necessary tool for many residents and commuters in the downtown area. Its on-demand plotting of optimal commute routes and its ability to provide goods and services options significantly reduces the personal "to-do" lists of its users—creating a loyal fan base.

The AVF service provider is now able to convince the City Council to give them control over the traffic lights. They also propose a plan to evolve the AVF service into a Smart Transportation Grid, wherein all the public-transit systems and traffic-control applications will be managed through a single platform, running AI algorithms that are real-time adaptable to changes in transportation and emergency-services demand. The plan is enthusiastically accepted by the city.

2038: Something's Amiss

Roughly two years into the launch of the AVF's expanded-service offering, users of the standard service start complaining about longer commute times and fewer route options. They insist that the standard cars are taking more fortuitous routes then previously traversed, and traffic lights are red for a longer period of time on routes predominantly filled with standard AVF vehicles.. In contrast, routes predominantly filled with upgraded vehicles enjoy longer green lights—effectively making them priority routes. Further, the upgraded cars seem to monopolize the routes running through the high-end shopping district. This is all brushed off by the city council as increased demand for the service. However, during an interview with a financial news outlet, a consortium CEO discusses accelerated revenue growth potential—citing the increase of premium- and executive-level vehicles to one-third of the overall fleet size.

News articles relay such observations and question the wisdom of an artificially intelligent transportation grid. The AVF Consortium and City Council work to reassure the public, but public opinion of the AVF becomes more and more polarized.

Stories like this one appear in the local newspapers:

A single mom, wishing to remain anonymous, works for housekeeping services in a luxury hotel in the downtown banking district. She received a call from the school nurse this morning; her daughter had flu-like symptoms and needed to be picked up early. On a typical day, mom normally uses the AVF mini-bus service. Today, however, she wanted to get to her daughter quickly, so she requested a standard single-rider vehicle. Walking up to the curb next to her was one of the shift managers—using the AVF app to make a similar request. Five minutes later, the shift manager was driving away in an upgraded AVF vehicle. The single mom waited several more minutes for her ride. The school was 4.5 miles due east, but the car took her five blocks due south, then zigzagged in a north-by-east trajectory—seeming to go out of its way to avoid the route directly through the finance district—adding, according to her smartphone's maps app, an additional 10 minutes to her commute time. When she picked up her daughter, she again requested the standard single-rider vehicle option. During the ride home, her daughter complained of sharp stomach pains, so she requested a change of destination to urgent care. The vehicle pulled over and asked her to exit and make a new route request. She and her sick daughter again waited, this time for 7 minutes, before another car came to take her the two miles to the doctor's office. . . . The AVF spokeswoman assured the public that the issue was just a glitch, and it would be corrected shortly.

For every story of this kind, there are dozens if not hundreds that go unreported because only the AI algorithms are fully "aware" of the inequity of treatment. For example, compare the single mom's experience with that of the mayor's daughter, who had a "pressing" shopping engagement with her girlfriends. She waited only three minutes for a car and hit all green lights during her four-mile excursion to the shopping district. And a business executive and a bank manager working in the same high-rise building both requested a premium ride at about the same time, going to the same restaurant, for unrelated lunch appointments. The business executive received her car first because the AVF service generated more revenue from her transactions than it did from the bank manager's.

The AVF service, meant to make life better for everyone, appears to be the catalyst of a new class warfare conflict. This is a disheartening blow to the well-intended engineers involved. They immediately go to work on design updates and algorithm retraining, but the path forward to redesign and modify code will likely introduce other unforeseen side effects—not to mention a hefty price tag.

How can we prevent being the designers and architects of the next class warfare conflict? For starters, we have to promote and protect healthy competition—see the inset on open architectures in the section "2045: AVF Contract Up for Renewal." In addition, we can use the Privacy Playbook (below) and privacy mechanisms in Chapter 9 to better control the types of data available for use in training deep learning algorithms. And, data scientists and AI researchers in university environments are in the perfect settings to foster think tanks in which individuals from diverse expertise areas and backgrounds can work shoulder to shoulder with scientists and technologists to derive design specifications for smart cities. Doctors, scientists, technologists, politicians, philanthropists, historians, psychologists, and economists should all be involved. And, their recommendations should be made public so the public can weigh in.*

Pirate or Vigilante

Incidentally, behind a locked door in a windowless room littered with aluminum cans, a small number of highly caffeinated IT techs run the AVF data center and have admin access to all user accounts. Unbeknownst to other employees, this self-proclaimed vigilante group might, now and then, modify the profiles of politicians openly critical of issues they favor. Thus, the VIP service does not work as well for some politicians as it does for others.[*]

> *By employing common hacker techniques such as the installation of hidden admin accounts and the clearing of select log entries, the IT insiders can remove traces of their actions.*
>
> *What could be done to prevent this? Insider attacks are hard to prevent altogether, but the best defense is the separation of duties and role rotation combined with detailed audits. The IT administrators who have admin rights to the servers and other machines should not also have admin rights to the databases and applications—that is, keep the application and IT administrator groups separated (separation of duties). Additionally, the group that does backups and monitors the log files should be a third separate group, and all logging should be done to a separate off-site data center to reduce the amount of time a log file sits around unprotected and able to be modified. At least once a year, a detailed audit of the servers and the log files should be done by an independent group (role rotation).*

In the meantime, the city and AVF Consortium are hard at work to complete the final phase of their public–private partnership—the integration of the AVF and other public transportation systems into a fully integrated, real-time responding Smart Transportation Grid.

2040: City Implements Smart Transportation Grid

The 10-mile radius surrounding the downtown city center is now off limits to personal vehicles. The smart grid now manages all motorized forms of public transportation, including motorized scooters.[†] Commuters wishing to use any form of public transportation must purchase their reservation using the AVF app, now called MySmart.

The city has also decided to conduct all municipal-related transactions using the Consortium's blockchain cryptocurrency. In exchange, the Consortium gives Myopia a% of the revenue generated from all transaction fees. This is great news for city managers, providing much-needed money to the city coffers, which is, in turn, used for social programs. The residents of downtown begin to see more pedestrian areas and green belts sprouting up around the city, providing some invaluable good will for the Consortium and its champions.

The executive and premium services still get preferential treatment, but commute times improve for both the standard and premium services thanks to the synchronization of all public transit systems. Tensions are reduced somewhat, but the inequitable treatment is still an issue raised by concerned citizens. Some of these citizens become regular fixtures in the public spotlight, and they learn quickly to hitch rides from other people reserving transportation.[‡]

---

[*] Be honest. If you had that kind of power, wouldn't you be at least tempted to use it to inflict frustration or worse on a politician you disliked? See also the footnote in the section "2040: City Implements Smart Transportation Grid" for an explanation of how this could be implemented.

[†] Why not? They're becoming a national trend, so, of course, the AVF would want in on that action.

[‡] Don't be hasty with your skepticism. This is not hard to accomplish. For example, a database architect might add a service level called "special" to the list of available service levels (i.e., VIP, EXEC, Premium, Standard, special). The AI algorithms would have to recognize this level as well and know it gets lowest priority in the system. Thus, anyone with access to user profiles can change the level in a person's user preferences.

In 2041, the downtown Farmer's Market sponsors a "Protect Organic Farmers" rally. Would-be participants report problems with the MySmart app that morning, citing long wait times and illogical route options that add unnecessary time to the commute.* Needless to say, participation at the rally is low. The city blames the incident on a system glitch and apologizes for the inconvenience. Assurances are given to the public that the city will work with the Consortium to fix the problem. Conspiracy theorists, however, are quick to point out that a major fertilizer company, with a strong lobbying arm, is headquartered in Myopia. This catches the attention of our favorite vigilante IT techs who happen to be locally grown organics enthusiasts. They have a little fun at the expense of the CEO of the fertilizer company.

In 2042, the Teachers Union decides to strike. A protest is planned at the steps of City Hall on Saturday morning. Thousands of people are expected to attend. The protest is getting statewide coverage, and other teachers throughout the state are looking to follow suit. A large turnout will be bad publicity, however, for the influential Mayor who is considering a gubernatorial run. But, the Mayor is known for his problem-solving skills and contingency planning, and a "few" disgruntled teachers will not deny him his political aspirations. He intends to make sure this protest never happens.

Saturday morning, the MySmart app rejects all requests within a 10-block radius of City Hall. The app message states the following: "Due to unpermitted protesting in the area requested, your transportation reservation cannot be processed. We apologize for the inconvenience. Please try again later."

How can one self-serving low-level politician, who undervalues education, possibly manipulate a smart transportation system? Any ideas? Well, the easy way is to use, or in this case abuse, its emergency services sub-system. This sub-system will likely be designed based on guidelines and specifications provided by a national agency such as the US Department of Homeland Security. The emergency services sub-system will have priority override capabilities, enabling it to reroute traffic, generate priority routes for emergency vehicles only, and take control of various other sub-systems in the smart city. How should emergency sub-systems be governed? Using cryptographically signed authorization tokens that can only be generated by at least two authorized and independent state and city officials, and the tokens should be verified by the system software against hardware trust anchors that cannot be subverted or changed by an IT administrator.

Angry at the Mayor's ability to so effectively shut down their protest, several teachers sat in the open patio space of a street-facing downtown café. They consoled themselves over lattes and brainstormed a plan to get their revenge on the grid—and the mayor. They brought the plan to the Teachers' Union, and it was quickly distributed to all union members. The plan would involve driving personal vehicles into the 10-mile downtown area the next business day—using them to block AVF traffic throughout the downtown area.

In the early hours of daylight, teachers, their spouses and friends, and some all too enthusiastic teenager drivers all raced into central Myopia, determined to wage traffic jam mayhem— vaguely resembling something out of Mad Max and the Thunderdome. Chaos ensued—child's play for a kindergarten teacher, but frantic hysteria for downtown businesses trying to begin their daily operations. Riot control is called, vehicles are impounded, little children cry as

---

* I will explain one way this is possible in the next italicized inset.

their beloved teachers are seen getting arrested on the nightly news,* and nerdy kids cheer as they see their gym teacher being pounced on for resisting arrest. Almost all of the teachers are released with a warning and many pay a fine to retrieve their impounded cars. The ringleaders, however, are brought up on charges. Recordings of their conversation at the downtown café are played on the newsfeeds, courtesy of one of the smart city surveillance cameras on a light post next to the café's patio.

The city is in a state of shock and outrage: people denied access to transportation, private conversations being recorded in public venues?! This is not what the city signed up for when agreeing to the expansion of the AVF service. Litigations against the city and the Consortium follow, but deep within the bowels of the MySmart app's user agreement there is language covering the Consortium's actions. Whether the citizens of Myopia could do anything but agree to the terms of the agreement is immaterial—or is it? The courts will decide. Politicians at the federal level will weigh in, but effectual change will take not only years of litigation, but a Congress that is willing to reach across the isle.

2045: AVF Contract Up for Renewal

Myopia's citizens are demanding that the AVF smart transportation grid be dismantled, and they want additional autonomous fleet providers to have fair access to the downtown area. They also want assurances that low-income individuals will have fair access to commute routes.

Now, if the smart transportation grid is built from an open platform architecture with well-defined interfaces, then Myopia might be able to appease its citizens by enabling other competitive transportation options. But Myopia is aptly named and has dug a deep hole that will be difficult to climb out of.

> *Do you know how hard it would be to repurpose a customized AI system controlling an entire Smart Transportation Grid or Smart City? I don't think anyone really does—best avoid the situation, if possible. That's why open platform architectures with well-defined interfaces to AI algorithms, data analytics, and other system assets are beyond important. They will enable much-needed competition and invoke discipline in architecture.*

8.7 The Deepfake and IoT

The threat of deepfakes on political elections has become a concern to some of the top leaders of the United States and Europe, and an alliance has been formed to specifically take actions to mitigate it. In 2017, both Anders Fogh Rasmussen, the former prime minister of Denmark, and Michael Chertoff, former secretary of the US Department of Homeland Security (DHS), founded the Transatlantic Commission on Election Integrity under the Alliance of Democracies Foundation.† The commission recently funded the building of a tool that detects deepfakes.[16]

---

* Imagine if one or more of the AVF consortium members owned a media outlet? This should impress on us the importance of protecting and preserving independent and diverse media outlets and platforms.
† Retrieved from http://www.allianceofdemocracies.org/initiatives/the-campaign/

DARPA's* MediFor† program is also working the problem by funding universities and companies using deep learning in the detection of deepfakes.

Deepfake has become the vogue term of choice to describe the tampering of source images and video by superimposing fake images onto original images using an AI subset called deep learning and a large number of authentic source images. The end result can be a realistic fake video or image that depicts a person performing an action that they have never done in reality.

The vogue term originates from a Reddit user's anonymous pseudonym "Deepfakes," which was his or her clever play on the words—"deep" from deep learning and "fake" for the fake product produced.[17] Words are not the only thing the apparently very lonely and misguided Deepfakes played with. Deepfakes' crude ideas of art and entertainment were quickly banned from Reddit shortly after being posted. This is a PG-rated chapter, so I will not elaborate further. Unfortunately, Reddit didn't seem to remove Deepfakes' content fast enough to prevent the community at large from inadvertently (or perhaps intentionally) paying tribute to Deepfakes by naming the AI image-tampering technique after him or her.

It doesn't require great skill to do what Deepfakes did. Thanks to some slick software, it's becoming child's play,[18] and celebrities and politicians will not be the only victims. Brokenhearted teens, students locked in competition, scorned spouses, and competing businesses will all be able to produce realistic images of those whose reputations they wish to tarnish. AI-based image editing tools such as FakeApp, which uses Google's AI framework TensorFlow®,[16,18] make it attainable for just about anyone to produce deepfakes of various levels of quality. Adobe® is also using deep learning, in its Adobe® Sensei™ technology, to enhance its editing tools.‡ Of course, these tools weren't designed for malicious use but rather to empower novices to express themselves and create renderings much like a professional. However, "Misused they will be," as a famous little green guru from a far-away galaxy might say.

With the evolution of the IoT-powered smart city, video and audio surveillance units will be everywhere, providing the volumes of images and audio needed to derive convincing deepfakes. Without cryptographic integrity protections from source to storage, detection of such fakes will be time consuming and, possibly, inconclusive. Engineers designing these tools could incorporate watermarking, a type of cryptographic integrity protection, into images, videos, and audio to make detection of tampering easier, Truepic® Inc.,[19] as an example, produces an app that enables images and videos to be certified for authenticity, as well as an SDK for photo forensics (see www.truepic.com). As with all things cyber security–related, however, it is an escalating arms race, where the deepfake tools will be constantly evolving to evade detection, while the detection tools are constantly evolving to detect them.

8.8 Learning from Smart Appliances, Myopia, and Deepfakes

As communicated in the first scenario and throughout "The Tale of a Smart City," there are numerous benefits to moving forward with smart IoT systems and, equally, numerous risks.

---

* Defense Advanced Research Projects Agency.
† MediFor is short for Media Forensics.
‡ In an Adobe blogpost by Tatiana Mejia, posted on 9-28-2018, she indicated that Adobe® has a working group to "examine the impact of [Sensei] technology, evaluating and creating best practices." See https://theblog.adobe.com/making-sense-of-ai-what-adobe-sensei-means-for-you/. That's promising news, and I hope to see more on the working group's output in the near future.

Many risks can be reduced by ensuring healthy competition and generating regulations that are agile and nearly as smart as the systems being regulated.* But industry can't just lean on the government to address the challenges described in this scenario. Regulations can't detect deepfakes, outthink a deep learning algorithm, or limit an insider's ability to cause harm. But perhaps you, the engineer, can. If engineers and tech executives commit to understanding the risks and commit time and resources to addressing them, then smart IoT systems can be as beneficial to society as the cellphone was to reducing global poverty. However, if this industry is not careful, the ushering in of the fourth industrial revolution will lead to war—a class war, that is, with few avenues of escape.

The other authors of this book and I, collectively, know hundreds of engineers in this field, and we do not believe any one of them, or you, wants to be complicit in discrimination, denial of service, or class warfare. Awareness, knowledge, and planning will make all the difference in the future outcomes of smart IoT systems. The authors of this book have faith in the IoT-engineering collective—the gifted creators of the world's digital ecosystem. We are confident that the same creative genius used to design this ecosystem can find ways to protect society from the unintended consequences of it.

8.9 Privacy Playbook

Let's now turn our attention to the ways an architect can expose privacy concerns in their systems and then, once exposed, take steps to mitigate them.

The privacy playbook, in the inset, is a guide that can be used to expose privacy issues during the planning and architecture phases of the product life cycle. The goal of the playbook is to generate an IoT architecture with appropriate privacy mitigations, using minimal cost in resources while achieving time-to-market objectives.

Since the human brain tends to recall the unusual more readily than the mundane, the privacy playbook has a deliberate twist on phraseology. The steps of the privacy playbook are described in the following sections.

> Design-Phase Privacy Playbook:
> 1. Bring in the "Great White Shark"
> 2. Know the pirate line-up
> 3. Believe in the data afterlife
> 4. Defy fate
> 5. Obfuscate Waldo

8.9.1 Bring in the "Great White Shark"

Every new design or product feature needs to undergo "the great white shark evaluation."† This is where an individual(s), not affiliated with the design team, comes in and reviews the design from the eyes of a predator, "tearing" the design to shreds with the goal of exposing as many misuse and abuse scenarios as possible.

The key to the success of the great white shark review is the person(s) doing the review. Dave's privacy-expert friend always recommended that at least one of the sharks chosen for the review be a white hat hacker, preferably with both a breadth and depth of experience.

---

* That will require engineers in Congress.
† This is the phrase used by a privacy expert Dave once worked with at Intel—who likes her privacy.

Such individuals are not always easy to find because they tend to specialize in specific attack vectors and also have the tendency to stop the analysis after finding the low-hanging fruit. At one point in my career, I would have called this lazy, and perhaps this is the case for some, but I've since wizened up a bit. So I don't mean to ruffle any hacker feathers here; their behavior is more of a byproduct of the "do more with less" ideology in Corporate America. White hat hackers are as time crunched as you are. So it may take two or three such hacker experts with different focus areas in order for you to achieve the proper coverage of your product. If you don't have access to hackers within your company, try using a senior security architect not on your project, or perhaps you have access to a security researcher in your company. If all else fails, every organization has that person who lives to point out the flaws in other peoples' work and loves to argue. One of the smartest things I did in my entire career was to figure out how to use this person. I had to swallow my ego, but doing preliminary reviews with this engineer and giving him ample credit helped to bulletproof all my designs, and this arguer started arguing *for* my work, not against it. I have him, in no small part, to thank for a double promotion I received one year. So, in a nutshell, a fresh set of security savvy eyes that are not enamored with the "sheer brilliance" of your design will be able to point out problems you are blinded to.

What happens in a great white shark review? As the architect, you walk the shark(s) through your product or features, in a similar way to the security architecture review described in Chapter 5. The difference here is that privacy is the target, and user and administrator misuse are the primary attack vectors. The shark analyzes all the ways the product or business case can go awry to reveal parts of a user's privacy. Limitations to the business case or creative workarounds are needed to ensure that privacy is adequately protected and preserved.

It's going to likely hurt a bit, but would you rather have a little pain to your ego up front, which everyone forgets about in a week, or a lot of pain to your reputation two years down the road when some upstart university student publishes an exploit on your work?

8.9.2 Know the Pirate Lineup

The "Great White Shark" reviews are not effective unless you, the architect, have a comprehensive understanding of who the real adversaries are in the privacy game. Some are more conspicuous than others. Knowing the pirate lineup will assist your team in deriving the adversary personas during the privacy review phase of your design. You can use Chapter 3 and the discussion of the cybercriminal motivations to help you get started.

Again, the white hat hacker(s) can help you understand some of the more obvious pirates that would be drawn to your particular product, but it also helps to consult with people in your own human resources (HR) department and friends in law enforcement. Ask them what their pirate doppelganger might do if they had access to the types of data and algorithms you're creating. They can help expand your understanding of what a pirate truly is.

I do not include sex predators in the pirate lineup. There are more nauseating terms available to better describe such societal deviants. Give them their own category in your threat models. They have to be accounted for, and law enforcement can go a long way in helping your team understand how seemingly innocuous data can be used by these predators to target the innocent. Dave's privacy-expert friend would begin privacy presentations with a picture of a young woman posing in front of a house. She would ask her audience to tell her what they see in the picture, then she would tell them what a predator sees based on her years of work as an investigative reporter. It is quite disturbing how much information someone can obtain from a mere picture.

As I discovered from a fascinating traveling exhibit on pirate society, the pirates were a complex bunch, often ruthless to outsiders but fair and equitable within their own ranks. In reality, there is a little pirate in all of us. Remember the IoT Attacker Profiles in Table 2.1, which includes insiders? The insiders include you and me. If given access to an ecosystem that enables us to deal a blow to our rivals, even if we have to bend the rules a little, then we just might succumb to the temptation. Conclusion: One person should not have that kind of power. So why not build that into your design? Why not require multiple individuals to make changes to algorithms and system configurations? This would have prevented the Mayor of Myopia from inhibiting protests.

Arrr Matey, hackers are not the only pirates you need be concerned with. The pirates in an IoT world are potentially anyone and everyone who has access to user data and algorithms—even your rivals, and even you. Knowing this will go a long way in helping you to see the potential for abuse of the technologies you are creating.

8.9.3 Believe in the Data Afterlife

We may not share the same beliefs in the afterlife, but we can all agree that digital data lives far longer than most of us ever imagined. Data, long since deleted from our personal cloud-based accounts, has an active afterlife that may outlast our physical existence.

During my teenage years, I was a smart-mouthed argumentative know-it-all who my mother swore would make a fine trial attorney someday. I am sure many of you can relate, as these adolescent traits appear to be particularly frequent in those who end up choosing engineering as a profession. I wonder how embarrassed I would be if some of what I said back then was floating around in cyberspace somewhere—just waiting to haunt me. I have changed and evolved, and my brain has fully myelinated* since then. I would like to think I've acquired a bit more wisdom along the way, so I'm grateful that all my English assignments were turned in on paper and social media wasn't popular when I was young. Many of you have kids, or will soon enough. You don't want them missing employment opportunities because of something they put on social media in their college days. The IoT ecosystem is capturing data from cameras and microphones linked to many different devices, so the potential for capturing a careless word or action is high. None of us want to be immortalized by our worst day, and we don't want our kids to suffer that fate either.

Protection is afforded by adding data longevity options to any cloud-based service you architect. Set up software-enforced policies that prohibit the sale and redistribution of certain content types as well as any content generated by individuals whose brains haven't fully

---

* There are two types of neurons in the human body—myelinated and unmyelinated. Myelinated neurons have faster conduction speeds. Myelination occurs in fully developed neurons. The brains of adults and teens are different, and teens process using the amygdala—the emotional part of the brain—whereas adults process using the prefrontal cortex—the rational part of the brain. The teen brain switches when the frontal cortex is fully myelinated, around age 25. All this means is that the unmyelinated teen brain can't make good decisions—it's just anatomy (http://www.urmc.rochester.edu/encyclopedia/content.aspx?ContentTypeID=1&ContentID=3051)

myelinated. I know teens are a lucrative consumer base. I'm not asking anyone to give up targeted marketing to teenagers—monitor their clicks all you want. I am suggesting that we respect them and their future enough to let them, or better yet, their parents, control where their content goes, who they share it with, and how long it stays there. These are basic tenets being written into many laws, like GDPR, but they haven't yet made their way into every country, including the United States. And yes, I know we have COPPA.* Have you read it? It's a great example of why we need more engineers in Congress—the ones who have actually built software, run the clouds, and designed digital security features.

I brought up children in particular because we really should be advocates for change if our company's business model cannot support respecting child privacy. But, as an adult, I often wonder why the apps I consider using don't offer an option to store my content locally. Even the note-taking app I recently downloaded wants to store all my notes in the cloud. Is that really necessary? I have plenty of available memory on my phone.

If we recall the Safe Driver App scenario above, the app could be redesigned to limit data longevity and provide a layer of ambiguity without sacrificing the objectives of the app. Traffic-violation incidents, of course, need to be collected and saved for long durations, but non-relevant GPS date/time stamps and other telemetry data need only be collected and stored for a limited amount of time and should not be sold or shared with third-party affiliates. The collection of high-level metrics rather than specific information items, where possible, can also reduce the ability of third-party entities to correlate disparate data sources and derive specific user identities. For example, in the safe driver's app, for long-term analysis purposes, average commute miles per day could be collected rather than specific routes. And, the app's dongle could tally the number of high-risk intersections the driver traversed per month rather than the specific intersections themselves. This way, the insurance company gets the information they need to assess risks without exposing the driver's personal habits.

When building that next app or service, providing the user with local and cloud-based data-storage options, as well as longevity and backup options, will leave the data afterlife in the hands of the life it impacts.

8.9.4 Defy Fate

In this fourth industrial revolution, of which IoT is a central part, with the combined strengths of predictive analytics, micro-targeting, IoT wearables, smart cities, and interconnected every-thing, it is more than possible for a few elite individuals to control large population groups, predetermining their fate. If we want to leave the question of fate in the hands of philosophers and religion professors and out of the hands of politicians, corporations, and engineers, then we have to defy fate by empowering individual choice and providing legitimate user options.

As touched on in "Remember the Data Afterlife," architects should provide options for how long data is persisted and where it is stored (cloud vs. local drives), who has access to it, and if it gets encrypted. Let users determine whether frequent destinations are stored, loca-tions tracked, pictures labeled, or persons identified in images, and what metadata is stored for images. Enable users to take control of the personal information that is sent to predictive

---

* COPPA is the Child On-line Privacy Protection Act. It is unimpressive when you look at what is omit-ted rather than what is there, and it only covers kids up to 13 years old.

algorithms. Data encryption should include features that give the user control of who has the decryption key, preventing the re-forwarding of data so that data only goes where the user intends for it to go.

Consumers have been asleep at the wheel when it comes to protecting their digital footprint, but that is changing thanks to the steady stream of headlines, lawsuits, and congressional inquiries that all center around the misuse of user data. Just type "data misuse" into your favorite search engine and find a plethora of examples. As a result of this privacy awakening, last year the European Union slammed its Italian leather boot into the backsides of data-hungry juggernauts when it approved the GDPR[*] (see the GDPR discussion in Chapter 9).

So, with all the new enlightenment of consumers and governments at large, it might behoove us to give consumers a choice with regard to how their personal data is utilized. If you give your users choices over the storage and dissemination of their personal data, they will thank you for it and give you their loyalty. And, you'll be doing your part to defy fate.

8.9.5 Obfuscate Waldo

"Where's Waldo"[†] was an obsession for many a child, parent, and seek-and-find enthusiast out there in the early 1990s. I was no exception. Finding that distinguishing red- and white-striped shirt in a sea of cartoon people was entertainment that left one satisfied and victorious when Waldo was finally spotted. Ah, the good old days!

Well, now it's time for a role reversal so to speak: Instead of being the ones finding Waldo, we need to hide him, and hide him such that he is only found when he wants to be. Speaking plainly, architects need to look for ways to hide users' identities from data analytics engines wherever possible. Data can appear to be anonymized and it should be, but it is all too easy to de-anonymize data by cross-referencing with an external dataset. For example, public health studies and measuring teenage shopping trends do not require an associated user ID. Sure, you want to manage your customers, but you can still restrict how the customer ID is associated with the information being collected (see the anonymization discussion in Chapter 9).

Perhaps thinking about our parents getting swindled or our teen getting cyberbullied or our spouse being denied health coverage because an analytics tool determined their future cancer risk is high, might just motivate us to protect user identity when possible—which is more often than not, probable.

8.9.6 Playbook Wrap-up

The privacy mechanisms and methodologies discussed in Chapter 9 are valuable and necessary in reducing the societal risks discussed in this chapter, but they cannot possibly account for all the possible IoT ecosystem variants out there, nor provide all needed privacy mitigations. Therefore, the playbook is designed to catalyze a much-needed dialog within your architecture

---

[*] GDPR is the General Data Protection Regulation adopted by Europe and discussed in the next chapter.

[†] "Where's Waldo" is a popular seek-and-find picture book for kids of all ages, and was first published in the UK by Martin Handford under the name "Where's Wally"; https://www.latimes.com/archives/la-xpm-1997-nov-26-ls-57728-story.html

team, and throughout your company, to brainstorm privacy solutions that are tailored to your products and represent a minimal impact to costs and schedule in the product life cycle.

References

1. Hern, A. (2018, January). "Fitness Tracking App Strava Gives Away Location of Secret US Army Bases." Retrieved from https://www.theguardian.com/world/2018/jan/28/fitness-tracking-app-gives-away-location-of-secret-us-army-bases

2. Jensen, R. (2007, August). "The Digital Provide: Information (Technology), Market Performance and Welfare in the South Indian Fisheries Sector." Retrieved from https://www.economist.com/finance-and-economics/2007/05/10/to-do-with-the-price-of-fish

3. Carter, J. (2015, October). "How Cell Phones Are Helping to End Extreme Poverty." Retrieved from https://blog.acton.org/archives/82494-how-cell-phones-are-helping-to-end-extreme-poverty.html

4. Wikipedia. "Information Broker or Data Broker." Retrieved from https://en.wikipedia.org/wiki/Information_broker

5. Leetaru, K. (2018, April). "The Data Brokers So Powerful Even Facebook Bought Their Data—But They Got Me Wildly Wrong." Retrieved from https://www.forbes.com/sites/kalevleetaru/2018/04/05/the-data-brokers-so-powerful-even-facebook-bought-their-data-but-they-got-me-wildly-wrong/#111b29163107

6. Naylor, B. (2016, July). "Firms Are Buying, Sharing Your Online Info. What Can You Do About It?" Retrieved from https://www.npr.org/sections/alltechconsidered/2016/07/11/485571291/firms-are-buying-sharing-your-online-info-what-can-you-do-about-it

7. NG Data. (2016). "How to Implement More Effective Micro-Targeting Programs. Leveraging All Your Data to Improve Marketing Precision and Performance." Retrieved from https://www.ngdata.com/wp-content/uploads/2016/08/NGD-eBook-micro-targeting-2016.pdf

8. Centers for Medical & Medicaid Services. (2018). "NHE Fact Sheet. Historical NHE, 2017." Retrieved from https://www.cms.gov/Research-Statistics-Data-and-Systems/Statistics-Trends-and-Reports/NationalHealthExpendData/NHE-Fact-Sheet.html

9. Bresnick, J. (2017, December). "How the CVS, Aetna Deal Will Overhaul Healthcare Big Data Analytics." Retrieved from https://healthitanalytics.com/news/how-the-cvs-aetna-deal-will-overhaul-healthcare-big-data-analytics

10. Monica, K. (2018, January). "Google Study Uses Entire Patient EHR for Predictive Analytics." Retrieved from https://ehrintelligence.com/news/google-study-uses-entire-patient-ehr-for-predictive-analytics

11. Harvard T.H. Chan School of Public Health (2009, Winter). "Employer Health Incentives." Retrieved from https://www.hsph.harvard.edu/news/magazine/winter09healthincentives/

12. Ferris, R. (2019, February). "It Seems Impossible That Tesla Will Have Self-Driving Car Tech Ready in a Year, Says Analyst." Retrieved from https://www.cnbc.com/2019/02/20/industry-analyst-doesnt-buy-musk-claims-about-tesla-self-driving-tech.html

13. Chan, C. (2017, September). "Advancements, Prospects, and Impacts of Automated Driving Systems." *International Journal of Transportation Science and Technology,* vol. 6 no. 3, pp. 208–216. Retrieved from https://www.sciencedirect.com/science/article/pii/S2046043017300035

14. Cohen, T., Rabinovitch, A., and Lienert, P. (2017, March). "Intel's $15 Billion Purchase of Mobileye Shakes Up Driverless Car Sector." Retrieved from https://www.reuters.com/article/us-intel-mobileye-idUSKBN16K0ZP

15. Walker, J. (2018, December). "The Self-Driving Car Timeline—Predictions from the Top 11 Global Automakers." Retrieved from https://emerj.com/ai-adoption-timelines/self-driving-car-timeline-themselves-top-11-automakers/

16. Browne, R. (2018, December). "Anti-election meddling group makes A.I.-powered Trump impersonator to warn about 'deepfakes'." Retrieved from https://www.cnbc.com/2018/12/07/deepfake-ai-trump-impersonator-highlights-election-fake-news-threat.html

17. Retrieved from https://en.wikipedia.org/wiki/Deepfake

18. Schwartz, O. (2018, November). "You Thought Fake News Was Bad? Deepfakes Are Where Truth Goes To Die." Retrieved from https://www.theguardian.com/technology/2018/nov/12/deep-fakes-fake-news-truth

19. Constine, J. (2018, June). "Truepic raises $8M to Expose Deepfakes, Verify Photos for Reddit." Retrieved from https://techcrunch.com/2018/06/20/detect-deepfake/

Chapter 9

Privacy Controls in an Age of Ultra-Connectedness

We ought to regard the present state of the universe as the effect of its antecedent state and as the cause of the state that is to follow. An intelligence knowing all the forces acting in nature at a given instant, as well as the momentary positions of all things in the universe, would be able to comprehend in one single formula the motions of the largest bodies as well as the lightest atoms in the world, provided that its intellect were sufficiently powerful to subject all data to analysis; to it nothing would be uncertain, the future as well as the past would be present to its eyes.

— Pierre-Simon de Laplace, *French Mathematician (1749–1827)*[*]

9.1 Introduction

It was unfathomable in the 16th and 17th centuries to be able to predict the location or movements of the planets. Even well after Galileo, celestial bodies appeared to move through the sky under the influence of some unknown force. It wasn't until Pierre-Simon Laplace linked Sir Isaac Newton's theory of gravity to the mutual influence those bodies had on one another that the movements of the planets were mapped out like clockwork. Without good privacy controls on IoT data, our movements, intentions, and desires will be as easily mapped out by artificial intelligence engines as astronomers predict the movements of the planets.

After having looked at the evolving privacy challenges posed by IoT, one may wonder if privacy is a thing of bygone centuries. Artificial intelligence (AI) and big data are being used to map out our actions, preferences, and behaviors, much like we do the stars. Is there anything

---

[*] Laplace used Newton's theory of gravity to mathematically show the correlation between the recorded observations of planetary movements and their theoretically ideal orbits. He also demonstrated the use of probability to explain and clarify scientific data: https://www.britannica.com/biography/Pierre-Simon-marquis-de-Laplace. Quote retrieved from https://www.goodreads.com/quotes

that can be done to protect one's privacy concerns in an ultra-connected world? This chapter takes a more technical look at privacy, providing some definitions and categorizations of privacy and data, and then explores some workable strategies for designing and building privacy into IoT systems, offering a balanced perspective of definitions, policies, legal protections, and technical controls. We start with a discussion of the definition of privacy so that we can create a foundation upon which to have a technical discussion regarding what controls actually protect the type of privacy we have defined. We then talk briefly about the different groups that are concerned with privacy to highlight the fact that different perspectives have slightly different views of what privacy actually is. In Section 9.5, we get to the technical specifics and talk about the different types of privacy controls and cryptographic mechanisms available, providing you with a basic understanding of these controls so that you are armed with the types of protections you can put into your systems. And, finally, Section 9.6 covers some of the privacy laws with which all engineers should be familiar.

9.2 Defining Privacy and Information Privacy

Privacy is the "the interest that individuals have in sustaining a *personal space* free from interference by other people and organizations."[1] It is the right to be left alone at times of your own choosing—like in the shower. But this definition can cover both the physical and digital worlds. As applied to data or the digital world, **information privacy** is "the interest an individual has in controlling, or at least significantly influencing, the handling of data about themselves."[2] The collection or discovery of this data that an individual considers private covers data collected from both the physical and digital realms. A definition of information privacy that has been used in US legal cases is:

> *[informational privacy] encompasses any information that is identifiable to an individual. This includes both assigned information, such as a name, address, or social security number, and generated information, such as financial or credit card records, medical records, and phone logs. Personal information will be defined as any information, no matter how trivial, that can be traced or linked to an identifiable individual.*[2]

It is interesting to note the increasing interrelation between personal/physical privacy and information privacy. As an example, location information, which is really just data, when linked to some other piece of trivial data, such as 10 digits (i.e., a phone number), can be considered an integral part of physical privacy, since knowing where someone is can aid or allow personal intrusion as well as personal harm.

In many countries, privacy is enforced by law. But the laws and the specifics of their provisions very often differ between countries, and even between regions or states.* So, there isn't really a single definition. The exact types of data and knowledge about a person that is covered by privacy laws, and what exactly constitutes personal information, will vary by country and legal domain. In Europe, there is a strong push to unify such definitions. We discuss a little more about various privacy laws in Section 9.7.

---

* An interesting website is https://www.privacypolicies.com/blog/privacy-law-by-country/, which describes some privacy regulations for many different countries. However, some information is out of date.

So, privacy is freedom of interference in our personal space, and information privacy is the restriction or protection of data that identifies an individual.

9.3 A Better Definition of Personal Information and How That Becomes *Personal Knowledge*

The definitions of physical and information privacy above are only vaguely helpful. We need to better understand how data becomes personal information, and how personal information grows to become something more dangerous, which we refer to as *personal knowledge.*

Personal information can be described as any data that resolves uncertainty or provides the answer to a question of some kind with respect to a particular individual. Data is represented as value attributes or parameters that can be associated with a person, such as <weight, 180> or <hair, brown>. Personally Identifiable Information (PII) is a specific type of data that uniquely identifies a particular individual, or can significantly aid in uniquely identifying an individual. This is because a name, for example, John Smith, does not uniquely identify a person within the United State; other countries, likewise, have common names. Regardless, a name does significantly contribute to identifying an individual. But combine that name with additional information—<"Lives on", "Druery Lane">, <"Occupation", "Baker">—and all of a sudden this data collection becomes PII. Single data elements can be PII as well. PII includes things such as a social security number, a telephone number, a complete home address, or even a network IP address. Personal information, such as someone's appearance or weight, is one thing. But PII tends to narrow the focus down to a single person or a few people, for instance, a household.

Combining multiple data sets of personal information and PII together represents a different class of information that we call *personal knowledge* or, as in this chapter, just knowledge. Knowledge signifies an understanding of complex subjects or abstract concepts.

Perhaps attributes such as height, weight, and hair color do not seem like such a big privacy concern. But when that data set is collated with an hourly GPS position reading for the last month, and perhaps a third data set containing gender, eye color, ethnicity, and build, then privacy concerns begin to loom large. This becomes enough information to pick someone out of a crowd and understand where they live and work as well as the places they frequent. Knowledge is the accumulation of multiple data sets from different IoT and other systems into an information base that raises real concerns about how this information is going to be used and how to prevent it from being abused.

9.3.1 Data from a Fitness App Turns into Military Intelligence

As of this writing, San Francisco-based Strava provides an app that uses a mobile phone's GPS to track a subscriber's exercise activity. This is very useful to athletes and those desiring to be very physically fit for their jobs, for instance, police, fire fighters, and military personnel.

The company's review of 2017 showed a heatmap of all routes taken by its users across the world. It was released in November 2017. In January 2019, Australian student Nathan Ruser noticed that trails from Strava users in certain countries made it possible to identify clusters of military personnel from the United States and other nations.

Although the location of many military bases is generally well known, and satellite imagery can show the outline of buildings, the heatmap can reveal which of them are most used, or the routes taken by soldiers. In some cases, it identified groupings of soldiers, or perhaps intelligence operatives, that were associated with previously unknown installations. It was postulated that these could be intelligence operatives or so called "black sites." The location data included tracking of individuals outside of these base locations, which many supposed were commonly used exercise routes or patrolled roads.[3]

9.4 Who Cares about Privacy?

It is worthwhile to consider the different groups that care about privacy and the basis for their concerns.

1. **Corporations** and businesses care about data privacy for three primary reasons:
   a. They wish to acquire data to make their businesses more viable, to increase revenue, or to open up new market opportunities.
   b. They wish to protect data that gives them a necessary advantage in achieving their mission—for instance, protect intellectual property such as a fast, efficient algorithm for searching text or protect operational data that reflects the speed or efficiency of business operations.
   c. They wish to prevent harm to their customers, employees, or shareholders that could occur through unauthorized access or misuse of data the corporation holds or processes concerning those groups.
2. **Individuals** want to prevent access, theft, and usage of their personal data that could cause them criminal, financial, and psychological harm.
3. **Governments/nations** care about privacy in a similar way to corporations, but with potentially more severe consequences if their data is accessed by unwanted entities. For instance, it is highly undesirable to allow a foreign entity access to a country's top-secret military deployments or capabilities. Often, nations aren't just the defenders; nations also play the role of antagonists by attempting to access data belonging to other nations or private companies.
4. **Lawyers** and legislators who help to define what constitutes legal or illegal usage or access of data. They are also the ones charged with battling it out to resolve data privacy–related disputes.
5. **Cybercriminals** (and other threat agents, as described in Chapter 3) are the entities seeking to obtain data from or about parties in order to cause those parties harm. Their goals are usually profit, attention (sometimes as part of activism), or disruption.
6. **Privacy Advocacy Groups** inform lawmakers, governments, businesses, and individuals about the impact of privacy on our society and attempt to provide checks and balances to prevent the erosion of privacy.[*]

Why would you even care about this list? It is important to note that the definition of privacy is different for all of these groups. Even though they are using the same words, their

---

[*] One such organization is CDT (http://cdt.org), which is "a nonpartisan, nonprofit 501(c)(3) charitable organization dedicated to advancing the rights of the individual in the digital world."

intentions and goals are very different. This necessarily leads to miscommunication and conflict. As security engineers, we need to understand these differences and attempt to build bridges between these groups. After all, it is our products and our mechanisms that are often at the crux of these discussions.

9.5 Privacy Controls

As designers, architects, and advocates of the Internet of Things (IoT), we need to carefully think through the privacy impacts on the systems we design and offer into the marketplace. We must design controls that can protect our users and customers, as well as the companies that build these systems and the innocent citizen who happens to stumble past our devices but who elicits no measurable benefit from the IoT system. But we also have to remember that the data being collected must be used to drive business and operational processes. Therefore, we have to allow data to be collected with enough context and accuracy to allow such processes to operate correctly. In this section, we look at the types of controls that can be used to protect privacy, and the effectiveness of such controls.

9.5.1 Access Controls

Of course, the very first thing that should be considered when protecting privacy is access control. What do we mean by access control? The following mechanisms can be used to control access to private data:

- **Restrictive operating system permissions on files**—Files that contain sensitive data should always have restrictive permissions set for each of them. This means that "world" or "everyone" readable files should never be allowed. Permissions should be set to a single account or to a group with a minimum number of members.
- **Mandatory logins for systems and applications**—All accounts on the system should require a login with a password or other authentication verification mechanism (e.g., single sign-on, biometrics, etc.). Guest accounts or accounts without a password set should be disable or removed from the system. Ensure that the passwords are changed on a regular basis, even if only once per year.
- **Use of complex passwords for all accounts**—All passwords should use mixed case, digits, and special characters and be a minimum of 8 characters. A better length is 12 to 13 characters, but this normally requires human beings to use passphrases instead of passwords. A user-awareness campaign is important to make all users cognizant of the importance of strong passwords. This must be followed up by auditing, where the information security team runs password-cracking tools linked with password dictionaries to uncover accounts with weak passwords. Ensure that this type of scan includes all accounts on servers, including system services. As a designer, don't leave all this up to the customer, but instead build in the security management features that enable them to require password changes, restrict passwords to only strong forms, and perform the security audits.
- **Two-factor authentication for privileged accounts**—Consider using a two-factor authentication system for accounts that have access to applications or data files with

significant amounts of private data—database administrators, accounts for AI algorithms, etc. This includes using biometrics, secondary device verification such as tokens or even cellphones. Some will argue against using cellphones as an authentication mechanism, but some secondary verification is better than a single factor such as a password.

- **Audit accounts and delete old or unused logins**—Some systems tend to accumulate user accounts. Regardless, ensure that an annual review of accounts and permissions is performed, at the minimum. This should verify that anyone who has left the company or who has changed roles has had their permissions appropriately revoked or updated. Sounds like an operational control? What about a system service that automatically disables an account when it hasn't been used in four months? How about a system report that is automatically mailed to a system admin, including accounts that haven't been used in a while and accounts that have been dormant and suddenly used for the first time. Can you see where this kind of thinking improves the product and how these kinds of system features helps system administrators perform their job function?

9.5.2 Anonymization

Data anonymization is the process that removes PII from a data set. This can be done in the following ways[*]:

- **Deletion**—Remove the data from the data set. While effective, this can reduce the value of the data set. This approach is usually reasonable when sharing a copy of a data set with an affiliate.
- **Masking**—Replace the characters of fields containing PII data with a fill character, such as an asterisk or a space. This is also called redaction. Masking allows the recognition that a piece of data exists, and even perhaps how long that data is, but does not reveal the data itself. Be careful when revealing the length of a particular piece of PII, such as a password. More on this below.
- **Generalization**—Replace the fields containing PII data with a general notion of the content that does not reveal personal information. An example is replacing GPS coordinates with just the city name, or replace the specific name of a person with their nationality.
- **Tokenization**—Replace the fields containing the PII data with a substitute token value. This can be effective when sharing or allowing access to data by affiliates. Recurrence of the PII data can be identified, but the specifics are not revealed. It is possible to have a master record that maps the token to the real PII data, giving those with authorization access to the complete data set. This is particularly effective when you want to delete certain PII data but don't want to search through all databases and records and log files and backup tapes for that data—think conformance to the GDPR, as discussed in Section 9.6.1. If data is tokenized when collected, it is easy to delete the single record mapping the token to the PII data, and all the PII information is effectively removed.

The approaches above are examples of creating datasets with the **k-anonymity** property, meaning that the data cannot be distinguished among (*k-1*) other distinct records. The value

---

[*] These techniques were discussed in a presentation at IETF 104 in Prague by the Privacy Research Group.[4]

of k is used as a rough measure of the level of privacy afforded to the data set. Although anonymization works well within your own data set, it can be reversed. Using a different, potentially large dataset that has k similar but distinct records as compared with a majority of the records in the anonymous data set, it is possible to find corollaries between the anonymized data set and the second data set. This is often how data brokers increase the accuracy and breadth of their data sets over time. As an example of how this can work, consider the following actual incident. It was reported[5] that when New York City released anonymized data on taxi rides within the city, pictures from celebrity gossip sites were used to identify specific taxi rides (correlated by the pictures which identified pick-up and drop-off locations) of certain celebrities and then cross-referenced the information from those pictures to the data set released by the City. It was possible to identify specific rides by some celebrities, which was believed to reveal their residence and work locations. So just remember, anonymization is not a complete answer to privacy.

9.5.3 Differential Privacy

Differential privacy (DP) is a relatively new technique used to obfuscate a data set by introducing a certain amount of error into the data set while still allowing the entity querying the data set to make relevant statistical inferences. DP mandates that no single entity's private data has much effect on what the querier sees from queries on the data set. The formal definition of differential privacy is[6]:

For $\varepsilon \geq 0$, we say that a randomized mechanism $M: X^n \times Q \rightarrow Y$ is ε-differentially private if, for every pair of neighboring datasets $x \sim x' \in X^n$ and every query $q \in Q$, we have
$$\forall\, T \subseteq Y,\ Pr[\,M(x,q) \in T\,] \leq e^{\varepsilon} \times Pr[\,M(x',q) \in T\,]$$

What this definition says is that if we have two data sets that are different only by a single element—if we think about databases this means that the databases differ only by a single record—then the answers to the queries are the same, or close enough to the same (same subset T) so as not to make a measurable difference. Effectively, this means that an observer cannot tell the presence or absence of a single record, thereby preserving the privacy of the entity represented by that single record.

How does one implement differential privacy? It turns out to appear quite simple for the basic case of counting queries—a query that returns a count of records that satisfy a particular query—by introducing noise (technically a LaPlace mechanism), or an epsilon-error value, into the response. However, as the same query is repeated multiple times or multiple slightly different but related queries are posed, the noise can be effectively factored away; multiple colluding parties could work together to evaluate the epsilon-difference and remove it. Thus, the equations used to calculate the epsilon value added to a query must take into account the global effect of privacy over all queries. This makes differential privacy a useful tool, but not a silver bullet for all privacy issues.

One way to manage the noise level to preserve privacy is to maintain a privacy-sensitivity value for different subsets of the data and evaluate the impact of any new query on the privacy of the exposed subsets. If the DP falls below a certain acceptable level, the query is simply not answered.

DP is a useful tool, but one that requires a very carefully thought out mathematical basis. It is not a silver bullet for privacy concerns but should be considered in situations where the queries can be controlled.

9.5.4 Homomorphic Encryption

Homomorphic encryption (HE) is a special class of encryption algorithms that allow some types of computation to be performed on encrypted data without ever having to decrypt the data. Since data is always encrypted, even during use, privacy is preserved. The algorithmic details of homomorphic encryption are beyond our scope. However, details can be found in some cryptography texts.[6]

There are some interesting limitations to homomorphic encryption schemes. The first is that all data must be encrypted with the same key. This prevents operations between different parties, where each party want to maintain the privacy of their own data but join together in some computation. Another limitation is the operations that can be executed on HE-encrypted data; some HE algorithms are limited to only additions whereas others are limited to multiplication or a limited combination of the two operations. Further, as multiple computations are performed on the encrypted data, noise within the computed result grows. Eventually, the noise grows to the point where decryption would fail. This means that the amount of computation that can be performed on data is limited unless the data is decrypted and then re-encrypted after the error is removed. It can be extremely inconvenient to pass data back and forth between the data owner and function execution engine. Finally, HE algorithms are rather expensive operations and therefore do not perform well in practice, especially on large datasets.

Homomorphic encryption is an active area of research and will likely improve as new techniques are explored to address these problems.

9.5.5 Secure Multi-Party Computation

With the limitations of anonymization—DP and HE algorithms—we are forced to pause and consider what we want to accomplish in privacy. A useful perspective can be defined under multi-party computation. In multi-party computation, we want to create a system in which multiple different parties with ownership or authorized access to private data can release such data to other parties for the purpose of computing a function on that private data. In this construction, the parties owning the data get no knowledge of the function, and may not even get the final computed result, while the parties owning the functions get no knowledge of the private data *other than* the computed result. There are variations of this system. For example, the data owners could also be the function owners, but they own only a portion of the input data needed to compute the final result. In this situation, they cooperate with other data owners to compute some function, potentially passing partial results among all data owners until the final computation is complete. Passing partial results should not expose the private data of any owner, and, in some cases, the release of the final result must be guaranteed to all data owners so that no owner is cheated by allowing others to use their data but not gaining the benefit of having the computed result.

Secure multi-party protocols exist for performing such operations, but other solutions exist that leverage trusted execution environments (TEE). The TEE solution is interesting since it does not depend on specialized protocols. The TEE is a special computation container, usually built with special hardware, which prevents software outside the TEE (e.g., the operating system) from seeing the memory or executing software inside the TEE. The TEE can produce signed attestations that guarantee what code (i.e., functions) are executing inside the TEE.

Data enters the TEE encrypted and the TEE can decrypt it because it also holds the decryption key. The function to operate on the data runs inside the TEE completely protected from anything outside. Output results can be sent to authorized parties in encrypted form to ensure that data results are not leaked to unauthorized parties.

New initiatives around confidential cloud computing are beginning to appear which utilize this TEE method of secure multi-party computation. Such initiatives will allow companies to *rent out* their AI algorithms with confidence that the algorithms and ML models cannot be leaked, and customers can be assured that their data is private and won't be retained by the cloud service provider (CSP) or the algorithm owner.

9.5.6 Zero-Knowledge and Group Signatures

In the secure multi-party computation case, what happens to the identities of the parties contributing data? Oftentimes, with standard security protocols such as TLS, the identities of the communicating parties become known. This may not be desirable since that can leak sensitive information, but at the same time we need to authenticate the parties that we allow into our shared computation. Zero-knowledge protocols and group signatures are solutions to this problem.

Zero-knowledge is a combination of a protocol (an exchange between two parties) and a hard mathematical problem that allows one party (the claimant) to prove to another party (the verifier) that the claimant knows a particular secret answer to the hard mathematical problem without revealing the actual secret to the verifier. There are several well-known techniques for zero-knowledge, including the Fiat-Shamir, the Feige-Fiat-Shamir, Guillou-Quisquater, and the Chaum-Pederson zero-knowledge algorithms.[7,8] The hard mathematical problems include solving a square, square root, or inverse modulo a large composite of two primes, or solving a decisional Diffie–Hellman (DDH) problem.

These algorithms are useful for anonymous identification schemes using a trusted third party that issues zero-knowledge parameters to claimants. Each claimant can prove that they have been issued a secret parameter by the trusted third party without revealing that secret. A verifier can verify that the claimant is part of the trusted group but has no information to determine which individual within the group the claimant actually is. Multiple interactions with the same claimant result in different proofs due to randomized values in the protocol, meaning that verifiers cannot even correlate multiple interactions with the same claimant. This is in contrast to RSA, DSA, and Diffie–Hellman certificates, which leak the same public key value for each interaction, allowing the verifier to identify that they are talking to the same party, even if they do not know the party's actual identity.

Group signatures are a particular instance of these zero-knowledge proof systems. They allow the claimants to use parameters issued to them to sign messages, in the same way that RSA and DSA allows one to sign messages. The difference is that most zero-knowledge systems require multiple interactions between the claimant and verifier to *prove* knowledge, whereas the group signature schemes allow a single signature to prove a claimant is part of the group.

Clearly there are some drawbacks to this system. The group size must be large enough so that practical identity information does not leak. Additionally, as we stated previously, there are other identity characteristics that still leak even when using this type of mechanism, such as the IP address of the claimant, browser, or operating system characteristics due to the way the computer or server constructs network packets, the time of day of the interactions, and many other subtle clues. These techniques must all be considered as tools, not as silver bullets that completely solve the privacy problem.

9.5.7 Data Retention and Deletion Policy

As we have seen with all these techniques, although they provide a solution to part of the problem, they cannot solve the entire privacy conundrum. As designers and engineers of these systems, we have to be cognizant of the information we collect, or we allow our systems to collect, and manage that information carefully. This means that we need to agree to and implement policies in our systems that define what data is collected, what data is retained, and for how long.

The first thing to think about is what data your system collects, both intentionally and unintentionally. When your system logs the successful authorization of a particular party, that log entry includes a date–time stamp. Over time, these log entries can reveal the geographical region that party resides in—Europe, East Coast of the United States, or India. Over time, correlated with other events, particular cities can also be identified. So, something as harmless as a normal log entry that almost all systems perform and consider trivial, reveals something.

The solution is to condense these log entries over a short period of time, and then eventually delete the logs. Can the log entries be consolidated to identify the number of times an event occurred in a particular day or week instead of the actual times? After three months, how valuable is that log entry data? It is probably only useful if investigating a system breach, but if that information is based on an anonymous ZK or group signature algorithm, the identity information isn't there anyway so the information is useless to that investigation. We point this out because as we add privacy control and requirements to our system, we need to use the "Great White Shark" review we discussed in the last chapter to determine if what we have always done is still relevant and useful.

The takeaway here is to review *all data* elements against privacy, not just the PII elements. Evaluate them against what value they bring to the valid uses for the system, and the dangers to privacy. Then construct the system to implement proper retention and deletion actions based on the policy. Do not leave this for the system administrator to go in and manually perform a file delete on all the log files. Would you want to sit at a terminal for three hours clicking the delete key over a hundred thousand files, or be the one whose deletion script file accidentally removed some critical data of the system and crashed the company's moneymaking application? Step up your game and build proper data management capabilities into your system. You can easily justify the time spent as an effort to delight the customer. Your customers will agree!

9.6 Privacy Legislation

Let's switch gears for a moment and discuss some of the privacy legislation that we as engineers should be familiar with. This section, and the whole of this book, does not provide any legal advice. Let's make this perfectly clear at the outset. Furthermore, this section is not intended to be comprehensive, but we will discuss the major legislation that has crossed Dave's path over the past several decades of his career.

9.6.1 European Union Data Protection Directive

The EU Directive 95/46/EC is known as the Data Protection Directive and outlines the desire of the European Union (EU) to "protect the fundamental rights and freedoms of natural persons, and in particular their right to privacy with respect to the processing of personal data."[9] This regulation includes a good definition of personal data as

> *any information relating to an identified or identifiable natural person ('data subject'); an identifiable person is one who can be identified, directly or indirectly, in particular by reference to an identification number or to one or more factors specific to his physical, physiological, mental, economic, cultural or social identity.*[10]

This definition specifically outlines the correlation of multiple factors to construct an identity, and we can infer that such individual factors alone would not have provided such personal information had they not been correlated or combined together. This provides the link back to data processing.

Although this directive is interesting as it is one of the first comprehensive personal privacy regulations, it has been primarily replaced by the General Data Protection Regulation (GDPR).

9.6.2 General Data Protection Regulation

The General Data Protection Regulation (GDPR)[10] is the recent EU regulation that defines how personal information associated with natural persons (EU citizens) must be protected by companies that collect, hold, or process that information, and it outlines the penalties companies will face should they be responsible for a data breech of a EU citizen's personal data. Some of the interesting elements of the GDPR include the following:

- Controllers holding or processing personal data of a natural person must provide information regarding the holding/processing of the data in accordance with the GDPR *to the data subject in a concise, transparent, intelligible and easily accessible form, using clear and plain language, in particular for any information addressed specifically to a child.* (Chapter 3, Article 12)
- A controller must act on behalf of a request by a data subject *unless the controller demonstrates that it is not in a position to identify the data subject.* (Chapter 3, Article 12)
- *The data subject shall have the right to obtain from the controller without undue delay the rectification of inaccurate personal data concerning him or her.* (Chapter 3, Article 16)
- A data subject may request their data be erased, as long as certain conditions apply. (Chapter 3, Article 17)
- A data subject may request their information be removed from any processing, especially if such data cannot be completely deleted due to legal or administrative restrictions. (Chapter 3, Article 18)
- A data subject has the right to object to their data being used by a controller for direct marketing, for public interests, or the exercise of an official authority; such objections shall take into consideration the data subject's personal situation, with special considerations if the subject is a child. (Chapter 3, Article 21)

- Complaints lodged by an individual with an EU administrative authority must be responded to in three months or the individual may seek judicial relief. (Chapter 8, Article 78)
- Not-for-profit organizations active in the privacy arena may represent individuals in the courts. (Chapter 8, Article 80)
- Compensation for both material and non-material damages may be claimed and awarded to individuals who suffered damage from a controller or processing agent. (Chapter 8, Article 82)
- Where multiple controllers or processing agents are involved in a violation of the GDPR, each may be individually held liable for the full damages of the breech, and then may request compensation from the other controllers and processors to equalize the payment; this is specifically enacted to ensure that the individual harmed receives full payment for the damages. (Chapter 8, Article 82)
- An EU supervisory authority may also impose administrative fines that are dissuasive of the practices that led a data breach or violation of the GDPR. (Chapter 8, Article 83)
- EU member states may impose additional penalties for violations not covered by administrative fines that are *effective, proportionate, and dissuasive.* (Chapter 8, Article 84)

These elements of the GDPR highlight the need for administrative and management functions that allow the controller and processor to locate records and manage, amend, and delete records, or restrict their use for particular purposes. As time goes on, companies without such controls should expect to be levied fines and penalties.

9.6.3 California Consumer Privacy Act of 2018

The California Consumer Privacy Act (CCPA)[11] is applicable to citizens of the United States residing in California and was enacted to enhance the privacy protection of California residents. The act specifically outlines the following goals:

- Disclosure of the categories of personal information collected about a data subject and their children:
  - Contact must include a toll-free number to call and a website address.
  - Disclosure must include inference data generated from other data elements.
  - Data must include biometric data that includes but is not limited to DNA, iris, pictures of the face, gait information, voiceprint, typing patterns, and sleep or exercise information.
  - Personal information does not include public or de-identified information, but no definition of de-identified information exists in the Act.
- Information on the sale or disclosure of personal information.
- Right to require a business to stop selling the personal information of a data subject.
- Prevention of retaliation for exercising these rights by denying or modifying future services, or increasing charges for a service.
- Requirements on the proper protection of consumers' personal information.
- Businesses are held accountable for damages under violations of the Act, with minimum monetary compensation per consumer of $1000. Consideration for damages may take

into account *the nature and seriousness of the misconduct, the number of violations, the persistence of the misconduct, the length of time over which the misconduct occurred, the willfulness of the defendant's misconduct, and the defendant's assets, liabilities, and net worth.*[11]

Although this law is fairly comprehensive, the limitation on personal information to exclude de-personalized information remains a concern, given the weakness discussed on *k*-anonymity in Section 9.5 and the lack of any definition of de-personalization, including the appropriate values for *k* that provide sufficient privacy.

9.6.4 California Online Privacy Protection Act

The California Online Privacy Protection Act (CalOPPA) is the California law that requires websites to post a privacy policy and comply with what the company posts, including whether websites track users across their site and other party's websites. More information on California privacy laws can be found on the California Attorney General's website.[12]

9.6.5 Children's Online Privacy Protection Act of 1998

The Children's Online Privacy Protection Act (COPPA) is a US federal law that applies to websites that collect data from children under the age of 13. The Federal Trade Commission (FTC) is tasked with enforcement and published the Children's Online Privacy Protection Rule[13] (COPPA Rule) to define how COPPA is enforced by the FTC.

The COPPA Rule was amended in 2013 to update enforcement and certain requirements of the rule due the FTC's perception of changes in online technology since the Act's initial rollout in 2000.

COPPA covers information collected from or about a child, including passive collection of a child's online activities, or allowing a child to make public statements that could include personal information. The rule requires that parental consent must be obtained before such information can be collected and identifies disclosure rules by operators.

COPPA Rules define what constitutes an operator and includes those who do not directly run a website or web service, but for whom information is collected. This expansion of the term *operator* makes those who use such collected information responsible, regardless of whether they were directly involved in the information collection.

9.6.6 Health Insurance Portability and Accountability Act of 1996

The Health Insurance Portability and Accountability Act of 1996 (HIPAA) is the United State's health privacy law and defines certain information as protected health information (PHI), mandating protections on that data. PHI includes *any oral or recorded information relating to any past, present, or future physical or mental health of an individual, provision of healthcare to the individual, or the payment for the healthcare of that individual.*[14] HIPPA defines 18 specific types of PHI that are classified specifically as individually identifiable health information (IIHI) that require special handling. HIPPA does not cover information that has been de-identified.

The following list comprises the 18 IIHI categories[15]:

1. Names
2. All geographical subdivisions smaller than a state
3. All elements of dates (except year) for dates directly related to an individual
4. Phone numbers
5. Fax numbers
6. E-mail addresses
7. Social Security numbers
8. Medical record numbers
9. Health plan beneficiary numbers
10. Account numbers
11. Certificate/license numbers
12. Vehicle identifiers and serial numbers, including license plate numbers
13. Device identifiers and serial numbers
14. Web Universal Resource Locators (URLs)
15. Internet Protocol (IP) addresses
16. Biometric identifiers, including finger and voiceprints
17. Full-face photographic images and any comparable images
18. Any other unique identifying number, characteristic, or code

Since HIPAA has been around for quite some time, it is interesting to view its effects and the problems that have arisen as a result of HIPAA. A study from 2007[16] identified the two biggest issues health providers saw as a result of HIPAA included barriers to legitimate usage of PHI and confusion between practitioners and IT management over the enforcement of HIPAA, including confusion experienced by the general public about how HIPAA applied or did not apply to certain situations. Finding qualified individuals to execute the technical controls was also considered a significant barrier. We expect that similar issues will result from new privacy regulations.

Although the conclusion of that report reflected a desire to have a better overall umbrella policy that solved the *understanding* problem, as technology evolves at a pace faster than legislation, that aspiration is probably unrealistic. Our recommendation and the important takeaway from this section on privacy regulations is to create a specific policy and set of guidelines, appropriately informed by your legal representative, which outlines valid goals and usage limitations for the use and handling of PII to avoid the issues identified with HIPAA.

9.7 The Future of Privacy Controls

If the recent trends in legislation and regulation are any indication, we can expect further governmental oversight of the Internet and IoT. The GDPR created a ripple effect, causing many legislative bodies to consider whether they were doing enough to protect their citizens. COPPA was the result of this thinking within the United States, and other states are considering their own equivalent regulations. Eventually this will bubble up to the federal level within the United States. However, the EU is not yet finished. An EU E-Privacy Regulation has been in draft form for a few years, but its rollout seems to be delayed due to concerns over the impact of additional regulations over and above GDPR. As citizens continue to lament the way their data is treated, and subsequently breeched, legislators may be forced to act and address loopholes such as the de-personalization gap discussed in GDPR, COPPA, and HIPAA.

Likewise, the sophistication and complexity of controls, such as DP and HE, will continue to evolve, but the most plausible short-term savior of data privacy is likely to be multi-party computation using TEEs. Staying informed and layering privacy controls to create robust defense-in-depth protections of data in IoT systems is the prudent approach to addressing customer needs, citizen's concerns, and regulatory mandates.

References

1. Clarke, R. (2010). "Introduction to Dataveillance and Information Privacy, and Definitions of Terms." Carnegie Mellon course on Introduction to Computer and Network Security. Retrieved from https://web2.qatar.cmu.edu/iliano/courses/10F-CMU-CS349/slides/privacy.pdf
2. Supreme Court of New Jersey. (2008, April 21). "STATE of New Jersey, Plaintiff-Appellant v. Shirley REID, Defendant-Respondent." Retrieved from https://caselaw.findlaw.com/nj-supreme-court/1377890.html
3. Hern, A. (2018, January 28). "Fitness Tracking App Strava Gives Away Location of Secret US Army Bases." Retrieved from https://www.theguardian.com/world/2018/jan/28/fitness-tracking-app-gives-away-location-of-secret-us-army-bases
4. Guest, R. (2019, March 14). "Log Data Privacy: Techniques for Data Privacy in Application Logs." Retrieved from https://datatracker.ietf.org/meeting/104/materials/slides-104-pearg-ryan-log-data-privacy-00
5. Winton Group, Ltd. (2018, September 4). "Using Differential Privacy to Protect Personal Data." Retrieved from https://www.winton.com/research/using-differential-privacy-to-protect-personal-data
6. Vadhan, S. (2017). "The Complexity of Differential Privacy." In *Tutorials on the Foundations of Cryptography*, edited by Y. Lindell (Chapt. 7). Cham (Switzerland): Springer International Publishing.
7. Trappe, W. and Washington, L. (2002). *Introduction to Cryptography with Coding Theory*. Upper Saddle River (NJ): Prentice Hall.
8. Menezes, A., Oorschot, P., and Vanstone, S. (1997). *Handbook of Applied Cryptography*. Boca Raton (FL): CRC Press.
9. EU. (1995). "Directive 95/46/EC of the European Parliament and of the Council of 24 October 1995 on the Protection of Individuals with Regard to the Processing of Personal Data and on the Free Movement of Such Data." Official Journal of the European Communities of 23 November 1995, no. L. 281, pp. 31–50.
10. Intersoft Consulting. (2019). "The General Data Protection Regulation." Retrieved from https://gdpr-info.eu/
11. Online Archive of California. (2017, November 20). "The California Consumer Privacy Act of 2018." Retrieved from https://oag.ca.gov/system/files/initiatives/pdfs/17-0039%20%28Consumer%20Privacy%20V2%29.pdf
12. California Office of the Attorney General. (2019). "Privacy Laws." Retrieved from https://oag.ca.gov/privacy/privacy-laws#
13. Federal Register. (2013, January 17). "Children's Online Privacy Protection Rule; Final Rule." Vol. 78, No. 12. Retrieved from https://www.ftc.gov/system/files/2012-31341.pdf
14. Beaver, K. and Herold, R. (2004). *The Practical Guide to HIPAA Privacy and Security Compliance*. Boca Raton (FL): Auerbach Publications.
15. UC Berkeley Human Research Protection Program. (2019). "HIPAA PHI: List of 18 Identifiers and Definition of PHI." Retrieved from https://cphs.berkeley.edu/hipaa/hipaa18.html
16. Houser, S., Houser, H., and Shewchuk R. (2007, March 23). "Assessing the Effects of the HIPAA Privacy Rule on Release of Patient Information by Healthcare Facilities." *Perspectives in Health Information Management*, vol. 4, no. 1.

Chapter 10

Security Usability: Human, Computer, and Security Interaction

The test of the machine is the satisfaction it gives you. There isn't any other test. If the machine produces tranquility, it's right. If it disturbs you, it's wrong until either the machine or your mind is changed.

— Robert M. Pirsig[*]

10.1 Poor User Experience Design Isn't Just Inconvenient, It's Painful

It is Monday morning, and you are just about set to leave for the office. You will be leading an 8:30 am meeting, so you would like to get in 15 minutes early. Thankfully, it is only a 15-minute commute. To ensure that you get to the office on time, you have gobbled up your breakfast by 7:10 am, leaving you about 1 hour to get to the office. Now, in preparation for the morning's workout, you don a pair of large overalls that cover your work clothes. Afterwards, you grab your tool box and office bag, and then head for your car. Once in the car, you get to work. That is, the work before "work."

In your car's front-engine layout, the starter is mounted low down near the back of the engine. You grab your handheld electrical generator from your tool bag before leaning over the engine. Yes, your car's engine and its various components are clearly visible from the front seats of the car. With your handheld electrical generator, you apply current to the starter switch. Meanwhile, you deftly ensure that your overalls do not get caught in the generator. Beads of

---

[*] Retrieved from https://www.goodreads.com/quotes/125342-the-test-of-the-machine-is-the-satisfaction-it-gives

sweat form on your forehead. Thankfully, you used antiperspirant so your underarms shouldn't be too bad. Still, there is more to do.

The starter switch feeds current to the solenoid, and you hurriedly pull back your electrical generator to avoid dangerous sparking. Once the solenoid receives electricity, the electromagnet attracts an iron rod. The movement of the rod closes two heavy contacts, completing the circuit from the battery to the starter. The starter begins turning your engine. Immediately, you flick the return spring on the rod so that the starter switch stops feeding current to the solenoid, the contacts open, and the starter motor stops. You need to flick the return springs because the starter motor must not turn more than it has to in order to start the engine. Your car revs to life, as it starts. You sit back with a deep sigh. Now, it is time to drive this thing. With a determined expression, you get to the task; after all, time waits for no one.[1]

Having read the previous paragraphs, you might be wondering why anyone should have to go through such contortions to get a car to start. Well, the scenario just described is probably what would occur if a car were to be designed and built solely by mechanical engineers. It would be a good car, and perhaps it could be a great car. It would be highly functional, able to do all that a car ought to. But as you just saw, it would be scarcely usable. Yes, you are right, if that were the case, we should all just take the subway, the overground, the bus, anything! But what if those were designed in the same way? We would be in a proper kerfuffle, that's what. Finally, if you do not understand many terms in the last two paragraphs, never mind—we scarcely did either, until we did some research.

Do not get us wrong. We are not saying that mechanical engineers have a problem; and, for that matter, neither are we saying that any engineer has a problem. We, the authors, are engineers ourselves. Rather, what we are saying is that those who touch any product can be placed into two broad categories—the makers and the users. Engineers who build cars are in the maker category, and they are smart. People who use cars are also smart, but they are not necessarily interested in the "cool" engineering specifics of how the car gets going. Such folks have all sorts of things besides cars that they are interested in understanding inside-out. Most people appreciate the insight and work of those who built the car, but with the car built, they just want to use it.

> *All I gotta do is turn a key (or press a button), and it just runs great and doesn't blow me up? Great! Where can I get one?*

It is the very same with security engineering in IoT systems. IoT security architects and engineers must realize that the IoT system and its security mechanisms are being built for users who are largely uninterested in the engineering or security specifics of the system. They just expect it to work. As such, IoT systems not only need to have secure defaults, but the security mechanisms should not hinder the reasonable use of the system. In addition, any security configuration or security functionality that users must interact with must be designed to be usable as well as functional.

In this chapter, we shall frame the security usability problem in the context of IoT, explore the challenges of security usability, and then suggest principles for designing usable security as it pertains to IoT systems.

10.2 Nightmare at 40: When Too Many Convenient Devices Become Too Difficult to Manage

We hope you like stories and imagination as much as we do, since we have another scenario for you to step into. In this case, you own a home, and being tech savvy (you do not have to be an engineer to be tech savvy), you would like to make it a "smart" home. Over the years, you have been upgrading your home with IoT conveniences. At your last inventory, some of your smart items included a thermostat that controls heating and cooling; a home security system fitted with comprehensive video camera coverage; doors to the house and garage that can be remotely locked or unlocked; a fridge that detects when food is going bad; window blinds that open and shut at right time using their knowledge of your schedule and preferences; light bulbs that can be controlled with gestures and commands; a virtual voice assistant that can be used for many basic tasks, such as searching the web, playing music, and setting alarms or timers; a watch that monitors your heart rate and counts your steps; a pill box that ensures that your pills are taken at appropriate times; and the list goes on. Let us say that your total comes up to about 40 devices, and we mustn't forget your TV, computer, and smartphones.

Why 40? It is just a number. More importantly, did your head spin as you read through that list? Yet, the scenario we just illustrated is not fantasy; in fact, it is present reality. What's more, it is not inconceivable that many of these devices will not be from the same manufacturer. In addition, each device is likely to have its own website or mobile app for system administration. Even if all devices were made by the same manufacturer, and even if there was a central dashboard that covered all devices, there are still 40 devices to monitor and administer.

For most consumers, managing 40 devices is a chore. A corporation with a smart office system can afford to employ system administrators and information security specialists whose sole jobs will be to manage the IoT devices owned by the company. Even so, it would still take lots of work on the parts of the "chosen ones" to ensure that all the company's IoT systems are operational and secure. For the regular user, in a smart home, for instance, as long as the systems are operational, the security of those systems will be assumed.

IoT is unique in that it drastically increases the number of devices that need to be secured. In the home, we could pick a number such as 40, and that sufficiently illustrates the point. But in many other scenarios, the number of devices are even higher, and very much so. For instance, IoT devices in an electrical power grid easily number into the thousands and millions.

In other chapters of this book, we cover the consequences of security breaches in IoT systems. The point that must be made here is that if security features are designed such that they are inconvenient to use, they are unlikely to be used properly, if at all. Needless to say, that increases the risk of security breaches.

10.3 Challenges of IoT Security Usability

In this section, we will explore some of the challenges of security usability in IoT systems. We seek to answer the question, "Why is security usability hard?"

10.3.1 Security Doesn't Make Sense to the Regular User

We alluded to this point in the introductory sections of this chapter. We must point out that the regular user of an IoT system is not stupid. But the regular user is uninterested in the security mechanisms of an IoT system. He or she just wants the system to work. Let us consider an individual, Sandy, who uses a pacemaker to regulate her heartbeat. A pacemaker is a medical implant that uses electrical impulses, delivered by electrodes, which regulate the beating of the heart by contracting the heart muscles. The pacemaker in our example can be monitored remotely, and a remote administrator can perform diagnostics on the pacemaker via an Internet-connected hub that Sandy keeps with her at home or in the office. Sandy also has a mobile app on her smartphone (from the manufacturers of the pacemaker) that she uses to view stats on her heart rate and the strength of her heartbeat. The mobile phone connects to her pacemaker, via the hub previously mentioned. The pacemaker hub has certain configurations that Sandy has to set up to ensure that only her app can connect to the hub.

Sandy cares about her heart, and Sandy cares about her pacemaker. But Sandy isn't a security specialist. She does not understand the security implications of the different configurations on her app that are used to connect to the hub and restrict connections to the hub from other devices. Given Sandy's knowledge, either of two occurrences is likely to happen.

One of those occurrences is what is referred to as the Kruger–Dunning paradox,[2] which teaches that the less people know about a particular subject matter, the more likely they are to overestimate what they do know. This is because they do not know what they do not know. Try this: Tell an experienced security professional about the scenario we just walked through: a pacemaker that can be remotely accessed via the Internet. It is likely that that you will see their eyes widen like that of an owl or observe them wince in pain. You might even see some wet their pants. Why do the security professionals react so dramatically? They know all too well how easy it is for things to go wrong if the system is not set up right. In fact, it is fair to wager that a security professional will only connect an Internet-accessible device to their heart as a very last resort. Until today, some of our close friends and colleagues adamantly refuse to use any car that is equipped with Internet connectivity.

The other likely occurrence is that Sandy just assumes that the system must be secure or that the various configurations she can choose from must be secure. When people have little or no knowledge about a particular subject matter, they readily turn to experts and are happy to trust that the experts have it figured out. In Sandy's case, it is easy to imagine her thinking, "Well this thing is connected to my heart, so why would anyone will be so foolish as to build it with insecure configurations? It must be securely developed." In addition, she probably thinks, "The government regulates all these companies. They cannot allow them to develop insecure products and connect them to people's hearts!"

Contrary to the regular user, security professionals know enough about the Internet and software and hardware vulnerabilities to be cautious when deploying IoT systems. For instance, a dear friend of ours, a master Security Architect took the extra measure of placing his home security system (which is managed and accessible to the solution provider) on a separate network from all the other devices (his devices) in his home. Based on his knowledge, he was able to find out something that is unsurprising but perhaps unknown by many. The technician installing his home security system had very little understanding about how computer systems and networks work. The technician was savvy enough to connect one component to the other as taught, so that the system worked, but that was about it. He knew practically nothing about the security of networks, computers, and software. Was the technician at fault? No. He knew what he needed to know. The onus then lies on us, the designers and builders of IoT systems, to ensure that our solutions can be securely deployed without the installation technician or the user being a master Security Architect.

10.3.2 Security Is Not Interesting to the Regular User

If you are a security professional, your heart may have fallen just a little. But perk up, security pro, and pick your chin up; you are not alone in your travails for the safety of those you defend. A soldier does not go to war for praise but rather to defend the peace and safety of those they love (or at least that is the only mildly palatable reason for the carnage of war). In any case, we digress. Back to the point, let us consider an automobile and the seat belt.

The seat belt in automobiles was designed to reduce the risk of injuries or fatalities in car accidents. Without seat belts, the force of impact during a car accident could cause passengers to be violently thrown against the body of the car and, in many cases, ejected from the car through the glass windows. The seat belt keeps passengers strapped to the car, thus significantly reducing the risk of injury. A study conducted in the United States in 1984 revealed that the use of seat belts reduced fatalities for front seat passengers considerably, from 20% to 55%. The range of major injury was likewise reduced, from 25% to 60%.[3]

Yet, the United States and virtually all other nations had to pass laws to ensure that seat belts were used by passengers. If users innately care about safety, why do we need laws and enforcers of the law to ensure that passengers wear their seat belt? In recent times, society has managed to impress upon automobile users the virtues of the seat belt, and the immense risk of harm that exists without it. Historically, however, the focus for automobile users was getting to their destination, rather than the inconvenience of strapping some rope across their chest.

As mentioned earlier, the regular user of an IoT system does not understand the security intricacies of the IoT system or its consequences. As such, the regular user is uninterested. Once we realize that the user of an IoT system is most concerned with putting it to use for the purpose for which it was acquired, we are able to understand that security must work alongside the IoT system seamlessly, rather than be a hindrance.

10.3.3 Usable Security Is Not Demanded from Vendors

We just covered how the regular user is uninterested in the security specifics of their IoT system, how they do not really understand it, and how both of those are related. Thus, it is easy to understand why regular users are not demanding usable security from vendors.

On the other hand, we have organizations that deploy IoT systems. For instance, organizations operating in sectors such as smart cities, agriculture, nuclear plants, and power generation. Usually, such organizations are savvy enough to realize the need to manage their security risk responsibly. However, since they do not measure the time users spend on configuring or navigating security features, they are unaware of the cost of security. Therefore, they do not demand usable security from their technology vendors. As a result, the vendors have no incentive to supply it. The main goal of most technology vendors, vis-à-vis information security, is their profitability and the avoidance of any bad publicity that could adversely impact said profitability.[4]

This causes organizations—who use IoT systems—to vacillate between two states. In one state, they must expend lots of time and effort to enable robust security via the technology provided by their vendors. This often stifles their productivity and efficiency. In the second state, they take the easy route by only implementing minimal security controls, which is an open invitation to the bad guys.

10.3.4 Barriers to Necessary Workflow

If security is designed into a system in such a way that it impedes the objective of the system, such security becomes a barrier to the necessary usage of the system. This leads users to a necessary circumvention of security controls.[5] For instance, if an IoT sensor hub was designed to only identify and authenticate all sensors via X.509 certificates, it becomes impossible to use the sensor hub with any sensors that do not support Transport Layer Security (TLS). Currently, a good number of sensors do not have the compute power or memory to support secure identification or authentication via TLS. This is likely to lead users of the sensor hub to turn off the identification and authentication of sensors altogether, thus placing the system at risk of data corruption via rogue sensors.

For the architects of such a sensor hub, it is their responsibility to design a system of sensor identification and authentication of sensors that caters to both TLS- and non–TLS-enabled sensors. In addition, the system should either detect the supported authentication mode of sensors automatically (this is preferred) or make the choice of configuration and the associated implications clear to system administrators. A secure default should also ensure that the sensor hub will not communicate with sensors without the configuration of sensor identification and authentication.

10.3.5 Different Views of Security, from Executive to Architect to Implementer, Then the User

There is a popular saying, "Too many cooks spoil the broth." This saying means that when many conflicting views are involved in the development of a product, it often leads to an incoherent result. The saying omits the obvious, which is that if the cooks were to be in harmony, with proper vision and leadership, well-defined roles and responsibilities, and proper management, the soup should turn out fine. Of course, providing all that structure for the cooks can be expensive and is only justifiable in certain situations.

In systems development today, there are many cooks. In most companies, the senior executives usually have a vision regarding how the company can harness its talent pool of skilled people and its expertise with certain technologies to create certain products that will attract or retain customers. Architects are tasked with designing the products that realize the vision, while engineers work with the architect to implement the design.

Security architects play an important role in ensuring that the vision, design, and implementation of security features are in harmony. Our challenge to security architects is that they must also consider the usability of security, thereby harmonizing the views of the user as well. Without such harmony, security features are hard to use, at best, and wrongly implemented, at worst.

10.4 Principles for Designing Usable IoT Security Controls

Is it possible to design security controls that are simultaneously functional and convenient for users? Let us find out as we outline some nuggets of wisdom (from our experience and research) for designing security controls that do not cause users to throw their hands up in frustration.

Principle 1: Implement Secure Defaults

The default state of an IoT system, its hardware, software, communication, etc., should be secure. This means that the configuration that is in effect when a new user takes ownership of an IoT system should be secure from end to end.

Secure defaults cover everything that can be implemented into an IoT system to ensure that the risk of a security breach is reduced from the outset, for the system and its users.

One example of this is secure boot—a mechanism by which a device cryptographically verifies the signature of the software used to start the device. In this case, a cryptographic key that's stored in hardware (and is very difficult to tamper with) is used to verify that the software used to instantiate a device is signed by a trusted authority. The cryptographic key that is used to verify the device instantiation software must be securely built into the hardware by the device manufacturer. The manufacturer (and any partners) are also responsible for ensuring that the authentic software that is to be used to boot the device is appropriately signed before it is placed on the IoT device. If things are done properly, the user of the IoT device has nothing to worry about or configure. Any time the device starts up, it should be able to automatically check and verify that its startup software has not been tampered with. Secure boot is described in more detail in Chapter 6, Section 6.4.1.

Principle 2: Display Targeted Risk Communication Information for Configuration Options

Yes, we should implement secure defaults into IoT solutions. Yet, most creatives understand that one configuration rarely caters to all users or customers. This means that for our systems to be broadly deployable, it is necessary to support various configurations.

As mentioned before, the regular user of a security option or security configuration is usually not well versed in the security implications of a particular configuration. However, the "regular user" frame of reference is not reserved solely for the Joe Somebody who buys consumer IoT devices. It also applies to people such as software engineers or system administrators who might not be security experts. Thus, this principle strongly suggests that every selectable security configuration should be accompanied by a targeted message or help notice that outlines the security benefits or risks of that configuration. To go the laudable extra mile, the selection of a known "less-than-secure" configuration can also be accompanied by a pop-up with the risk highlighted.

For instance, the cloud applications architect for a modern farm may be tasked with configuring the mechanism by which the IoT gateway at the farm is authenticated to the cloud application provided by the vendor. The vendor allows two kinds of authentication, X.509 certificates and pre-shared symmetric keys. Most security architects will understand that pre-shared symmetric keys are less secure since the same cryptographic key, upon which authentication depends, has to be stored in multiple places, thus increasing the risk of compromise.* The documentation and the configuration interface should highlight the risks of pre-shared symmetric keys sufficiently.

Principle 3: Provide Buckets of Security Configuration Options with Different Security Levels

To reduce complexity for our users, we can create buckets of security configuration options that they can choose from. A crude example will be buckets of high, medium, and low security. The selection of any of those buckets should lead to the automatic selection of certain options applicable to the chosen level of security. For IoT, we can imagine a system with a cloud back-end and edge devices that can either communicate within a private, corporate intranet, or over the Internet. The risk of the compromise of a communication channel is usually greater for channels that go over the Internet than it is for those that exist within a corporate intranet. As such, different security levels can apply for the different network types, and different buckets of security configuration options can be used to simplify the user's experience.

This principle works in tandem with Principle 2, since it would require that the security benefits or risks of any bucket are clearly articulated. It should also be possible for users to do any of the following: select security configuration options manually, select a security level that automatically selects associated security configuration options, or combine the selection of a security level with manual modification of auto-selected configuration options. Indeed, insight and skill are required to implement this principle for any given use case, but everyone would build usable security if it were effortless.

Principle 4: Design to Scale

Given the large number of devices typically involved in IoT deployments, the scalability of deployed solutions is often essential. It is no different for security. When designing security solutions for IoT devices, you must ask the question, "How does this scale?" Depending on the situation, your solution might need to scale to hundreds, thousands, millions, or even billions of devices.

Examples of this were provided in Chapter 6, Section 6.4.2, where we explored IoT device provisioning. Just-in-time registration allows new IoT devices to be automatically registered and activated with the cloud as they start up via their built-in device identification X.509 certificates. We also presented the certificate vending machine mode, which allows IoT system owners to securely and automatically transmit device identification certificates to devices that could not be preinstalled with certificates during manufacturing.

---

* Retrieved from https://www.digicert.com/ssl-cryptography.htm

Principle 5: Enable Active Authorization That Requires Consent of the Grantor

This principle applies to user authorization. For instance, an IoT system administrator must consent to the granting of any authority to another user.[6] Such a system should also allow a user to seamlessly request authority.

Principle 6: Allow the System Administrator to Revoke Previously Granted Authority

System administrators must be able to revoke previously assigned authority. This applies to users as well as devices or subsystems that have been granted authority to access resources.

As an example of usable revocability, let us consider an IoT cloud that is connected to five thousand edge gateways. The gateway devices are authenticated and authorized to transmit data via X.509 certificates that are installed on them. For usable revocability, the IoT cloud application should be able to use device health checks (covered in Chapter 6, Section 6.4.1) and machine learning to detect compromised edge gateways. This detection should happen on the fly, with alerts and reports sent to designated system administrators. The IoT cloud application should also allow system administrators to revoke or invalidate the X.509 certificate for the edge gateway that is suspected of compromise. This will ensure that the compromised edge gateway cannot send rogue data to the cloud while the IoT solution owner investigates the situation.

Principle 7: Store Secrets in a Way That Significantly Hinders Compromise

It is a fair argument that this principle belongs among principles of secure design. We believe it belongs here as well. We believe that secure design is particularly so if the security it produces is usable. In addition, we also believe that secret management is so difficult to do well that it is best to leave as little as possible in the hands of the user, be it a regular user or a system administrator.

In conjunction with Principle 1, this principle recommends that secure storage and management of secrets should occur out of the box, with little interference required by the system user. Examples of secrets are cryptographic keys, digital certificates, passwords, and usernames. Means of secure storage mechanisms include hardware-based Trusted Platform Modules (TPMs), software-based secret managers or vaults, restricted operating system directories, among others. Please visit Chapter 5 for details on secure storage and management of secrets.

Principle 8: Monitor Circumvention of Security Controls

Studies show that regular, "good" users often circumvent security controls.[5] The referenced resource describes different scenarios of security circumvention, as well as the reasons that necessitated circumvention. It shows that, as was mentioned in the previous section, the primary reason for security circumvention is that security is impeding necessary workflow, making life difficult for those who need to use the system.

Technology makers usually build in telemetry* that enables them to monitor—preferably anonymously—the statistics regarding user behavior, as well as overall system behavior and health. This principle recommends that system monitoring should also cover the usage, lack of usage, or circumvention of security controls. The data produced will be helpful in creating security controls that better fit user needs.

10.5 The Cause of Usable Security Belongs to All of Us

There is an existential tension between usability and good security. The purpose of security usability is the alleviation of that tension. If we strive for the utmost security, the system becomes unusable. If we are too lax with security for the benefit of the greatest usability, the system is soon compromised and everyone is up in arms. We must find a balance.

As technology moves in its relentless tide, and new products and security controls are built, the "right" balance will necessarily have to be refined. It behooves system designers to not only keep security at the forefront of their minds as they innovate but also watch and ensure that the security that is implemented is usable.

In this chapter, we have outlined some of the challenges in the provision of usable IoT security controls. We have also presented eight principles that help you address the security usability challenges we outlined. We do not frame the eight principles as an exhaustive list, but rather as a foundation that provides you with the necessary insight and starter fuel required to build usable IoT security. As usual, we invite the reader to extend our list. Now, Go Innovate!

References

1. "How the Starting System Works." Retrieved from https://www.howacarworks.com/basics/how-the-starting-system-works
2. Kruger, J. and Dunning, D. (1999, December). "Unskilled and Unaware of It: How Difficulties in Recognizing One's Own Incompetence Lead to Inflated Self-Assessments." Retrieved from https://www.ncbi.nlm.nih.gov/pubmed/10626367
3. National Highway Traffic Safety Administration, Plans and Programs, Office of Planning and Analysis (1984, July 11). "Passenger Car Front Seat Occupant Protection." Retrieved from https://crashstats.nhtsa.dot.gov/Api/Public/ViewPublication/806572
4. Lampson, B (2009, November). "Usable Security—How to Get It." *Communications of the ACM,* vol. 52, no. 11, pp. 25–27. Retrieved from https://cacm.acm.org/magazines/2009/11/48419-usable-security-how-to-get-it/fulltext
5. Blythe, J., Koppel, R., and Smith, S. (2013, October). "Circumvention of Security: Good Users Do Bad Things." *IEEE Security & Privacy,* vol. 11, no. 5, September–October 2013. Retrieved from https://www.cs.dartmouth.edu/~sws/pubs/bks13.pdf
6. Payne, B. and Edwards, W. (2008, June). "A Brief Introduction to Usable Security." Retrieved from https://www.cc.gatech.edu/~keith/pubs/ieee-intro-usable-security.pdf

---

* Retrieved from https://whatis.techtarget.com/definition/telemetry

Part Three

Chapter 11

Earth 2040—Peeking at the Future

> *We are made wise not by the recollections of our past,*
> *but by the responsibilities of our future.*
>
> — George Bernard Shaw[*]

11.1 Whacking at the Future of IoT

Aaron Fechter is an engineering entrepreneur who enjoys a good laugh as much as anyone else. In 1974, he invented an arcade game that is as funny as it is annoying. He called it Whack-a-Mole. A typical Whack-A-Mole machine consists of a waist-level cabinet with five holes on the top and a large, soft, black mallet. Each hole contains a single plastic mole and the machinery necessary to move it up and down. Once the game starts, the moles begin to pop up from their holes at random. The objective of the game is for the player to force the individual moles back into their holes by hitting them directly on the head with the mallet, thereby adding to the player's score. No sooner does the player manage a hit than the grinning mole pops ups in a random spot.

Imagine if we were to increase the number of holes to hundreds or perhaps thousands, with the same lone mole popping up randomly, a mallet that could be shortened or lengthened with a special switch, and the player in the middle surrounded by the endless holes.

For us mere mortals, the prior scenario is akin to predicting a future event, especially the further we attempt to project into the future. History shows us that life often presents limitless possibilities, seemingly random. Even if those possibilities are not random, they might as well be, as the creation of links or connections between events and the impacts of those connections are just as difficult to predict.

---

[*] Retrieved from https://archive.org/stream/backtomethuselah13084gut/13084.txt

In his fascinating book, *Sapiens: A Brief History of Humankind*, Yuval Noah Harari describes why it is usually impossible for most dictators to predict a revolution. If a fastidious dictator were to hire the best statisticians to monitor the level of satisfaction of his or her fellow citizens and compare the results with public sentiments and events that have led to prior or current revolutions in other countries, the findings would undoubtedly influence the future actions of the dictator. Most dictators, upon learning of intense national dissatisfaction that has the makings of a revolution, will either embark on measures that present them in a better light, such as tax reductions, or use force to subjugate the malcontents. Thus, the predicted revolution does not materialize until it does, unexpectedly and unpredicted.[1]

It is no wonder that many brilliant minds have proclaimed that, "the best way to predict the future is to create it."[2]

Still, some might wonder how it is that savvy investors are able to make good bets on companies and stocks. The answer . . . trends. Investors and market analysis algorithms analyze historical data and current circumstances to make best-guess predictions of a future event. A sleek investor, sometimes with the aid of fancy algorithms or software, is able to predict (or actually guess) that a company X will do well or perform poorly in the near future, given the state of the market (i.e., supply and demand in a particular sector or multiple sectors of the economy); the leadership, products, and revenue of company X; and the results of similar situations in years past. No worthy investor will predict that company X will be valued at price W on day Y in year Z. The chances of getting all those right are very small to begin with, but they get even smaller, the further into the future such an investor attempts to predict.

In this chapter, we will appropriate the approach of the savvy investor. Using an analysis of current innovation, research, and the deployment of IoT technology, we will attempt to predict the future state of IoT in different sectors of the economy.

11.2 The Fascination of Technology Innovation

Technology innovation continues to evolve apace. Research and discoveries in materials science and computing continue to impact technology innovation, as small, yet powerful, computing devices are applied to various problems. Sociopolitical developments in human interaction, partly motivated by technology, also continue to influence changes in user expectations of any technology. For instance, it just so happens that many of today's technology aficionados are largely concerned with reducing waste and any negative impacts of technology on the environment.

Generally, it is unsurprising that, as technology changes, the problems we care about also change. Often, such changes require new strategies for solving new problems (or old problems in new situations) with newer technology. As we have described in prior chapters, a culmination of events in technology and, indeed, human innovation has led us to what we might call the burgeoning age of the Internet of Things (IoT). In our peek at the future of IoT, we are most concerned with the information and computer security problems birthed, and in other cases, accentuated by the proliferation of IoT.

The term "proliferation" is often used generously in reference to IoT, and rightly so. According to predictions by many industry experts, we can expect IoT to connect at least 25 "smart" things by 2020.[3] Considering such exponential growth of IoT devices and solutions, along with the related IoT security concerns that we began to explore in Chapter 2, it is unsurprising that we are fascinated with the future.

What might the future of IoT be two decades from now? How could innovative IoT solutions affect our lives via industries or sectors such as healthcare, agriculture, transportation, homes, and our cities? What are the security implications of such solutions?

Join us as we put on our frontier hats and boots and venture off into the future of IoT-enabled technology and solutions.

11.2.1 Clairvoyance or Science?

Most people cannot help wondering about the future and the possibilities it might hold. As though trying to capture and hold the morning dew, we strain and stretch, while wondering if there is any sense in such actions, but cannot seem to stop.

In centuries past, and to a much lesser extent today, such curiosity made it possible for the practice of fortune telling to flourish. That practice often involved the prediction of events regarding an individual's life. A popular tool used by clairvoyants is a glass or crystal ball, which was said to induce visions as the user stared into it. Largely, the patronization of fortune tellers is driven by our fear of the unknown. Clients of fortune tellers often seek resolution of conflict in tough decision making, and the affirmation or comfort of positive future outcomes.

Scientists and technologists also reach for the future. But unlike clairvoyants, scientists and technologists use existing evidence to explore and study current as well as future occurrences. For instance, considering the introduction of computers for the management, protection, and monitoring of homes, what security problems exist with the current technologies and the ways they are deployed? And what problems might come to light if computers become the *de facto* way that all facilities in a home are managed and monitored, locally and remotely?

As IoT aficionados and technologists, we are interested in the future. But we must leverage the evidence outlined by current trends and technological advancements to envision future trends and their impact on IoT solutions and their users.

11.2.2 Now

As we showed in prior chapters, innovations in sensors and miniaturized (yet powerful) computers have introduced IoT "smarts" or automated computing assistance into different sectors of the economy and our lives. In IoT, "smarts" involves the use of technology to detect, analyze, and respond to changes in a physical environment. Currently, sensors are able to detect the presence of chemical or biological components, temperature, pressure, and more. Powerful Internet or cloud-hosted compute infrastructures are then used to analyze the large numbers of events picked up by the sensors.

Current Examples of IoT "Smarts"

IoT is no longer novel. This is exciting. The possibilities IoT presents are being investigated and indeed deployed, actively.

Here's a brief review of present IoT innovation in a few sectors of the global economy.

Agriculture:

Smart watering and fertilizing systems are able to sprinkle water and dispense fertilizer intelligently, using information about the weather and soil composition.

Healthcare:

- In hospitals, advanced sensors monitor skin color, breathing, and temperature, and alert nurses of any changes.
- Computerized pillboxes open at preconfigured times for patients to take their drugs, and they also send messages to people such as a family member or nurse, confirming that the drugs were taken.
- Pacemakers are medical device implants that use electrical impulses, delivered by electrodes, which regulate the beating of the heart by contracting the heart muscles.

Energy:

- Sensor-enabled power lines and pipelines can be analyzed to detect and isolate maintenance problems.
- The smart power grid is being designed and implemented to receive data information about the time-of-day usage of transmitted power. In the future, it is theorized that power will be dynamically priced (via smart grids) based on demand. The demand from a household, for instance, will be transmitted in real time to smart meters, thermostats, and appliances so that they can draw the power they need at off-peak times, when it's cheapest.[4]

Manufacturing:

Microsensors embedded throughout the manufacturing and assembly process are helping manufacturers to drastically reduce product defects.

Home and Office:

The poster child for smart homes and offices is the Nest® Learning thermostat—a device that is designed to control the temperature in a living space via the manipulation of heaters and air conditioners, but that's not all. Using machine learning algorithms, it continuously observes and learns the schedule and preferences of the users of the space—that is, when and how they use the heater or air conditioner. Using that knowledge, it automatically modifies its own behavior regarding the regulation of the temperature of the living space.

Transportation:

- Sensors in an aircraft's engines can now detect and isolate developing problems—in part by measuring the temperature of a jet engine's exhaust—and communicate these problems to both pilots and ground crews while the plane is still in the air.

- The latest cars from Tesla® include an autopilot feature that uses a camera, radar, and 360-degree sonar sensors to automatically drive on open roads and in stop-and-go traffic, and to not only find, but back into, parallel parking spots.[5]
  Autopilot Is Not Equivalent to Full Autonomy: It is noteworthy that at the time of writing, the Tesla Autopilot is still quite a long way from being a fully autonomous car. Drivers still need to be aware and are advised to have both hands on the steering wheel, as many situations could arise that autopilot might not handle as well as a human would.[6]
  On May 7, 2016, a man died in the first known fatal car accident involving the use of Autopilot. Postaccident investigation indicated that the man had his hands off the wheel while the car drove on via autopilot, and, supposedly, he ignored seven warnings given by the car. According to the automaker, Tesla, "Neither Autopilot nor the driver noticed the white side of the tractor trailer against a brightly lit sky, so the brake was not applied," and the car drove under the trailer.[7]

Cities:

- Solar-powered garbage cans crush waste and send a message to a dispatcher requesting pickup when they are full, resulting in a significant reduction in the number of weekly garbage-collecting shifts and associated maintenance and labor costs.
- Stoplights with embedded video sensors can adjust their greens and reds according to where the cars are and the time of day, thus reducing traffic congestion.
- Smart LED streetlights turn on only when a pedestrian or vehicle approaches, saving lots of power.

Some of the developments described above are still being piloted. We will explore them in greater detail soon, as we delve into our vision of the future.

11.2.3 The Major Types of Change Introduced by IoT

Having considered some of the major IoT innovations that are already underway, let us go on to explore how those changes might evolve. Two decades from now, what environments, problems, and solutions will be produced by the increasing "connectedness" of machines among their own kind, as well as machines with humans?

To provide a framework through which we can explore the future, we will start by considering the various change types introduced by IoT.

Changes in Products

A product can be defined as anything, tangible or intangible, which can be offered to a market to satisfy a want or need. Markets are mediums of exchange. Products can be exchanged for other products, or for money.

As we mentioned in the previous section, miniaturized computers capable of sensing their environment have led to the creation of new solutions and products that provide some form of value to consumers.

Generally, advances in technology give rise to new solutions to problems, and sometimes create new problems. Developments in IoT have produced a demand for products that combine environmental sensing with automation to create "smart" solutions or environments.

Changes in Lifestyle and Work Style

In 1908, Henry Ford, founder of the Ford® Motor Company, introduced what is widely regarded as the world's first affordable automobile, the Ford Model T™. Ford's invention was not the first car, but it was priced such that it was affordable to the middle-class or average buyer. The reduced cost of the car was mostly due to increased production efficiency via assembly line production and reduced labor costs. In 1913, Ford built a moving assembly line, which greatly improved production efficiency even more. As a result, cars were rapidly adopted in the United States, where they replaced horse-drawn carriages and carts as the primary means of transportation. The Model T was named the car of the 20th century.[8]

It is fair to say that technology is often engineered to impact how we live. Such impact occurs in varying degrees, from the major impact of the Model T to what we might consider the lesser impact of a child's toy. Impact must exist for technology to be considered useful.

The impact of IoT on our lifestyle and work style is brought about by the change it produces in our interaction with our environment. Since IoT solutions can sense our environments and analyze what they detect, they have a sizable effect on how we behave in any environment that encompasses the technology.

Sticking with the car theme, the Model T greatly impacted road transportation. It effectively started the trend in general usage of the car as a means for transport. In 2015, there were about 1.3 billion cars in use, worldwide.[9] With IoT, fully autonomous and thus driverless cars are likely to be the next major wave of transportation innovation. This could lead to a considerable reduction in car usage, particularly in more industrialized countries, where cars, which are currently not in use and parked 90% of the time, could be replaced with public self-driving taxis.

Changes in Technology Support Infrastructure

Take a moment to consider broad IoT deployments at any large office building. One can imagine solutions such as facial recognition—and other biometrics—for employee access and intrusion detection, smart thermostats for regulating the temperature, or auto-stocked vending machines for food and office appliances.

Such an organization could easily find itself loaded with hundreds and perhaps thousands of IoT devices. Very few organizations are able or willing to take on the task of managing all those devices.

Today, most large organizations contract out the supply, management, and repair of IT devices such as computers, computer peripherals, and networking equipment. This approach is likely to be replicated with major IoT deployments.

IoT devices, including edge devices such as smart thermostats, run software. The deployment and management of the software on those devices is critical to the success of the entire solution. Similar to the last point, the overhead in deploying and managing IoT software is likely to give rise to middleware-oriented businesses geared toward the provision and management of software applications on IoT devices.

Currently, smartphones are powered by two major operating systems—Google® Android™ and Apple® iOS®. Third-party software applications (not bundled with the operating system) are generally deployed via digital distribution platforms, which allow any software developer to publish software that smartphone users can download onto their devices. A similar system is likely to arise for IoT systems, as software authors and IoT users seek to extend or customize the basic software solutions set that most IoT vendors can be expected to deploy onto their creations.

Changes to the Economy

It is relatively easy to judge that the various IoT solutions and prospects already explored in this chapter will impact the economy. Autonomous cars and smart power grids are just two examples that can be expected to have far-reaching effects. Please refer to Chapter 3 for our detailed exploration of the regular economics of IoT and how cyber criminals subvert that economy to make money by compromising IoT systems.

Changes to Security

As we have seen in the preceding chapters of this book, security risks are inherent to the nature of IoT systems. This is due to factors such as their connectivity, the sensitive data they process, and the compute power that is available via IoT devices. We have also seen that those risks can be greatly reduced with secure system design and development.

This chapter's exposé on the future heralded by IoT will describe the imminent security risks in that future.

11.3 The Evolving Cyber Threat Landscape

In Chapter 2, Section 2.3.1, the current threat landscape was analyzed in great detail. We sought to understand who might want to attack our IoT systems and their motives for attack.

In Table 11.1, we take things further, as we explore the evolution of the threat landscape in the coming decades.

11.3.1 Threat Agents and Cyberattackers of the Future: AI and ML

We expect that artificial intelligence (AI) and machine learning (ML) systems will emerge as a new form of attacker.

AI refers to the field of computer science that is involved in the development of machines that mimic the "cognitive" functions that humans associate with human minds, such as

Table 11.1 The Evolution of the Threat Landscape

Threat Agent	Description	Vision 2040
Cybercriminals	Perpetrators of organized crime for money, with computers and networks as both tools and targets	A detailed analysis of cybercrime in IoT is covered in Chapter 3—The IoT Security Economy
Nation states	Country- or government-backed cyberattackers	Heightened tension among industrialized nation states as the cyber capabilities become akin to nuclear capabilities
Indstrial spies	An organization's homegrown or hired attackers who wish to steal from or otherwise harm the competition	Unlikely to change*
Hacktivists	Attackers with a political agenda	Will require greater skill, as governments will beef up security as a matter of course
Security researchers	Skilled security experts seeking unknown or undiscovered flaws in existing systems and protocols	Unlikely to change*
Malicious insiders	Disgruntled employees or contractors	Unlikely to change*
Script kiddies	Hobby hackers	The ability to make any significant impact will further reduce as defenses and legal penalties increase

* Unlikely to change: Tools and techniques might change, but the attacker type and attributes will be relatively unchanged.

complex and adaptable "learning" and "problem solving." ML is a related, but much more specific, field that seeks to give computer systems the ability to "learn" (in this case, by progressively improving performance on a specific task in a specific problem domain) with data, without being explicitly programmed.

Today, ML systems are actively used to build defensive systems that can identify anomalous behavior in a given environment. For instance, network intrusion detection systems aim to spot attackers seeking to compromise a network via traffic patterns and attack tool signatures.

Cybersecurity is a cat-and-mouse game that pitches "bad" guys against "good" guys in a timeless battle for supremacy. Hence, we can expect that if defenders use ML, attackers will use such tools and approaches as well. AI systems are yet to be publicly or commonly deployed for attack or defense, although it is not unlikely that the deep-pocketed nation states already have such capabilities.

As cybersecurity expert, Bruce Schneier, put it: "Today's Nation state secret techniques are tomorrow's PhD theses and the following day's cybercrime attack tools."[10]

Three future AI/ML attack types are as follows:

1. AI used by attackers to create systems that can intelligently develop evasive malware—that is, machines creating bad software.
2. AI used to build systems that intelligently discover vulnerabilities in computer systems and networks. Currently, automated tools are used to scan and infect vulnerable systems via the Internet. AI will take that further, creating an ability to intelligently discover

vulnerable patterns or flaws that even the most skilled security researchers find hard to spot. Closer, perhaps less-scary versions of this are ML systems used to learn vulnerable patterns in a specific domain. We can expect that such ML systems are in use already.

3. ML-based attacks will be deployed to subvert cyber defenses. ML-based cybersecurity systems are designed to classify certain things as good and others as bad—that is, good network traffic and bad network traffic, good software and bad software (malware). An attack that causes such systems to classify bad as good is a hit for the bad guys. ML systems are heavily dependent on the data used to train classification models and the data they have to classify. ML systems often learn on the fly—that is, using the data processed in real time to update their understanding of normal and abnormal. An attack that is able to supply lots of bad data, undetected, will cause the system to learn incorrectly.

Garbage In, Garbage Out . . . AI Style

In 2016, Microsoft® created "Tay," an AI chatbot modeled to speak like a teenage girl. Tay was developed to improve customer service for Microsoft's voice recognition software. Tay was designed to learn from its interactions with users. Embarrassingly for Microsoft, Internet trolls (people who sow discord on the Internet by starting arguments or upsetting people, or by posting inflammatory, extraneous, or off-topic messages in an online community) taught Tay to swear and utter the most shocking profanities in less than 24 hours. It had to be taken off the web.[11]

Returning to our examination of the attributes of threat agents, ML systems are not independent agents, since they are not designed to "think" on their own. It becomes evident that any threat agent attributes of an ML system will be those of its creator or user (trigger puller), who is the actual threat agent.

In the near future, the same will be mostly true for AI. Today's AI is able to already able to learn in new ways or learn new things, but if or when AI becomes truly successful—that is, smart and autonomous enough to function, think, and create independently of humans, in a sustainable manner—that changes.

Currently, and for the next few years (five to ten), the attributes of an AI agent are are follows:

- **Goals:** Depends on the creator or user
- **Risk tolerance:** Depends on the creator or user
- **Work factor:** High to extreme
- **Methods:** Very sophisticated and very unique

It will take much skill and effort to create AI capable of intelligently infiltrating systems. But it is also expected that just as many cyber attack tools, such as the Mirai botnet, are released free of charge on the Internet, advanced AI attack tools will be released as well. In addition, it could be possible to repurpose advanced AI designed to address non–security-related problems, in order to use it for discovery and exploitation of security vulnerabilities. Such redesign will not take nearly as much skill as the creation of AI from scratch, thus lowering the bar for the average cybercriminal.

11.4 A Vision of 2040

Now, it is time for us to step into a time machine and travel two decades into the future.

In this section, we will explore the future of IoT, covering different economic sectors as we go along. Our voyage will take us from healthcare to agriculture, smart cities, energy, and more.

For each sector, we will:

1. Seek to understand the sector's objective and the current usage of IoT.
2. Ask the question, "What's next?," as we describe future scenarios.
3. Consider the technical and architectural implications of future scenarios.
4. Wrap up a detailed analysis of the security implications and concerns from scenarios described.

11.4.1 Healthcare

Health is wealth is an all-time favorite truism that all teachers and parents seem to love drumming into their kids, from the United States to Nigeria. They are right. After all, nothing is really enjoyable—even the most intriguing code wonking or systems hacking—if one is in poor health. We all care about healthcare, which caters to the improvement or maintenance of health.

The field of healthcare or medicine has never been slow on the uptake of technology innovation. Thankfully, the days when healthcare professionals hacked away at their patients like butchers carving up beef are long gone. Now, advances such as microsurgery allow doctors to perform surgical operations with needles tiny enough to pass through a human hair—yes, you read that correctly, a surgical needle through a human hair. In recent years, many top hospitals have been quick to try out and incorporate IoT solutions and machinery.

Hospitals in technologically advanced countries are beginning to adopt beds fitted with sensors, allowing hospital administrators to receive auto-notifications once beds are freed, reducing wait times for new patients by hours.[12] Smart beds also monitor patients' signs such as heart rate, breathing, and sleep restfulness.

Healthcare equipment such as MRIs are also getting their sensor fix. This allows hospitals to monitor the health of their equipment and reduce failures or outages. For patients, there is continued innovation in technologies that make use of sensors to monitor vital signs such as temperature, breathing, heartbeat rhythm, blood sugar, etc. Traditionally, such signs or symptoms helped healthcare professionals diagnose diseases and other medical problems. For instance, the disease referred to as diabetes is caused by having too much sugar in the blood stream. Sensors (and devices) that measure blood sugar keep patients well informed, allowing them to adapt smartly. Sensors for patients are usually made either in the form of external patches affixed to the skin or as subcutaneous implants.

The sensors described above are usually connected to an analytics system that crunches the data they process before presenting alerts and suggestions to healthcare professionals or patients.

It is evident that IoT has the potential to increase healthcare efficiency by allowing healthcare professionals real-time access to the vitals of patients, the status of hospital equipment, or the availability of wards or doctors. This can greatly reduce noise in an often hectic domain, placing healthcare professionals in a better position to leverage their meticulously acquired skills to ascertain the health (or lack of it) of their patients, the health of their equipment, and the health of the overall hospital process flow.

What's Next?

IoT will play a central role in the running of major hospitals in most industrialized countries by 2040. Undoubtedly, this will affect the lives of many.

For starters, I expect that many patients will be transformed into cyborgs—humans with both organic and mechanical body parts. Although a good number of people might not have or need robotic organs or appendages, biological sensors will be widely adopted. This is important.

A system of biological sensors connected—via edge gateways and public networks—to cloud analytics systems means that healthcare professionals will be able to monitor the health, symptoms, and disease progression or regression of patients, in real time. In most cases, patients will also have access to that information, providing them with clarity and education concerning their condition and proactive steps that ought to be taken to aid healing as well as the medical procedure(s) or drugs that they have been or will be administered. For instance, a diabetic patient with an insulin patch/sensor (either on the skin or via an implant) will be able to receive real-time alerts on blood sugar levels, based off predefined thresholds that can vary, depending on the patient. Ingestible sensors in pills will also let the patient and physician monitor how well the patient adheres to the required schedule for using any given pill.

The hospital will change drastically as well. Remember those long wait times for access to a healthcare professional? Gone! Many appointments will be scheduled upfront, but that is the case already in technologically advanced economies. A major leap forward will be the automatic scheduling of medical appointments with the aid of biological sensors that enable the generation of opt-in appointment alerts.

Similar sensor capabilities and solutions will eliminate waste and inefficiencies in hospital administration of facilities such as beds, wards, and operating theaters. To illustrate further, the occupancy or expected occupancy of either bed or operating theater, based on disease or surgical procedure, will be monitored and predicted, directly influencing auto-scheduling of such facilities.

The scenarios just described involve lots of data processing, with alerts and notifications generated and sent off, based on the data. Data processing will either be handled by hospitals or contractors hired by the hospitals. One thing is clear though, regardless of how it is done; hospitals will become data analytics and data warehousing companies. This means that some hospitals will deploy their own data centers, some will use third-party cloud data center offerings to store and crunch data, and others will turn over data crunching to third-party agencies almost completely.

Data, very private data, will be everywhere, often transmitted by public networks:

- Hospitals to remote patients and vice versa
- Hospitals to third-party cloud providers, software/middleware providers, data analysts, or IT administrators
- Hospitals sharing data as the patients switches medical providers
- Hospitals and insurance agencies

Technical Implications

By 2040, many hospitals will be data aggregators. Dashboards will become key, so as to enable doctors and patients to make sense of all the data flying around. Most dashboards will be built as mobile and web applications. This will make such applications central to how we monitor, examine, and understand health. Dashboards will exist for both patients and hospital personnel. Patients will use dashboards to keep tabs on their health, and doctors and hospital personnel will be able to stay informed about the health of their patients and the health or status of the hospital.

We can expect that multiple sensor types will be connected to central dashboards. The sensors will support different networking protocols such as Bluetooth®, WiFi, ZigBee®, and Z-Wave®, as well as different messaging protocols such as MQTT, XMPP, and AMQP. Please check Chapter 7 for an in-depth study of IoT communication protocols.

Currently, networking and messaging protocols are being actively researched to improve existing technologies and identify new ones, in order to reduce the power and processor compute drain on IoT edge devices, including sensors. It is likely that new protocols will continue to emerge, even as it is evident that compatibility or standardization of IoT networking or messaging protocols is a major need that will simplify the adoption and maintenance of IoT solutions. For instance, can a Z-Wave gateway talk to a ZigBee sensor? Or can an AMQP messaging server process MQTT clients?

Data privacy and security will also be a major concern, as very personal health data traverses public networks—the Internet. At the edge, in transit, and at the back-end services that crunch data and present dashboards, strong encryption will be required to protect the personal data of patients.

Currently, back-end services that store medical data must be HIPPA compliant. However, will that suffice when those services also store real-time data about the activities and health of a patient?[13]

Dashboards are not limited to data presentation. It is to be expected that hospital administrators or healthcare professionals will have a means by which they can send commands to IoT devices and update them. This command and control functionality gives great power to any users who can access the tool in the hospital, and as such, access must be restricted. Username/password based–authentication will not suffice. Two- or three-factor authentication must be enabled.[14] Access control to the configuration mechanisms of the sensors and gateways with patients must also be strictly enforced.

Back-end services often require a connection to the Internet, if they are to receive data from remotely deployed IoT devices. This is the same whether back-end services are hosted locally by the hospital or via cloud providers. This increases the attack surface of the hospital and its patients.

Few hospitals will have the technical capacity to manage the hundreds to thousands of IoT devices on premise, and potentially tens of thousands and more that will be deployed with patients outside the hospital. Enter middleware-oriented businesses who will partner with hospitals and IoT solutions vendors to provide a platform that enables seamless management and updates of software applications resident on the IoT devices.

IoT devices tend to have a lifetime of about a decade.[15] Lifetime refers to the expected usage term. The lifetime expectation of a fridge (even if it is "smart") is much longer than that of consumer products such as laptops and smartphones, which usually have a lifetime of about 18 to 24 months. Hence, the components in IoT solutions and devices will have to be maintained and updated for quite a while. The update mechanism of any software on IoT devices must be hitch free. Otherwise, hospital solutions (and many others) will carry security vulnerabilities for years. IoT device manufacturers can also find themselves in the unenviable position of paying huge sums of money to recall and replace flawed devices. Even then, not every device that has been sold comes back in a recall. Many stay in the wild, gleefully vulnerable, inviting the next hacker to an easy prize.

Security Concerns

Dashboards might seem like a tiny piece of an IoT solution, but there is no getting away from them. They are the portal through which the lay person unlocks the wizardry—or shall we say the passionate engineering—of an IoT solution. We kicked off the previous section by indicating that dashboards will continue to be mostly mobile and web application based. Such applications have been extensively analyzed and attacked by security researchers for years. This will not change. Although circumventing security controls in web and mobile applications is not novel, it is important to note that dashboards present a juicy and well-known attack vector for threat agents seeking to compromise healthcare systems. The threat agents need not be very skilled either. Script kiddies with access to a web or mobile dashboard can be expected to try run-of-the-mill hacking tools and techniques used for those platforms. Dashboards for patient usage are likely to be easily accessed by script kiddies.

Another well-known attack vector is the network. Such attacks are actually older than web and mobile attacks. As hospitals get connected, locally or within the hospital, to cloud back ends, and in some cases to patient's home devices (i.e., if a hospital were to host a service locally that patients' client devices can access), they become increasingly susceptible to network intrusion and distributed denial-of-service (DDoS) attacks. The risk here is exacerbated by the fact that in the near future, much hospital equipment will be connected—that is, not just computers but beds and MRIs. A compromise of any device on the network potentially places other devices at risk. It goes without saying that attacks, including DDoS, which make a hospital's computers or equipment unusable (even for a few minutes), will place the lives of many at risk.

'US hospital pays $55,000 to hackers after ransomware attack'

In January 2018, Hancock Health revealed that cyberattackers had infected their IT systems with the strain of malware known as ransomware. Their files, including health records, were encrypted and inaccessible. The cyberattackers requested four Bitcoins in payment, the equivalent of $55,000 at the time. They were given seven days to pay up, before the files became permanently encrypted. After a few days, Hancock health paid up.[16]

An intriguing and often dramatized component of information security in healthcare is that successful attacks in this space can directly put lives at risk. Interestingly, even though we are yet to see a trend in attacks actively perpetuated in an attempt to harm an individual, we know the stakes are high. If it is possible to harm one life, that is a problem. It is also evident that the ability to target and harm an individual can only increase as patients and, indeed, all humans get physically connected to the Internet.

The theft and sale of medical data is a well-known cybersecurity concern in healthcare. That is not going to change. Cybercriminals are notorious for selling medical data for a pretty penny on the dark web.[17] Please refer to Chapter 3 for a detailed exploration of the dark web and how cyber criminals use it for nefarious gains.

As mentioned earlier, robust yet seamless authentication and fine-grained access control must be deployed to restrict access to the central systems and dashboards used to control edge devices. In an environment such as a hospital or a healthcare facility, people just want technology that works. Authentication (and other information security controls) that is difficult to use will either be misused or turned off. In addition, there is a need for proactive account management. For instance, healthcare professionals who have left the hospital must be unable to access the IoT systems, starting immediately after their very last moment in the office.

Data is king now and will continue to hold its royal place. Homomorphic encryption[18] could be the holy grail that allows the sharing of data with external parties who are able to use that data without accessing the potentially private information it contains. Homomorphic encryption allows computation on encrypted data without the need to decrypt that data. This could be a security solution that allows hospitals to securely share data with entities that need to understand patterns in the data without understanding any necessary specifics of said data. Currently, this problem is tackled by data anonymization processes that must be painfully tailored to each instance of data sharing, and the data involved.[19]

Besides the privacy of data shared by users, there is a larger, although related, security concern regarding the privacy of the users or individuals themselves. This is not difficult to imagine considering the expected proliferation of sensors and IoT edge devices that can pick up the location, actions, exertion, utterances, etc., of their users. Due to such concerns, it becomes essential to allow users to control what is shared about them and whom it is shared with. This is very similar to the current permissions model for today's mobile applications and smartphones.[20]

Possible Consequences of Successful Attacks

Data theft, fraud, unavailability of healthcare when needed, and physical harm to individuals.

· · · · ·

As we just saw, the future of IoT impacts healthcare through the introduction of new products and solutions, along with changes to our lifestyle, as the administration and delivery of healthcare evolves; it introduces new requirements for technological support infrastructure and directly impacts the economy as well as the security of hospitals and patients.

11.4.2 Agriculture

It is very unlikely that we would be writing this book if we were starving, and it is just as unlikely that you would be reading it, if you were. Needless to say, agriculture keeps us here, providing the fuel for all other pursuits.

For most of us, farming is rarely the first or second thing that comes to mind when we consider technology. But contrary to what most of us might think, farmers have long been ardent supporters of technology. Their profit margins can be tight, depending on their crops or livestock. This is exacerbated by the unpredictability of environmental conditions such as temperature, rain, or pests. As a result, farmers are often willing to consider any technology that can help to reduce operating costs and thus increase profit, depending on their budget and the cost of the technology, of course.

Today, IoT technology is having a very positive impact on agriculture, in areas such as precision farming, livestock monitoring, pest control, and self-driving tractors. In precision farming, sensors are used to acquire a wealth of data concerning soil composition, humidity, temperature, air quality, and about enabling farmers to make smarter decisions. The sensors mentioned are often connected to automated fertilizer and sprinkler systems, achieving efficiency and reducing waste. To control pests, solutions have been developed that are able to sense the presence of large amounts of certain pests, leading to the dissemination of phero-mones that disrupt their mating process.[21]

Livestock-monitoring solutions allow the remote observation of the health, behavior, and location of livestock. One innovative solution from Cattle Watch involves the deployment of solar-powered hub collars on 2% of cattle, while other cattle don collars or ear tags powered by lithium batteries. The hub collar is responsible for the transmission of data to a remote satellite network.

Finally, self-driving tractors have been operational in agriculture for years. This should not be surprising considering that the operation of a self-driving tractor on a farm is far less complex and entails far fewer variables, when compared with a car that has to navigate hundreds of other cars and pedestrians. The tractors on the market are not fully autonomous and require a human behind the wheel.

What's Next?

As mentioned earlier, agriculturists have long been avid technologists. By 2040, there will be fully autonomous tractors responsible for tilling the soil, planting, and fertilizing the crops. Robust and varied sensor solutions will be used to monitor the health of the farm. This will include the monitoring of weather conditions, soil composition, and crop maturity. Fertilization and irrigation will be fully automated, using data provided by the sensors.

The data harnessed from the agricultural farm just described will provide its owners with invaluable data that influences what crops they plant, when they plant, when they harvest, and when they put cattle to pasture. The automation of farms also means that the data produced by multiple smart farms can be stored and accessed via centralized databases.

Currently, food databases on the web are sourced via a virtually manual process that incorporates a few hundred farms. In the future, a large number of automated farms will supply data seamlessly. This will facilitate the creation of a global and open dataset for food production in different areas. That data will be fed into ML systems, enabling governments and private sector entities to predict crop yield and crop needs, and to forecast any need for intervention in specific locations.

However, we do not expect that the technologies just described will be universally available on all farms by 2040. Rather, they will be in operation in the most industrialized countries, and in most advanced farming operations.

Technical Implications

In Chapter 4, we mentioned that the core architecture of all IoT systems is pretty much the same—that is, of sensors and actuators connected to a central gateway that relays data to and from a back-end system. The back-end system crunches data, providing insight that either enables system owners and administrators to make informed decisions, or automatically influences a related action elsewhere in the system. On a farm, such automated actions could be the dispersal of water or fertilizer, noisy sounds to frighten predators, pheromones to confuse pests, or even alerts to system administrators.

The main technical implications of IoT-enabled agriculture are similar to what we expect in healthcare. A major difference between both environments is data sensitivity. We can assume that data processed by IoT solutions on the farm is proprietary to the owners of the farm. But it is not as private as the information that is regularly accessed and manipulated in healthcare.

It is also fair to say that a cyberattack on agricultural IoT has less of a direct impact on human lives when compared with healthcare. However, there is indirect and considerable impact on the food produced, since the solutions we have described are responsible for the dispersal of fertilizers and other agents that affect the food produced. But it is unlikely that any produce will be sent for public consumption without prior quality analysis.

This indicates that although agricultural IoT will require a very good level of security, it does not required the extremely high level of security controls necessary in healthcare.

Security Concerns

The threat model for agriculture comprises three primary threat agents or malefactors:

1. **Malicious Insiders:** System administrators with bad intent
2. **Cyber Espionage/Industrial Spies:** Entities seeking to steal confidential information from a competitor, so as to gain market advantage
3. **Cybercriminals:** Individuals expected to try either the deployment of ransomware or the hijacking of the computers as part of a Botnet.

All three threat agents enumerated above indicate that, perhaps unsurprisingly, data and compute power are the assets—at least at a high level—of an agricultural IoT solution. This leads us to the basic requirement that all usage and configuration access to each component of the system must employ appropriately robust authentication. In addition, data should always be transmitted over a secure channel. None of this is novel, and is actually quite routine.

However, depending on the preferences of the system owners, data sent between equipment on the farm might not be sent via a secure channel. This is due to the low sensitivity of the data being passed, coupled with the compute overhead for devices handling encryption. The decision concerning the encryption of every communication among IoT components on the farm is made by management. This decision should be based on risk analysis that is informed by a comprehensive threat model. All communication exiting the farm must be encrypted.

Functional safety is also a security concern for any self-driving or fully autonomous vehicle, including tractors. Whereas we can describe safety as freedom from unacceptable risk of physical injury or of damage to the health of people, functional safety in a mechanical device is the part of the overall safety that depends on a system or equipment operating correctly in response to its inputs. It involves the detection of a potentially dangerous condition, resulting in the activation of a protective or corrective device or mechanism to prevent hazardous events from arising or providing mitigation to reduce the consequence of the hazard.[22] Although functional safety does not seek to address the prevention or mitigation of cyberattacks, security professionals and business leaders must consider that the safety of the system is related to its security. For instance, can a cyberattack or malicious data cause a functional safety failure?

Possible Consequences of Successful Attacks

Intellectual property theft, hijack of farm equipment as part of botnet armies, financial losses to farm owners, and a less pertinent risk of bad farm produce.

· · · · ·

The discipline of farming continues to undergo promising and exciting changes. This is influenced by the combination of embedded computing, which has existed as a field and practice for many years, and the ultra-connectedness ushered in by IoT. By 2040, advanced automation, data gathering, and AI promises to drive down operating costs while preserving production efficiency and increasing the quality of farm produce.

As we have seen, industries and customers seeking to adopt IoT on a large scale will require a major technical support infrastructure to deploy and maintain the IoT solutions. Currently, manufacturers of technologies such as sensors and gateways are the entities that are primarily responsible for the deployment and maintenance of their products. But, for connected farms to flourish, a pluggable architecture is desirable. Pluggable architectures refer to systems and components designed such that sensors from one manufacturer must be able to interact with a gateway from another manufacturer. In turn, such a gateway ought to be able to connect to any cell tower or satellite network, regardless of the manufacturer, wherever necessary.

11.4.3 Cities and Homes, Energy, and Autonomous Transportation

We hope you have found our time travel as intriguing and fun as we have. And there is more to come. On this final leg of our trip, we shall attempt to be even braver in our exploration. We will combine our analysis of cities and homes of the future with the projected effects of technologies such as autonomous driving and alternative energy.

By 2050, it is estimated that about 60% of the world's population will live in cities or urban areas.[23] Cities can be most simply described as human settlements with a high-population density and numerous man-made infrastructures.

Most cities provide infrastructures or services such as housing (homes), transportation, water, sanitation, electricity, security (law and order), entertainment, sewage, and governance (leadership and administration). It is easy to see that the infrastructure provided by the city is central to the existence of city dwellers. Due to the density of cities, they are often the powerhouses of most states or countries. As such, they tend to be the focal point for most business and government institutions, thereby also affecting the lives of those who live outside the city.

Inevitably, any changes to how a city is run or how various services or infrastructures are provided affect the lives of its inhabitants.

As we discovered in the earlier "Now" section, before our time travel began, a number of innovative city administrators and occupants have begun to build and pilot solutions for their cities. For instance, smart street lights that only turn on when cars are around, connected waste management disposal bins, traffic lights that analyze traffic congestion, autonomous driving, pay-as-you-go decongested driving lanes, etc. Last but not least, and well worthy of many books of its own, is the smart power grid, which is still evolving. The goals of the smart grid include enabling users to monitor their power usage in real time, enabling end-user appliances to intelligently request the amount of power they need at a particular time, effectively integrating alternative energy sources such as wind and solar, and many more.

In homes, current IoT solutions are used to provide improved security, energy-usage management, and ambience automation. This is not surprising, as the priorities at home for most persons are security, control, and comfort.

What's Next?

About two decades from now, many homes, particularly those in cities, will be remotely accessible. Sensors controlling everything from the locks on the doors, to temperature and lighting, will be remotely accessible via the Internet. Smart power meters will be connected to smart grids, enabling smart devices and user appliances in the home to track peak power usage periods and current costs of electricity, and to adjust their demand for electricity or even shut it off. Similar to what we expect in agriculture, many a lawn will be mowed by autonomous robots and automatically sprinkled by systems that monitor humidity.

Cities have the potential to change drastically, depending on how much connectedness is embraced. First, fully autonomous cars will relegate most drivers to the passenger seat. As a result, I expect accident rates will plummet, parking problems will just about vanish, streets will narrow, and commuting by automobile will become a mere extension of sleep, work, and recreation. Vehicles, including cars, buses, and trucks, will be able to pick up multiple passengers, verified via a city or national database, thus reducing the need or requirement for individuals to own an automobile. We can expect that traffic congestion will eventually become a thing of history.

Crime will drop with the inclusion of sensors and cameras in vehicles, road signs, traffic lights, street lights, and buildings.

City dwellers can expect better health as well. Fewer vehicles in operation means less pollution. In addition, pollution will be actively monitored and reliably forecasted via sensors strategically placed around the city, aiding a continued pursuit of reliance on green energy.

Cities can also expect to see significant savings in the administration and maintenance of infrastructure such as power lines and water and gas pipelines. All such facilities will feature sensors connected to a central monitoring system, alerting system administrators to leaks, predicting problems in the lines, and enabling administrators to proactively monitor the health of power lines and pipelines across the city.

For years, embedded computing has existed in various forms in some of the systems we just described. The beginning of IoT. A little sensing here, some monitoring there. But it is a different world when all those computers are connected to the Internet, proactively sharing or learning about ongoing events.

Technical Implications

The infrastructure and solution sets required to bring all that was just described to life are quite extensive. As we described in Chapter 1, the vision of IoT is computers everywhere, augmenting and improving our ability to interact with and react to our environment. In a smart city, that certainly holds true.

Hence, in a smart city or smart home, the following major elements must work together in a variety of interconnected systems:

- Sensors and cameras that pick up data.
- Network communication media, wired and wireless, that transfer data.
- ML systems that grind out patterns from data, creating actionable knowledge. Some of such systems will exist on devices in the city, such cars, and will also be hosted in remote systems in the cloud.
- Humans or technology that receive and act on raw data or knowledge distilled by ML.

To further illustrate, although one autonomous car must have all the aforementioned systems functioning, it must also be able to analyze what it can see about hundreds of other cars that have similar systems. Additionally, devices such as smart traffic lights will also be observing cars and vice versa.

A very complex system.

To help us consider the high-level technical implications of such a system, the following list comprises some of the major components required for large, complex environments such as smart cities, smart homes, and autonomous vehicles:

1. **Processors:** IoT means we have processors in different environments and devices that have different constraints, such as performance, security, and reliability. The requirements for a processor in an autonomous car are very different from those we find in a street light, and both of these are quite different from the requirements of a fitness tracking armband.

   As an example, the processor in a driverless or autonomous car must be powerful enough to process and analyze data locally, in real time.

2. **Edge Devices and Local Infrastructure:** As discussed in Chapter 1, edge devices such as sensors and gateways get things going in the "smart" IoT life cycle by performing data gathering. Local infrastructure such as the network in an autonomous car plays a central role, in that it passes on data to the back end or receives control information from the back end, for the edge device. Such devices and all local infrastructure must be designed quite securely.

3. **Critical Infrastructure:** This refers to all processes, systems, facilities, technologies, networks, assets, and services that are essential to the health, safety, security or economic well-being of a community people and the effective functioning of their government.[24]

Our definition includes networks and technologies as part of critical infrastructure. However, in a smart city of the future, large scale IoT deployments will be used to automate, manage, and improve processes, facilities, and services such as power delivery, water delivery, transportation, and waste management.

As a result, connected technology—to a local network and sometimes to the Internet—not only becomes an element of critical infrastructure but also a major part of all critical infrastructure.

With IoT, critical infrastructure becomes a very dynamic attack vector.

4. **Communication Infrastructure:** Here, we are referring to basic communication facilities required to make IoT solutions possible. For a city to comprise publicly utilized IoT solutions, including autonomous cars (as previously described), neighborhood or citywide communication infrastructures (such as fiber optic cables, cell towers, or phone lines) must be available. This is why IoT solutions such as smart cities, homes, or autonomous vehicles will first come to maturity in industrialized cities that already possess a reliable communication infrastructure.

It can also be expected that not all vendors of IoT solutions will have the funds required or indeed care to reserve funds for the communications infrastructure required to connect the different pieces of their IoT solutions. This means that, similar to what exists today, where large telecommunications providers lease their infrastructure to smaller providers serving niche markets, large IoT communications providers (including today's telecommunications and Internet access providers) will deploy and lease the communications infrastructure required to connect the different IoT components in a smart city, including numerous homes and autonomous cars.

5. **Cloud Back-End Platforms:** Large IoT solutions providers will develop cloud back ends catering to various IoT needs. This will arise quite naturally from the need to receive, store, and process data received from IoT edge gateways and sensors deployed in the field. Multiple cloud-based services will spring up, allowing access to shared compute infrastructure—for example, Infrastructure as a Service (IaaS)—which exposes databases, knowledge, or inferences derived from ML, web applications or services that run over the underlying data, and alert or notification frameworks. Please see Chapter 6: Securing the IoT Cloud, Section 6.2, Table 6.1, for the different cloud-service models, including IaaS and SaaS.

Similar to communications infrastructures, cloud back ends will be shared. In a simplistic example, we can consider a generic SaaS agricultural IoT back-end solution, which allows innovators in specialized agriculture such as fish farming to leverage the data and back-end infrastructure put together by the larger SaaS agricultural IoT solution.

6. **Application Enablement (and IoT Management) Platforms:** This is a type of cloud back-end platform, but we considered it worthy of being presented separately from the previous point. This is because, rather allowing shared access to shared databases, or software and inferences based on shared data, this category of broad IoT deployment provides IoT companies with the ability to use IoT solutions and devices while outsourcing the headache of managing and monitoring the devices in the field.

SaaS of this kind, some of which are already in the field today, will be essential for any form of broad IoT deployment, enabling businesses to administer and monitor their devices in real time. Device management could also include firmware and software updates, when supported by the underlying device.

For instance, a city government with thousands of street lights and stop-lights deployed will have a hard time creating technology for monitoring and control-ling all of them, especially since technology management is not their expertise. The vendors of certain IoT solutions for smart street lights and stoplights might also provide management and monitoring tools, but what happens when a city deploys multiple IoT solutions and devices from different vendors? That is where Application Enablement and IoT Management Platforms come into play. At the time of writing, examples of such platforms are Jasper™, Cumulocity™, and 2lemetry™.[25]

Security Concerns

From our discussion thus far, it is evident that in a smart city, many lives are simultaneously impacted or perhaps at risk due to IoT.

Our overview of security concerns will cover smart cities, smart homes, and autonomous cars, in turn.

1. **Smart Cities and Energy:** A defining characteristic of smart cities is that critical infra-structure will be connected. This provides a wide attack vector through which a wily attacker could directly affect the lives of many. The smart city of the future is one in which energy grids (electricity and alternate sources), road traffic control, water provi-sion, and waste disposal are connected to sensor networks and the cloud. Successful attacks via DDoS, admin systems, and dashboards (both in energy-generation facilities or in the cloud) could ensure that inhabitants of one or more sections of a city do not have access to basic amenities. In a smart city with IoT-enabled critical infrastructures, defensive software and technology, secure defaults, and fail-safes are non-negotiable.

   In Chapter 2, we described a DDoS attack that caused a building in Finland to lose heating in winter. Not to join the doomsayers, but can you imagine such an attack on an entire city? How about the disorder that would necessarily ensue if smart traffic lights or stoplights were to be hacked?

   The evaporation of privacy is also a major concern, as was shown in Chap-ter 8. Increasingly, many security and privacy experts believe that the existence of digital privacy is an illusion. Daily, many of us use computers and smartphones, with location-aware apps that are often on, constantly monitoring our location. Very few people perform the due diligence required to ensure that their privacy is protected by the technologies that consume information about them. Still, is it fair to wonder whether the government of a city controlling sensors and cameras all over the city could be the official death knell to the privacy of the individual?

   Possible Consequences of Successful Attacks: *Physical harm to many indi-viduals, acts of war or holding a city (or nation) ransom, large scale nation-state espionage, and data theft.*

2. **Smart Homes:** Automation of access via smart locks and remote management or moni-toring is often marketed as providing improved security and control to the homeowner.

Never mind that unless such solutions are intelligently built with strong end-to-end security technologies and protocols, a happy hacker[26] in the neighborhood or employees of the IoT solutions vendor could easily access or monitor the home.

Possible Consequences of Successful Attacks: Hijack of home devices as part of botnet armies, financial losses to homeowners through cyberattack-induced robberies, and data theft (including spying).

3. **Autonomous Driving:** A rather obvious concern with autonomous vehicles is that a successful hack places the lives of individuals at risk of direct physical harm. For instance, a terrorist who is able to hack into an autonomous car is not only able to hurt or kidnap targeted persons, the hacked vehicle can also be used to run over pedestrians. Autonomous vehicles can also be expected to contain some sensitive information about their users, such as daily routes, home address, and biometrics (for access to autonomous public transport systems). Such information becomes dangerous ammunition in the wrong hands. In addition, vehicles, like homes, are private spaces. Any security holes in an autonomous vehicle places the privacy of commuters at risk. An additional privacy concern is the potential access and usage of video and audio feeds by manufacturers or owners of the autonomous vehicle.

Possible Consequences of Successful Attacks: Harm to individuals and data theft, which includes spying.

· · · · ·

From autonomous cars, to IoT-enabled stoplights and street lights, as well as robotic lawn-mowers, we can see that the functional safety of these devices, their software, and networks is an immediate concern. The objective of functional safety is freedom from unacceptable risk of physical injury or of damage to the health of people, either directly or indirectly.

Privacy, on the other hand, seems to evaporate. That is a concern. With sensors everywhere in a smart city, are the residents always monitored by the authorities? What happens if nation state attackers were able to silently infiltrate the networks that connect those sensors and the facilities that support critical infrastructure? Does this dawn a new age for blatant yet "unrisky" espionage?

Many questions. As we consider them, we find ourselves getting into the realm of the security of the State. It is evident that we must be serious about building and deploying robustly secure IoT systems, if those systems are intended to safely run our cities and our lives.

11.5 The Emergent Future of Cloud Computing

In Chapter 6, we delved into the cloud back end of IoT solutions and explored how the cloud enables and scales IoT security.

Now, we will spend some time examining four emergent trends in cloud computing and their security implications. Spoiler alert! Most of the security concerns we unpack here were addressed in Section 6.6 of Chapter 6, as we designed a secure IoT cloud. But at that time, we were succinct about the specific mitigation steps necessary to create a secure cloud. Here, we will take a broader look at the cloud features before providing a brief security summary that should shed more light on the relevant mitigations that were applied in Chapter 6.

As you well know, cloud computing services are based on virtual infrastructure. This means that, by design, they are plastic—easy to deploy, modify, scale, and tear down. Any infrastructure architect or data engineer will tell you that this is not nearly the case with "physical" infrastructure. The plastic nature of cloud computing is huge for IoT, as it provides the flexibility required to support thousands, millions, or even billions of IoT devices.

Furthermore, cloud computing, particularly public cloud computing infrastructure, has become a cornerstone for deploying tech solutions without the need for an up-front cost of capital investment in local infrastructure. It also eliminates the waste in unused local compute, due to the pay-as-you-go or pay-when-you-use cost structure of public cloud computing providers.

Unsurprisingly, the points discussed above have created sustained incentive for cloud computing providers to innovatively explore ways to reduce the overhead for creating, deploying, and maintaining cloud computing–based solutions.

11.5.1 Infrastructure as Code

At its core, this simply means using code files or code templates to provision and manage cloud infrastructure, as opposed to the more traditional usage of physical tools or interactive configuration software tools.

This is a huge efficiency and security win. Cloud infrastructure management is effectively reduced from hours to minutes, and human error (which usually results in security flaws) is also greatly reduced. Further advantages include:

- Reduction in the cost required to administer cloud infrastructure
- Increased speed of cloud infrastructure deployment and updates
- Reduction of the risk of insecure infrastructure due to the use of automation to create, validate, and audit the code that generates the cloud infrastructure

Security Brief

Creating and maintaining cloud computing infrastructure in code means that security assessments of the configuration can be automated, and known good (and secure) configuration can be saved for the future, as well as for multiple use. This applies to most threats that are pertinent to an IoT cloud.

11.5.2 Serverless Architecture

This approach to cloud computing is also referred to as Functions as a Service (FaaS). It involves enabling developers to create cloud computing solutions without having to worry about server management or capacity planning. This feat is accomplished by allowing software developers to write and deploy solutions through event-driven code functions, while the cloud computing provider takes care of setting up the infrastructure, scaling it, and tearing it down, as required.

Please check Chapter 6, Section 6.2, for the traditional cloud service models, Infrastructure as a Service (IaaS), Platform as a Service (PaaS), and Software as a Service (SaaS).

FaaS or serverless computing is the latest evolution of cloud service models, and it fits between PaaS and SaaS, since users are abstracted away from most administrative details of the platform but are still able to write and deploy new solutions. This is a very attractive choice for software architects or developers, start-ups, and small businesses that are thus empowered to focus on writing code for their shiny new solution without worrying about infrastructure management or about hiring infrastructure architects. This can significantly reduce their development and maintenance costs.

The event-driven nature of serverless functions means that the code they contain is not always running. Thus, in an IoT example, a function that parses and stores temperature data, only runs and is billed when temperature data is received from a field sensor. If there is not a need to collect data constantly, a fastidious solutions architect could store the data for specified periods on the edge gateway before sending them to the cloud in batches. At this juncture, it becomes necessary to add that serverless computing providers usually allow solution architects or developers to specify the kind of hardware upon which their code functions should run. If IoT architects expect to perform complex processing on a large amount of data, they can select higher-grade hardware while setting up their functions and still only pay for the time their code runs.

Security Brief

As you have probably intuited, a major security attraction for going serverless is that a bulk of the security considerations for the cloud environment are handled by the cloud provider.

For the most part, this helps to narrow the security focus of solutions architects to the following:

- Writing good, well-tested FaaS code with attributes such as:
  - Robust input validation or filtering. No secrets (such as credentials, authorization tokens, keys, or passwords) stored in the code.
  - Modular functions that are designed to achieve a small, clear objective.
  - Functions that do not run for long periods, and timeout in the case that an error or other unexpected event causes them to run for longer than allowed.
- Creating (or using, where one is provided) an API gateway to prevent direct caller access to the code. Cloud providers who provide easy-to-customize and deploy API gateway services will be the victors here.
- Applying the principle of least privilege when assigning permissions to functions so they can only access functionality in the cloud environment that they need to access.
- Manage user permissions using solutions provided by the cloud provider.

Table 11.2 shows the foremost FaaS solutions and vendors today.

Table 11.2 Serverless Computing Solutions, Platforms, and Vendors

Serverless Computing Solution	Cloud Computing Platform	Vendor
AWS Lambda™	Amazon Web Services™	Amazon®
Azure Functions™	Microsoft® Azure®	Microsoft®
Google Cloud Functions™	Google Cloud Platform™	Google®

11.5.3 Elastic Container–Based Cloud

Software containers are products of operating system–level virtualization or containerization.[27] OS-level virtualization refers to an operating system feature in which the kernel allows the existence of multiple isolated user-space instances. A computer program running on an ordinary operating system can see all resources (connected devices, files and folders, network shares, CPU power, quantifiable hardware capabilities) of that computer. However, programs running inside a container can only see the container's contents and devices assigned to the container. In addition to isolation mechanisms, the kernel often provides resource-management features to limit the impact of one container's activities on other containers.

Currently, one of the most well-known container technologies is Docker™.* Docker is used to package applications into containers that can be run in disparate environments with little or no overhead. The Docker Engine makes use of OS-level virtualization to create user-space instances that can contain a software application's code, configuration, and all the dependencies that it requires to run. This allows Docker users to create and deploy their applications without having to worry about supporting different operating systems or environments. As a result, Docker allows developers to build, test, and deploy applications with environmental consistency and minimal operational overhead. A single software solution could comprise multiple containers that work together via a central orchestrator.

Elastic container–based cloud computing services are designed to manage the provisioning and maintenance of servers or server clusters that run user-supplied containers. Examples are Google Kubernetes® Engine and AWS Elastic Container Service™.

Security Brief

Automated container management and deployment services such as those we just mentioned greatly reduce the attack surface that cloud solutions architects and software developers need to worry about.

For the most part, security concerns in such environments can be placed into two large buckets:

- **Application security:** This involves the security of the software and configuration that will run inside containers.
- **Container configuration security:** Ensuring that the containers are configured to protect the apps inside the container from compromise (or reduce the risk of compromise) by processes running outside the container. A related concern is the need to protect the underlying operating system from compromise, even if the container were to be compromised.

11.5.4 Autoscaling

This is fast becoming a standard feature of cloud computing. It involves the ability of a cloud provider to automatically increase provisioned computing resources, based on the load on the

---

* https://www.docker.com/

service or application that is using those resources. For instance, increasing the number of servers that support a web service during periods of high demand.

Security Brief

Autoscaling is a short-term mitigation to DDoS attacks. A DDoS is a large-scale DoS attack where the perpetrator uses more than one unique IP address, and often thousands of them. Since the incoming traffic flooding the victim originates from many different sources, it is impossible to stop the attack simply by using ingress filtering. It also makes it very difficult to distinguish legitimate user traffic from attack traffic when spread across so many points of origin. Services with autoscaling infrastructure have a greater chance of staying up, while the system administrators address the DDoS attack.[28]

It is a mitigation that comes at a price (as most do), since the increase in utilized computing resources is billed. However, the incurred costs are usually agreeable to vendors who could incur much higher losses if their service was to go down, even for a short period. Still, it is advisable to configure limits on scaling to avoid exorbitant charges. Once the attack is over or thwarted, autoscaling automatically scales down the number of servers.

11.5.5 Summarizing the Security Advantages of Emergent Trends in Cloud Computing

Table 11.3 summarizes the security wins to be found in the emergent cloud computing trends just covered. These advantages are relevant to individuals or entities who are consumers of a cloud computing offering but are not responsible for running and maintaining the underlying cloud computing infrastructure—for example, a software development company using AWS, Google Cloud Compute, or Microsoft Azure.

Table 11.3 Security Implications of Emergent Cloud Computing Trends

Trend	Advantages
Infrastructure as code	• Enables automated testing of cloud infrastructure configuration. • Known good configuration can be deployed broadly.
Serverless architecture	• Narrows the security scope for solution architects and developers since efforts required to secure cloud environments (servers and networks) are greatly reduced. • Allows developers and architects to focus on writing good, secure code and managing permissions assigned to users and software components such as cloud functions.
Elastic container–based cloud	• Allows architects and developers to focus most of their attention on writing good, secure code and configuring secure software containers to house their apps.
Autoscaling	• Provides short-term and cost-effective mitigation for DDoS attacks.

11.6 Do the Right Thing and the Future Will Take Care of Itself

IoT will change how we interact with ourselves and our environment. Daily routines, as we currently know them, will change. New and old technologies will be connected in different environments for different purposes, creating new experiences for the daily breather and thus introducing new (even when they are old) security concerns that we must address.

It is no small feat to predict the future, and we do not expect our forecast of 2040 to be accurate to the last letter. But that is hardly the point. Inevitably, the future has in its making, the ingredients from our decisions and our actions today. If a world permeated by IoT is to be secure, we all have a part to play. Let us do the right things right.

References

1. Harari, Y. (2015, February 10). *Sapiens: A Brief History of Humankind.* Harper.
2. "We Cannot Predict the Future, But We Can Invent It." Retrieved from https://quoteinvestigator.com/2012/09/27/invent-the-future/
3. Nordrum, A. (2016, August 18). "Popular Internet of Things Forecast of 50 Billion Devices by 2020 Is Outdated." Retrieved from https://spectrum.ieee.org/tech-talk/telecom/internet/popular-internet-of-things-forecast-of-50-billion-devices-by-2020-is-outdated
4. "What Is the Smart Grid?" Retrieved from https://www.smartgrid.gov/the_smart_grid/smart_grid.html
5. Matousek, M. (2018, January 29). "The Most Impressive Things Tesla's Cars Can Do in Autopilot." Retrieved from http://www.businessinsider.com/tesla-autopilot-functions-and-technology-2017-12#when-driving-on-the-highway-enhanced-autopilot-can-control-the-vehicles-speed-based-on-the-traffic-around-it-determine-whether-to-stay-in-or-change-lanes-move-between-freeways
6. Stewart, J. (2018, January 25). "Why Tesla's Autopilot Can't See a Stopped Fire Truck." Retrieved from https://www.wired.com/story/tesla-autopilot-why-crash-radar/
7. Davies, A. (2016, June 30). "Tesla's Autopilot Has Had Its First Deadly Crash." Retrieved from https://www.wired.com/2016/06/teslas-autopilot-first-deadly-crash/
8. Cobb, J. (1999, December 24). "This Just In: Model T Gets Award." Retrieved from http://www.nytimes.com/1999/12/24/automobiles/this-just-in-model-t-gets-award.html.
9. Statista. "Number of Passenger Cars and Commercial Vehicles in Use Worldwide from 2006 to 2015 in (1,000 units)." Retrieved from https://www.statista.com/statistics/281134/number-of-vehicles-in-use-worldwide/
10. Schneier, B. (2015, March 2). "The Democratization of Cyberattack." Retrieved from https://www.schneier.com/blog/archives/2015/03/the_democratiza_1.html
11. Reese, H. (2016, March 24). "Why Microsoft's 'Tay' AI Bot Went Wrong." Retrieved from https://www.techrepublic.com/article/why-microsofts-tay-ai-bot-went-wrong/
12. Today's Hospitalist. (2012, August). "A Smart Bed." Retrieved from https://www.todayshospitalist.com/a-smart-bed/
13. U.S. Department of Education. (2013, July 26). "Summary of the HIPAA Security Rule." Retrieved from https://www.hhs.gov/hipaa/for-professionals/security/laws-regulations/index.html
14. National Institute of Science and Technology. (2016, June 28). "Back to Basics: Multi-Factor Authentication (MFA)." Retrieved from https://www.nist.gov/itl/tig/back-basics-multi-factor-authentication
15. Lantronix. (2017, March 11). "Component Lifespan Considerations for Industrial IoT OEMs." Retrieved from https://www.lantronix.com/blog/component-lifespan-considerations-for-industrial-iot-oems/
16. Osborne, C. (2018, January 17)." US Hospital Pays $55,000 to Hackers After Ransomware Attack." Retrieved from: http://www.zdnet.com/article/us-hospital-pays-55000-to-ransomware-operators/
17. TrendMicro. (2016, June 30). "Healthcare under Attack: What Happens to Stolen Medical Records." Retrieved from https://www.trendmicro.com/vinfo/us/security/news/cyber-attacks/healthcare-under-attack-stolen-medical-records
18. Green, M. (2012, January 2). "A Very Casual Introduction to Fully Homomorphic Encryption." Retrieved from https://blog.cryptographyengineering.com/2012/01/02/very-casual-introduction-to-fully/
19. El Emam, K., Rodgers, S., and Malin, B. (2015, March 20). "Anonymising and Sharing Individual Patient Data." Retrieved from https://www.bmj.com/content/350/bmj.h1139
20. Hoffman, C. (2017, June 8). "How to Manage App Permissions on Android." How-To Geek. Retrieved from https://www.howtogeek.com/230683/how-to-manage-app-permissions-on-android-6.0/

21. Jasper, J. (2014, December 7). "Surprise: Agriculture Is Doing More with IoT Innovation Than Most Other Industries." Retrieved from https://venturebeat.com/2014/12/07/surprise-agriculture-is-doing-more-with-iot-innovation-than-most-other-industries/

22. International Electrotechnical Commission. "Functional Safety." Retrieved from http://www.iec.ch/functionalsafety/explained/

23. Herald Globe. (2014, July 12). "City Population to Reach 6.4bn by 2050." Retrieved from http://www.heraldglobe.com/news/223727231/city-population-to-reach-64bn-by-2050

24. "Critical Infrastructure Sectors." Retrieved from https://www.dhs.gov/critical-infrastructure-sectors

25. Casey, K. (2015, August 4). "10 Leaders in Internet of Things Infrastructure." Retrieved from https://www.networkcomputing.com/internet-things/10-leaders-internet-things-infrastructure/1612927605

26. Lemos, R. "Script Kiddies: The Net's Cybergangs." Retrieved from http://www.zdnet.com/article/script-kiddies-the-nets-cybergangs/

27. Hogg, S. (2014, May 26). "Software Containers: Used More Frequently Than Most Realize." Retrieved from https://www.networkworld.com/article/2226996/cisco-subnet/software-containers--used-more-frequently-than-most-realize.html

28. Soltanian, M. and Amiri, I. (2016). *Theoretical and Experimental Methods for Defending Against DDoS Attacks*. Waltham (MA): Syngress Publishing Inc.

Epilogue

As we mentioned in Chapter 1, IoT was birthed in 1988 when Mark Weiser coined the term *ubiquitous computing*, and it has continued to develop and mature until now. This book is neither the beginning nor the end of that journey. So where would we recommend you go from here?

First, because IoT is such a broad domain, there are multiple areas that we could not provide comprehensive coverage for—sensor architecture; cloud architecture; and the detailed architecture of different IoT verticals, such as healthcare, agriculture, automotive, oil and gas, and too many others to list here. We recommend further investigation and reading in these areas to broaden your IoT knowledge. We intend to add such material to our blogs—both material we compose and our recommendations and pointers to other's great books and articles. Look for our blog URLs under our bios at the front of the book.

We are gladdened by the possibility that you and other practitioners will find our work suitable to build upon. This could involve going deeper into any of the domains we have covered, outlining how to build secure IoT systems using newer or different security solutions, or innovating new security solutions and technologies. Throughout the book, we identified several hard problems for which improved or better solutions are still in high demand: robust separation of privilege in IoT devices for secrets management (Section 5.4.4–5.4.8); architecting for end-to-end encryption in IIoT (Section 5.5.4); architecting for longevity in cryptographic algorithms, especially those that are resistant to quantum attacks (Section 5.5.5); efficient re-provisioning of already provisioned IoT devices (Section 6.4.2); and architecting for privacy in AI systems (Chapters 8 and 9). Your contributions in adding to the literature of IoT, providing solutions through standards, and constructing well-architected products that solve these problems will benefit all of us and society as a whole.

Second, considering the realities birthed by IoT, wherein the Internet connects and powers cities, transportation systems, hospitals, and even people, we want to strongly encourage the continued push for standardization and open design in IoT to enable the interoperability that is necessary for reliable and resilient systems. Too many systems still do not interoperate in a way that realizes the goal of machine-to-machine (M2M) interaction.

Finally, we believe that systems and security architecture (including IoT architecture) should be taught to computer engineering and computer science students in universities and colleges, not just those students specializing in cybersecurity or information assurance. Some might argue that the curriculum is already up to date. But as is evident from what we have relayed through these pages, a lack of knowledge regarding security architecture, threat analysis, protocol analysis, privacy, and usability is having a detrimental impact on the IoT systems being built today. Therefore, we believe that computer engineering and computer science students entering the field of computing need a rigorous introduction to security architecture, security and privacy of end-to-end systems, and the secure design of systems, such as those found in IoT.

Thank you for helping us complete this work by reading it. If you like this book, would you do us a favor by posting a review on amazon.com. We would really appreciate it.

You can read more on "Dave's blog" at http://crypto-corner.typepad.com and "Damilare's blog" at https://tech.edgeofus.com.

— David M. Wheeler & Damilare D. Fagbemi

Index

Printed in the United States
by Baker & Taylor Publisher Services